# Integrated Video & Study Guide Workbook

*to accompany*

# Intermediate Algebra

## Fifth Edition

### Julie Miller
*Daytona State College—Daytona Beach*

### Molly O'Neill
*Daytona State College—Daytona Beach*

### Nancy Hyde
*Broward College*

*Prepared by*
### WiseWire LLC

INTEGRATED VIDEO & STUDY GUIDE WORKBOOK TO ACCOMPANY

INTERMEDIATE ALGEBRA, FIFTH EDITION

Published by McGraw-Hill Education, 2 Penn Plaza, New York, NY 10121. Copyright © 2018 by McGraw-Hill Education. All rights reserved. Printed in the United States of America. No part of this publication may be reproduced or distributed in any form or by any means, or stored in a database or retrieval system, without the prior written consent of McGraw-Hill Education, including, but not limited to, in any network or other electronic storage or transmission, or broadcast for distance learning.

Some ancillaries, including electronic and print components, may not be available to customers outside the United States.

This book is printed on acid-free paper.

1 2 3 4 5 6   QVS  21 20 19 18 17

ISBN 978-1-259-94906-7
MHID 1-259-94906-0

All credits appearing on page or at the end of the book are considered to be an extension of the copyright page.

The Internet addresses listed in the text were accurate at the time of publication. The inclusion of a website does not indicate an endorsement by the authors or McGraw-Hill Education, and McGraw-Hill Education does not guarantee the accuracy of the information presented at these sites.

mheducation.com/highered

# Contents

# Chapter 6    Radicals and Complex Numbers

# Chapter 7    Quadratic Equations, Functions, and Inequalities

# Chapter 8    Exponential and Logarithmic Functions and Applications

Video and Study Guide to accompany Intermediate Algebra, 5th ed., Miller/O'Neill/Hyde
Copyright © 2018 McGraw-Hill Education

# A Note from the Authors

Dear Students,

We know how busy you are and want you to be successful in the most efficient way possible. For this reason, we have developed this Integrated Video and Study Workbook to help you consolidate your notes as you work through the material, which will help you succeed in your course.

Each one of the videos that accompanies your text was created by us, with you in mind. We understand that you want help when you are struggling, on the topic you are struggling with—not an hour-long lecture over the whole chapter. But do you ever find that you aren't exactly sure what you should be learning from the videos? That's why we created this study guide for you!

As you watch the videos for each topic, we provide written examples to reinforce what you watch, along with written exercises for extra practice. You also have a place to take notes. The point of this is that after each video, this guide will help you process and apply what you learned in the video. This guide also has concise notes outlining the main concepts for each section in your textbook, and room for you to take notes given by your instructor.

At the end of the semester, this guide will be a robust notebook and portfolio showcasing your hard work throughout the semester. That is something to be proud of!

All the best for a successful semester,

Julie Miller, *Daytona State College*

Molly O'Neill, *Daytona State College*

Nancy Hyde, *Broward College*

Video and Study Guide to accompany Intermediate Algebra, 5ᵗʰ ed., Miller/O'Neill/Hyde
Copyright © 2018 McGraw-Hill Education

# Frequently Asked Questions (FAQs)

## How does this Integrated Video and Study Workbook enhance the experience of watching videos?

Often, watching educational videos is a passive experience. With this study guide, you work along with the videos by actively entering notes and solving related exercises to test your understanding.

## How is this Integrated Video and Study Workbook related to the textbook?

The videos are numbered in the same order as the textbook. Additionally, since we wrote the book and did the videos ourselves, the language and approach is the same. This makes it very easy to use the videos, the study guide, and the textbook together.

## Why should I use this Integrated Video and Study Workbook?

- With our own students, we find that they often get confused by all the items that are available for study help. This guide pulls it all together—your class notes, your reading/video notes, and your study guide.

- It makes study time much more efficient.

- You can easily jump right into practicing what you learn in the video.

- Flexibility—you can use this guide in class, in a lab, or at home.

- It promotes good study skills that you can apply in other classes.

- It's a great way to prepare for your tests!

**How do I use this Integrated Video and Study Workbook?**

Watch the videos in the order given in the study guide. The example used in each video is printed on the page in the study guide. In some cases, you are asked to work along with the author. In other cases, you are asked to work the example first, and then play the video to check your answer. After each video topic, read any related notes provided in the study guide and then answer the related exercises.

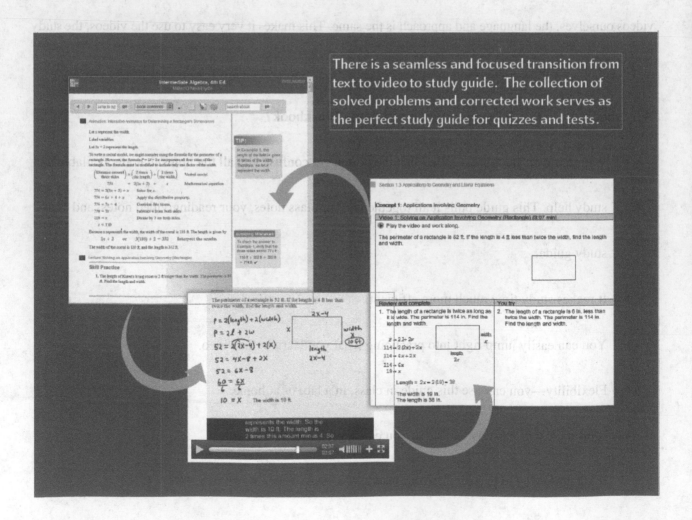

There is a seamless and focused transition from text to video to study guide. The collection of solved problems and corrected work serves as the perfect study guide for quizzes and tests.

**Concept 1**: The Set of Real Numbers

| Video 1: Introduction to Sets (2:26 min) |
|---|

▶ Play the video and work along.

Write a set for each of the following.

Natural Numbers:

Whole Numbers:

Integers:

| Video 2: Rational and Irrational Numbers (2:47 min) |
|---|

▶ Play the video and work along.

**Definition of Rational Numbers**

The set of **rational numbers**:

Examples:

**Definition of Irrational Numbers**

The set of **irrational numbers** is a subset of the real numbers whose elements cannot be written as a ratio of two integers.

*Note*: An irrational number cannot be written as a terminating decimal or as a repeating decimal.

Examples:

## Review and complete

1. Check the set to which each characteristic belongs.

|  | Rational Numbers | Irrational Numbers |
|---|---|---|
| Repeating decimals |  |  |
| Terminating decimals |  |  |
| Non-terminating decimals |  |  |

## Video 3: Identifying Sets to Which Numbers Belong (3:55 min)

■ Before starting the video, try this yourself.

Identify the set to which each number belongs.

|  | $\frac{2}{9}$ | $\sqrt{3}$ | $0.\overline{6}$ | 0.25 | −5 | 0 | 7 |
|---|---|---|---|---|---|---|---|
| Natural numbers |  |  |  |  |  |  |  |
| Whole numbers |  |  |  |  |  |  |  |
| Integers |  |  |  |  |  |  |  |
| Rational numbers |  |  |  |  |  |  |  |
| Irrational numbers |  |  |  |  |  |  |  |
| Real numbers |  |  |  |  |  |  |  |

▶ Play the video and check your answers.

| Review and complete | You try |
|---|---|

2. The set of real numbers consists of both the _____ numbers and the _____ numbers.

3. Place each of the following numbers in the correct box with in diagram.

$$0, \frac{4}{2}, \sqrt{5}, 1.75, \sqrt{9}, 0.3\overline{2}, \frac{4}{9}, -7$$

Since 0 is a whole number but not a natural number, we put a 0 under whole numbers. Next we have $\frac{4}{2} = 2$

which is a natural number, so we put 2 under natural numbers. Now you can complete the rest.

## Real Numbers

| Rational Numbers | Irrational Numbers |
|---|---|

Integers

Whole numbers
0

Natural numbers
2

**Concept 2**: Inequalities

▶ Play the video and work along.

Fill in the blank with the appropriate inequality symbol: < or >

a.  $-4$ _____ $-1$

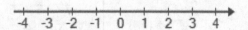

⏸ Pause the video and try this yourself.

b.  $\dfrac{4}{9}$ _____ $\dfrac{3}{7}$

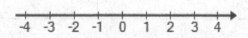

c.  $-2.66$ _____ $-2.\overline{6}$

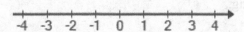

Play the video and check your answer.

| Review and complete | You try |
|---|---|

4.  Fill in each blank on the chart that summarizes the relational operators that compare two real numbers $a$ and $b$.

| Mathematical Expression | Translation |
|---|---|
|  | $a$ is less than $b$ |
|  | $a$ is greater than $b$ |
|  | $a$ is less than or equal to $b$ |
|  | $a$ is greater than or equal to $b$ |
| $a = b$ |  |
| $a \neq b$ |  |
| $a \approx b$ |  |

5.  Fill in the blank with the appropriate inequality symbol: $<$ or $>$

$$0 \underline{\hphantom{aaa}} -6$$

$0 > -6$    On the number line, note that $0$ lies to the

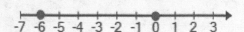

6.  Fill in the blank with the appropriate inequality symbol: $<$ or $>$

$$-5 \underline{\hphantom{aaa}} -3$$

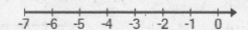

**Concept 3:** Interval Notation

| Video 5: Introduction to Interval Notation (3:24 min) |
|---|

▶ Play the video and work along.

$\{x \mid x > 2\}$

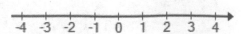

$\{x \mid x \geq 2\}$

---

## Video 6: Practice with Interval Notation (3:32 min)

■ Before starting the video, try this yourself.

Write the set in interval notation.

a.  $\{x \mid x \leq 0\}$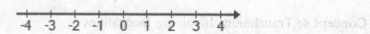

b.  $\{x \mid -4 < x\}$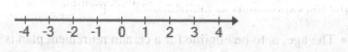

c.  $\{x \mid -1 \leq x < 2.5\}$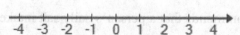

▶ Play the video and check your answers.

| Review and complete | You try |
|---|---|

**Using Interval Notation**
- The endpoints used in interval notation are always written from left to right. In other words, the smaller number is written first, followed by a comma, followed by the larger number.
- Parentheses ) or ( indicate that an endpoint is *excluded* from the set.
- Square brackets ] or [ indicate that an endpoint is *included* in the set.
- Parentheses are always used with $\infty$ or $-\infty$.

7. Graph the set, $\{x \mid x \le 1\}$ on the number line and then express it in interval notation.

Graph:

-4 -3 -2 -1 0 1 2 3 4

Interval Notation:    $(-\infty, 1]$

8. Graph the set, $\{x \mid x > -3\}$ on the number line and then express it in interval notation.

Graph:

-4 -3 -2 -1 0 1 2 3 4

Interval Notation:

**Concept 4:** Translations Involving Inequalities

**Video 7: Translations Involving Inequalities (2:07 min)**

▶ Play the video and work along.

Write the expressions as mathematical inequalities.

- The age, *a*, to be eligible for a certain retirement plan is at least 59 yr.

- A doctor recommends that a person's daily salt intake, *s*, is at most 2400 mg.

⏸ Pause the video and try these yourself.

- The total weight, *w*, in an elevator must not exceed 2000 lb.

- The average number of calories, $c$, that Jen takes in each day is between 2000 and 2400, inclusive.

▶ Play the video and check your answer.

| Review and complete | You try |
|---|---|
| 9. The total rainfall, $r$, for 3 consecutive months was no more than 4.5 in.<br><br>    $r \leq 4.5$ | 10. The number of hours, $h$, that Kaitlyn spent studying was for no less than 40 hr. |

**Concept 1:** Opposite and Absolute Value

### Video 1: The Opposite of a Real Number (1:16 min)

▶ Play the video and work along.

The opposite of 4 is _____

The opposite of $-\dfrac{5}{2}$ is _____

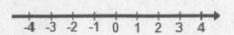

Simplify.   $-(12) =$ _____

⏸ Pause the video and try this yourself.

Simplify.   $-\left(-\dfrac{8}{5}\right) =$ _____

▶ Play the video and check your answer.

| Review and complete | You try |
|---|---|
| 1.  Symbolically, the opposite of a real number $a$ is _____. | |
| 2. Find the opposite of $2.5$.<br><br>The opposite of $2.5$ is $-2.5$.<br><br>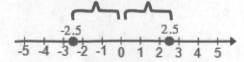 | 3. Find the opposite of $-\dfrac{2}{7}$. |

4. Simplify $-(-2.5)$.

$-(-2.5) = 2.5$    The expression $-(-2.5)$ represents the opposite of $-2.5$.

5. Simplify $-\left(\dfrac{9}{10}\right)$.

## Video 2: Practice with Absolute Value and Opposite (2:55 min)

▶ Play the video and work along.

Simplify the expressions.

$$|-9| \qquad\qquad -|-14| \qquad\qquad -(8)$$

$$-(-20)$$

⏸ Pause the video and try these yourself.

Determine if the statements are true or false.

$$|-2| \overset{?}{\le} 2 \qquad\qquad |-15| \overset{?}{<} -|15| \qquad\qquad -|10| \overset{?}{=} -(10)$$

▶ Play the video and check your answers.

| Review and complete | You try |
|---|---|
| Simplify each expression. | Simplify each expression. |
| 6. $-|-2|$ | 8. $-|-13|$ |
| ⎡This negative is outside of the absolute value.⎤ | |
| $-|-2| = -2$   The absolute value of -2 is 2, and the opposite of that is -2. | |
| 7. $-(-3)$ | 9. $-(-13)$ |
| $-(-3) = 3$   The opposite of -3 is 3 | |

Video and Study Guide to accompany Intermediate Algebra, 5ᵗʰ ed., Miller/O'Neill/Hyde

Determine if the statement is true or false.

10. $-1 < |-1|$

$-1 < 1$    Simplify each side.

The statement is _____.

Determine if the statement is true or false.

11. $|-5| \le -|5|$

---

## Video 3: The Absolute Value of a Real Number (2:35 min)

▶ Play the video and work along.

> **Definition of Absolute Value of a Real Number (informal)**
>
> The **absolute value** of a real number $a$, denoted $|a|$, is the distance between $a$ and 0 on the number line.
>
> *Note:* The absolute value of any real number is positive or zero.

$|-4|$          $|4|$

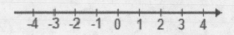

⏸ Pause the video and try these yourself.

$\left|-\dfrac{6}{7}\right|$          $|13.2|$          $|0|$          $|7|$          $|-12|$

▶ Play the video and check your answers.

## Review and complete

12. The absolute value of any real number is never _____.

**Concept 2**: Addition and Subtraction of Real Numbers

---

## Video 4: Addition of Real Numbers (3:58 min)

▶ Play the video and work along.

Add.

$$-4 + (-2)$$

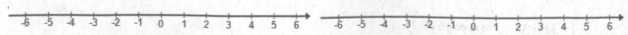

---

Video and Study Guide to accompany Intermediate Algebra, 5th ed., Miller/O'Neill/Hyde

$-4+2$

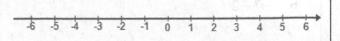

$4+(-2)$

 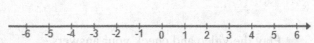

**⏸ Pause the video and try these yourself.**

$-15+(-5)$                    $-3.72+1.51$                    $\dfrac{7}{8}+\left(-\dfrac{1}{6}\right)$

**▶ Play the video and check your answers.**

| Review and complete | You try |
|---|---|
| **Adding Signed Numbers** | |

- To add two numbers with the **same** sign, add their absolute values and apply the common sign.
- To add two numbers with **different** signs, subtract the smaller absolute value from the larger absolute value. Then apply the sign with the larger absolute value.

| | |
|---|---|
| 13. Add each of the following. | 14. Add each of the following. |

**13.** 

a. $-3+(-2)=-(3+2)=-5$

   Both numbers have the same sign: add their absolute values and take the common sign.

b. $2+(-9) \Rightarrow 9-2=7 \Rightarrow$

   The answer is negative because the number with the greater absolute value is $-9$.

   Therefore, $2+(-9)=-7$.

**14.**

a. $-4+(-6)$

b. $5+(-8)$

**Video 5: Definition of Subtraction of Real Numbers (3:00 min)**

| ▶ Play the video and work along. | ⏸ Pause the video and try these yourself. |
|---|---|

Subtract.

a. $6-(-4)$                    c. $-12-(-10)$

b.  $-7-2$          d.  $8-20$

▶ Play the video and check your answers.

## Video 6: Practice with Subtraction of Real Numbers (1:44 min)

■ Before starting the video, try this yourself.

Subtract.

$$\frac{7}{6}-\left(-\frac{1}{4}\right) \qquad\qquad -3.783-4.92$$

▶ Play the video and check your answers.

| Review and complete | You try |
|---|---|
| 15. If $a$ and $b$ are real numbers, then $a-b=a+\left(\quad\right)$. | |
| 16. Subtract each of the following.<br><br>a.  $-2-3 = -2+\left(-3\right) = -5$<br><br>Since both numbers have the same sign, add their absolute values and take the common sign.<br><br>b.  $-15-\left(-5\right) = -15+\left(5\right) = -10$<br><br>Since the numbers have different signs, subtract the smaller absolute value from the larger absolute value, and then apply the sign with the larger absolute value. | 17.  Subtract each of the following.<br><br>a.  $-5-7$<br><br><br>b.  $-12-\left(-3\right)$ |

**Concept 3:** Multiplication and Division of Real Numbers

## Video 7: Multiplication of Real Numbers (2:37 min)

▶ Play the video and work along.

Multiply.

$$(-8)(-4.1) \qquad \left(-\frac{4}{7}\right)\left(\frac{21}{8}\right) \qquad -\frac{1}{9}(-9) \qquad (-4)(-3)(-2)$$

⏸ Pause the video and try this yourself.

$$(-1)(-2)(-5)(-6)$$

▶ Play the video and check your answer.

| Review and complete | You try |
|---|---|
| 18. The product of two real numbers with the same sign is _____. The product of two real numbers with different signs is _____. | |
| 19. Multiplying an odd number of negative factors is (positive/negative). Multiplying an even number of negative factors is (positive/negative). | |
| 20. The product of any real number and zero is _____. | |

**21. Multiply.**

a. $8\left(-\dfrac{3}{5}\right)$

$8\left(-\dfrac{3}{5}\right) = -\dfrac{24}{5}$    Since the factors have different signs, the product is **negative**.

b. $(-3)(-1)(-4)(2)$

$= -24$    Multiplying an odd number of negative factors is **negative**.

Multiplying from left to right, 3 times 1 is 3, times 4 is 12, times 2 is 24.

**22. Multiply.**

a. $\dfrac{4}{5}(-3)$

b. $(-1)(-2)(3)(-5)$

**Video 8: Practice Dividing Real Numbers (2:32 min)**

▶Play the video and work along.

Simplify.

a.  $-36 \div (-3)$

b.  $\dfrac{-100}{4}$

c.  $14 \div (-2.8)$

d.  $\dfrac{-11}{-3}$

e.  $\dfrac{0}{-7}$

f.  $-7 \div 0$

| Review and complete | You try |
|---|---|

23. When dividing real numbers with the **same sign**, the answer is always (positive/negative).

24. When dividing real numbers with the **different signs**, the answer is always (positive/negative).

| | |
|---|---|
| 25. Simplify $\dfrac{-25}{5}$.<br><br>$\dfrac{-25}{5} = -5$    Different signs: the quotient is **negative**. | 26. Simplify $\dfrac{-35}{5}$. |

### Video 9: Dividing Real Numbers Involving Fractions (1:24 min)

■ Before starting the video, try this yourself.

Simplify.

a.   $-18 \div (-12)$          b.   $-\dfrac{3}{10} \div \dfrac{6}{25}$

▶ Play the video and check your answers.

| Review and complete | You try |
|---|---|
| 27. Simplify $-\dfrac{4}{9} \div \dfrac{8}{15}$.<br><br>$= -\dfrac{4}{9} \cdot \dfrac{15}{8}$    Rewrite as multiplication.<br><br>$= -\dfrac{\overset{1}{\cancel{4}}}{\underset{3}{\cancel{9}}} \cdot \dfrac{\overset{5}{\cancel{15}}}{\underset{2}{\cancel{8}}}$    Divide out common factors in the numerator and denominator.<br><br>$= -\dfrac{5}{6}$    Multiply. | 28. Simplify $-\dfrac{10}{27} \div \dfrac{5}{18}$. |

**Concept 4:** Exponential Expressions

### Video 10: Simplifying Expressions Involving Exponents (3:08 min)

▶ Play the video and work along.

a.  $(-3)^2$

b.  $-3^2$

⏸ Pause the video and try these yourself.

c.  $\left(-\dfrac{1}{4}\right)^3$

d.  $-0.2^3$

▶ Play the video and check your answers.

| Review and complete | You try |
|---|---|
| 29.  Simplify $-4^2$.<br><br>Square the base which is 4.<br>The opposite of 16 is -16.<br><br>$-\left(4^2\right)=-16$ | 30.  Simplify $(-5)^2$. |

**Concept 5:** Square Roots

**Video 11: Evaluating Square Roots (2:38 min)**

▶ Play the video and work along.

Simplify the expressions, if possible.

$\sqrt{25}$          $\sqrt{9}$          $\sqrt{\dfrac{49}{16}}$          $\sqrt{-49}$          $\sqrt{-36}$

| Review and complete | You try |
|---|---|
| 31. The square root of a negative number (is / is not) a real number. | |
| 32. Simplify the expressions, if possible.<br><br>  a.   $\sqrt{144}=12$ because $(12)^2=144$.<br><br>  b.   $\sqrt{-16}=$ not a real number<br><br>  This is not a real number, because there is no real number that when squared equals -16. | 33. Simplify the expressions, if possible.<br><br>  a.   $\sqrt{36}$<br><br><br>  b.   $\sqrt{-25}$ |

**Concept 6:** Order of Operations

**Video 12: Applying the Order of Operations (2:16 min)**

Video and Study Guide to accompany Intermediate Algebra, 5th ed., Miller/O'Neill/Hyde

■ Before starting the video, try this yourself.

Simplify.

$$\frac{12-3\left[-4+\left(2-6\right)^2\right]}{-2\left|-8+2\right|}$$

▶ Play the video and check your answer.

| Review and complete | You try |
|---|---|
| 34. Simplify: | 35. Simplify: |
| $$\frac{\sqrt{6^2-3^2-2}-1\left[-1+\left(3-8\right)^2\right]}{-3\left|-7+4\right|}$$ | $$\frac{9-1\left[-4+\left(3-6\right)^2\right]-\sqrt{13^2-12^2}}{-1\left|-6+4\right|}$$ |
| $$=\frac{\sqrt{36-9-2}-1\left[-1+\left(-5\right)^2\right]}{-3\left|-3\right|}$$  Simplify inside the parentheses and the other grouping symbols. | |
| $$=\frac{\sqrt{25}-1\left[-1+25\right]}{-3\left|-3\right|}=\frac{\sqrt{25}-1\left[24\right]}{-3\left|-3\right|}$$ | |
| $$=\frac{5-24}{-3\left(3\right)}$$  Multiply before adding and subtracting. | |
| $$=\frac{-19}{-9}$$ | |
| $$=\frac{19}{9}$$ | |

**Concept 7**: Evaluating Formulas

**Video 13: Evaluating an Algebraic Expression (2:11 min)**

■ Before starting the video, try this yourself.

Evaluate the expression for the given values of the variables.

$$\sqrt{\left(x_2 - x_1\right)^2 + \left(y_2 - y_1\right)^2} \quad \text{for } x_1 = -9, y_1 = 4, x_2 = 3, y_2 = -1$$

▶ Play the video and check your answer.

| Review and complete | You try |
|---|---|
| 36. Evaluate the expression for the given values of the variables.<br><br>$$\sqrt{\left(x_2 - x_1\right)^2 + \left(y_2 - y_1\right)^2} \quad \text{for}$$<br>$$x_1 = -2, y_1 = 3, x_2 = 2, y_2 = 6$$<br><br>$$= \sqrt{\left((2) - (-2)\right)^2 + \left((6) - (3)\right)^2}$$<br>$$= \sqrt{(4)^2 + (3)^2}$$<br>$$= \sqrt{16 + 9}$$<br>$$= \sqrt{25}$$<br>$$= 5$$ | 37. Evaluate the expression for the given values of the variables.<br><br>$$\sqrt{\left(x_2 - x_1\right)^2 + \left(y_2 - y_1\right)^2} \quad \text{for}$$<br>$$x_1 = -2, y_1 = -9, x_2 = 3, y_2 = 3$$ |

**Concept 1**: Recognizing Terms, Factors, and Coefficients

| Video 1: Algebraic Expressions and Vocabulary (2:38 min) |
|---|

▶ Play the video and work along.

State the definition or give an example of each word.

- A term:

- An algebraic expression:

- Like terms:

- Unlike terms:

| Review and complete | You try |
|---|---|
| The **coefficient** of a term is the numerical factor of the term. | |
| 1. Identify the coefficient of the term $x^3 y^5$. <br><br> The coefficient is 1. | 3. Identify the coefficient of the term $5xy^5$. |
| 2. Identify the like terms: $6x^2$, $6y$, $x$, $x^2$ <br><br> The like terms are $6x^2$ and $x^2$. | 4. Identify the like terms: $5xy$, $6xy$, $y$, $x^2$ |

**Concept 2**: Properties of Real Numbers

| Video 2: Cumulative and Associative Properties of Real Numbers (3:54 min) ||

▶ Play the video and work along.

| Property | Examples |
|---|---|
| Commutative property of addition<br><br>$a+b=b+a$ | |
| Commutative property of multiplication<br><br>$a \cdot b = b \cdot a$ | |

| Property | Examples |
|---|---|
| Associative property of addition<br><br>$(a+b)+c=a+(b+c)$ | |
| Associative property of multiplication<br><br>$(a \cdot b) \cdot c = a \cdot (b \cdot c)$ | |

| Review and complete | You try |
|---|---|
| 5. Subtraction and division (are/are not) commutative or associative. ||
| 6. State the property that is illustrated.<br><br>a.  $(4+2)+3=4+(2+3)$<br><br>Associative property of addition<br><br><br>b.  $2 \cdot 3 = 3 \cdot 2$<br><br>Commutative property of multiplication | 7.  State the property that is illustrated.<br><br>a.  $(2 \cdot 3) \cdot 4 = 2 \cdot (3 \cdot 4)$<br><br><br><br>b.  $5+2=2+5$ |

## Video 3: The Identity and Inverse Properties of Addition and Multiplication (2:53 min)

▶ Play the video and work along.

| Property | Examples |
|---|---|
| Identity property of addition <br><br> $0+a=$ _____          $a+0=$ _____ | |
| Identity property of multiplication <br><br> $1 \cdot a =$ _____ | |
| Inverse property of addition <br><br> $a+(-a)=$ _____ | |

| Property | Examples |
|---|---|
| Inverse property of multiplication <br><br> $a \cdot \dfrac{1}{a} =$ _____ | |

| Review and complete | You try |
|---|---|
| 8. The number _____ is called the **identity element of addition**. <br><br> 9. The number _____ is called the **identity element of multiplication**. <br><br> 10. For any real number, $a$, the **additive inverse** of $a$ is _____. <br><br> 11. For any real number, $a$, the **multiplicative inverse** of $a$ is _____. | |
| 12. State the property illustrated. <br><br>   a.  $1 \cdot 7 = 7$ <br><br>     Identity property of multiplication <br><br>   b.  $5+0=5$ <br><br>     Identity property of addition | 13. State the property illustrated. <br><br>   a.  $12+(-12)=0$ <br><br><br>   b.  $4 \cdot \dfrac{1}{4} = 1$ |

## Video 4: The Distributive Property of Multiplication over Addition (2:28 min)

▶ Play the video and work along.

| Property | Examples |
|---|---|
| Distributive property of multiplication over addition<br><br>$a(b+c) = $ _____<br><br>$(b+c)a = $ _____ | |

Apply the distributive property.             $3(5w+4y+7)$

## Video 5: Practice Applying the Distributive Property (4:41 min)

▶ Play the video and work along.

Apply the distributive property of multiplication over addition.

$2(6x-9)$

⏸ Pause the video and try these yourself.

$-(9a-2b+7c)$

$-5(2-8x)$

$$-8\left(-4x-3y+5\right)$$

▶ Play the video and check your answers.

| Review and complete | You try |
|---|---|
| 14.  Apply the distributive property.<br><br>$6\left(-9x+8y-12z\right)$<br><br>$=6\left[-9x+8y+\left(-12z\right)\right]$<br><br>$=6\left(-9x\right)+6\left(8y\right)+6\left(-12z\right)$<br><br>$=-54x+48y-72z$ | 15.  Apply the distributive property.<br><br>$5\left(a-2b-3\right)$ |
| 16.  Apply the distributive property.<br><br>$-\left(-6x+3y\right)$<br>$=-1\left(6x+3y\right)$<br>$=-1\left(-6x\right)+\left(-1\right)\left(3y\right)$<br>$=\boxed{\phantom{xxxxxxx}}$ | 17.  Apply the distributive property.<br><br>$-\left(a-2b\right)$ |
| **Video 6: Combining Like Terms (1) (2:22 min)** | |

▶ Play the video and work along.

Simplify the expressions by combining like terms.

$8x + 3x$

⏸ Pause the video and try these yourself.

$-11y + 8y - y$                                    $6a + 5$

▶ Play the video and check your answers.

**Video 7: Combining Like Terms (2) (1:31 min)**

⏹ Before starting the video, try this yourself.

Simplify the expressions by combining like terms.

$8ab - 7 - 3ab + 12$                          $3.4x^4 + 7.1x^4$

▶ Play the video and check your answers.

| Review and complete | You try |
|---|---|
| 18. Simplify by combining like terms. | 19. Simplify by combining like terms. |
| $12 - 6x^2 - 8 + x^2$ | $5 + a^3 - 2 + 6a^3$ |
| $= -6x^2 + x^2 + 12 - 8$    Group like terms. | |
| $= -5x^2 + 4$    Combine like terms. | |

**Concept 3:** Simplifying Expressions

| Video 8: Clearing Parentheses and Combining Like Terms (2:18 min) |
| --- |

▶ Play the video and work along.

Simplify the expressions by clearing parentheses combining like terms.

$9 - 5(7x + 9)$

| Review and complete | You try |
| --- | --- |
| 20. Clear parentheses and combine like terms. | 21. Clear parentheses and combine like terms. |

20. $3(-5y + 2) + 7(y - 4)$
    $= -15y + 6 + 7y - 28$    Clear parentheses.
    $= -15y + 7y + 6 - 28$    Group like terms.
    $= -8y - 22$                Combine like terms

21. $-2(5x - 8) - 3(4 - x)$

| Video 9: Simplifying Expressions (3:43 min) |
| --- |

▶ Play the video and work along.

Simplify the expressions by clearing parentheses combining like terms.

$$\frac{1}{3}(12x+6)-\frac{1}{5}(5x+2)$$

⏸ Pause the video and try this yourself.

$$-7(7x-6y)-2(3x+9y)+5y$$

▶ Play the video and check your answer.

**Video 10: Simplifying an Expression with Nested Parentheses (1:39 min)**

⏹ Before starting the video, try this yourself.

Simplify the expression by clearing parentheses combining like terms.

$$-12x-2\big[4x-3(x+5)\big]-9$$

▶ Play the video and check your answer.

| Review and complete | You try |
|---|---|
| 22. Clear parentheses and combine like terms. | 23. Clear parentheses and combine like terms. |

22. Clear parentheses and combine like terms.

$$5x - 3\left[x + 2(3x - 6)\right]$$

$$= 5x - 3(x + 6x - 12) \qquad \text{Clear inner ( ).}$$

$$= 5x - 3(7x - 12) \qquad \text{Combine like terms.}$$

$$= 5x - 21x + 36 \qquad \text{Clear parentheses.}$$

$$= \boxed{\phantom{xxxxxxxxx}}$$

23. Clear parentheses and combine like terms.

$$3y - 2\left[4y - (5x + 1)\right]$$

**Answers:** Section R.2

1.

|  | Rational Numbers | Irrational Numbers |
|---|---|---|
| Repeating decimals | x |  |
| Terminating decimals | x |  |
| Non-terminating, non-repeating decimals |  | x |

2. rational, irrational

3.

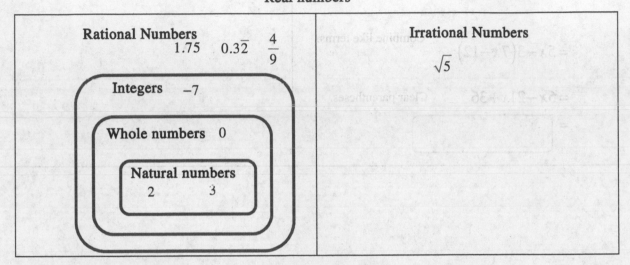

**Real numbers**

Rational Numbers    1.75    $0.3\overline{2}$    $\dfrac{4}{9}$

Integers    −7

Whole numbers    0

Natural numbers    2    3

Irrational Numbers    $\sqrt{5}$

4.

| Mathematical Expression | Translation |
|---|---|
| $a < b$ | $a$ is less than $b$ |
| $a > b$ | $a$ is greater than $b$ |
| $a \leq b$ | $a$ is less than or equal to $b$ |
| $a \geq b$ | $a$ is greater than or equal to $b$ |
| $a = b$ | $a$ is equal to $b$ |
| $a \neq b$ | $a$ is not equal to $b$ |
| $a \approx b$ | $a$ is approximately equal to $b$ |

6. <

8.    $(-3, \infty)$

10. $h \geq 40$

**Answers**: Section R.3

1. $-a$          3. $\dfrac{2}{7}$          5. $-\dfrac{9}{10}$          8. $-13$          9. $13$

10. true          11. False          12. negative          14a. $-10$          14b. $-3$

15. $-b$          17a. $-12$          17b. $-9$          18. positive; negative

19. negative; positive          20. zero          22a. $-\dfrac{12}{5}$          22b. $-30$

23. positive          24. negative          26. $-7$          28. $-\dfrac{4}{3}$          30. $25$

31. is not          33a. $6$          33b. Not a real number          35. $\dfrac{1}{2}$          37. $13$

**Answers**: Section R.4

3. $5$          4. $5xy,\ 6xy$          5. are not

7a. Associative property of multiplication     7b. Commutative property of addition

8. $0$          9. $1$          10. $-a$          11. $\dfrac{1}{a}$

13a. Inverse property of addition          13b. Inverse property of multiplication

15. $5a-10b-15$      16. $6x-3y$      17. $-a+2b$      19. $7a^3+3$

21. $-7x+4$      22. $-16x+36$      23. $10x-5y+2$

**Concept 1**: Linear Equations in One Variable

| Video 1: Introduction to Linear Equations in One Variable (3:04 min) |
|---|

▶ Play the video and work along.

| Equation | Solution | Check |
|---|---|---|
| | $x = -2$ | |
| $3x + 6 = 0$ | | |
| | $x = 4$ , $x = -4$ | |
| $x^2 = 16$ | | |

⏸ Pause the video and try this yourself.

$$-\frac{12}{x} = 4 \qquad x = -3$$

▶ Play the video and check your answer.

▶ Play the video and work along.

Determine if the following equations are linear or nonlinear.

$$x^2 = 16 \qquad\qquad -\frac{12}{x}$$

$$3x + 6 = 0$$

| Review and complete | You try |
|---|---|

1. Let $a$, $b$, and $c$ be real numbers such that $a \neq 0$. A **linear equation in one variable** is an equation that can be written in the form _____.

| 2. Label the equations as linear or nonlinear. | 3. Label the equations as linear or nonlinear. |
|---|---|
| a. $x^2 = 16$ <br><br> This is **nonlinear** because the exponent on $x$ is not 1. <br><br> b. $6x = 7$ <br><br> This is **linear** because <br> • It has only 1 variable <br> • Variable is raised to the first power only <br> • Variable is in the numerator | a. $3x^3 = 81$ <br><br><br> b. $x = \dfrac{7}{3}$ |

**Concept 2**: Solving Linear Equations

---

**Video 2: Addition and Subtraction Properties of Equality (2:34 min)**

▶ Play the video and work along.

Solve the equations.

$$-5 + y = 11 \qquad\qquad\qquad 10.6 = t + 4.1$$

Check: $-5 + y = 11$            Check:   $10.6 = t + 4.1$

---

| Review and complete | You try |
|---|---|
| 4. Addition Property of Equality: If $a = b$, then _____. <br><br> 5. Subtraction Property of Equality: If $a = b$, then _____. | |
| 6. Solve the equation and check the solution. <br><br> $\qquad -6 + x = 12$ <br><br> $\underline{-6 + 6 + x = 12 + 6}$    To isolate $x$, add 6 to both sides of the equation. | 7. Solve the equation and check the solution. <br><br> $\qquad -3 + y = 27$ |

$$x = 18$$

$$\{18\}$$

Don't forget to <u>check</u>.

$$-6 + x = 12$$

$$-6 + (18) \stackrel{?}{=} 12 \quad \text{Plug in 18 for x.}$$

$$12 = 12 \checkmark \quad \text{True.}$$

## Video 3: Multiplication and Division Properties of Equality (3:19 min)

▶ Play the video and work along.

Solve the equations.

$$-\frac{2}{5}p = \frac{3}{10}$$

$$8.5 = \frac{d}{6.2}$$

$$-y = 9$$

Check: $-\frac{2}{5}p = \frac{3}{10}$

Check:   $8.5 = \frac{d}{6.2}$

Check:  $-y = 9$

| Review and complete | You try |
|---|---|

8. Multiplication Property of Equality: If $a = b$, then _____.

9. Division Property of Equality: If $a = b$ and $c \neq 0$, then _____.

| | |
|---|---|
| 10. Solve the equation and check the solution. | 11. Solve the equation and check the solution. |

**10.**

$$-\frac{12}{13} = \frac{4}{3}b$$

$$-\frac{12}{13} \cdot \left(\frac{3}{4}\right) = \left(\frac{4}{3}b\right) \cdot \left(\frac{3}{4}\right)$$

To isolate $b$, multiply both sides by the reciprocal of 4/3.

$$-\frac{9}{13} = b$$

$$\left\{-\frac{9}{13}\right\}$$

Check:

$$-\frac{12}{13} \overset{?}{=} \frac{4}{3}\left(-\frac{9}{13}\right)$$  Plug in $-\frac{9}{13}$ for $b$.

$$-\frac{12}{13} \overset{?}{=} -\frac{12}{13} \ \checkmark$$

**11.**

$$-\frac{15}{11} = \frac{3}{2}b$$

12. Solve the equation and check the solution.

$$-y = 9$$

$$-1y = 9$$

$$\frac{-1y}{-1} = \frac{9}{-1}$$

$$y = -9$$

$$\{-9\}$$

Check:

$$-(-9) = 9 \qquad \text{Plug in } -9 \text{ for } y.$$

$$9 = 9 \ \sqrt{} \qquad \text{True.}$$

13. Solve the equation and check the solution.

$$-w = -5$$

## Video 4: Solving a Linear Equation Requiring Multiple Steps (1:33 min)

▶ Play the video and work along.

Solve the equation.

$$-3x + 4 = 19$$

Check: $-3x + 4 = 19$

## Video 5: Guidelines for Solving a Linear Equation in One Variable (2:19 min)

▶ Play the video and work along.

Solve the equation. $\quad 8 - 3 + y = 3(y + 4) + 15$

Check: $8 - 3 + y = 3(y + 4) + 15$

### Review and complete

Procedure for **Solving Linear Equations in One Variable**

1. Simplify both sides of the equation.
2. Collect all variable terms on one side of the equation.
3. Collect all constant terms on the other side of the equation.
4. Use the multiplication or division property of equality to obtain a coefficient of 1 for the variable.
5. Check the potential solution in the original equation.

### Video 6: Solving a Linear Equation in One Variable (2:11 min)

■ Before starting the video, try this yourself.

Solve the equation.

$$1 - 2x - 4 = x - 5(x - 7) + 4$$

<u>Check</u>: $1 - 2x - 4 = x - 5(x - 7) + 4$

▶ Play the video and check your answer.

| Review and complete | You try |
| --- | --- |

14. Solve the equation and check the solution.

$$1 - 5(p+2) = 2(p+13)$$

$$1 - 5p - 10 = 2p + 26 \qquad \text{Clear parentheses.}$$

$$-5p - 2p = 26 + 10 - 1 \qquad \text{Collect variables terms on one side and}$$
$$-7p = 35 \qquad\qquad \text{constant terms on the other.}$$

$$p = -5$$

$$\{-5\}$$

Check:

$$1 - 5((-5)+2) = 2((-5)+13) \qquad \text{Plug in -5 for } p.$$
$$16 = 16 \ \checkmark \qquad\qquad \text{True.}$$

15. Solve the equation and check the solution.

$$3 - 5(9-w) = 2(w+4) + 1$$

**Concept 3:** Clearing Fractions and Decimals

| **Video 7: Solving an Equation by First Clearing Fractions (2:49 min)** |
|---|

▶ Play the video and work along.

Solve the equation.

$$\frac{2}{5}x - \frac{3}{10} = \frac{1}{2}$$

Check: $\dfrac{2}{5}x - \dfrac{3}{10} = \dfrac{1}{2}$

| **Video 8: Solving a Linear Equation Containing Fractions and Parentheses (3:59 min)** |
|---|

▶ Play the video and work along.

Solve the equation.

$$\frac{1}{2}(x-3)+\frac{1}{7}(x+1)=1$$

<u>Check</u>:  $\frac{1}{2}(x-3)+\frac{1}{7}(x+1)=1$

| Review and complete | You try |
|---|---|
| 16. Solve the equation. | 17. Solve the equation. |

16.

$$\frac{1}{2}(x-4)-\frac{3}{4}(x+2)=\frac{1}{4}$$

$$(4)\cdot\frac{1}{2}(x-4)-(4)\cdot\frac{3}{4}(x+2)=(4)\cdot\frac{1}{4}$$

$2(x-4)-3(x+2)=1$    Clear parentheses.

$2x-8-3x-6=1$    Combine like terms.

$-x-14=1$

$-x=15$

$x=-15$

$\{-15\}$

17.

$$\frac{1}{3}(x+3)+\frac{1}{2}(x-4)=4$$

**Video 9: Practice Clearing Fractions, Numerator has two Terms (2:14 min)**

Video and Study Guide to accompany Intermediate Algebra, 5<sup>th</sup> ed., Miller/O'Neill/Hyde
Copyright © 2018 McGraw-Hill Education

● Before starting the video, try this yourself.

Solve the equation. $\dfrac{x+3}{3} - \dfrac{3x-1}{2} = 1$

| Review and complete | You try |
|---|---|
| 18. Solve the equation. | 19. Solve the equation. |

18.

$$\frac{x-1}{6} - \frac{x-3}{4} = 6$$

$$12 \cdot \left( \frac{x-1}{6} - \frac{x-3}{4} \right) = 12 \cdot (6) \quad \text{Multiply both sides by the LCD.}$$

$$12 \cdot \left( \frac{x-1}{6} \right) - 12 \cdot \left( \frac{x-3}{4} \right) = 12 \cdot (6)$$

$$\overset{2}{\cancel{12}} \cdot \left( \frac{x-1}{\cancel{6}_1} \right) - \overset{3}{\cancel{12}} \cdot \left( \frac{x-3}{\cancel{4}_1} \right) = 12 \cdot (6)$$

$$2(x-1) - 3(x-3) = 72 \quad \text{Clear fractions.}$$

$$2x - 2 - 3x + 9 = 72 \quad \text{Clear parentheses.}$$

19. Solve the equation.

$$\frac{x+2}{4} - \frac{x-1}{3} = 2$$

$$x = -65$$

$$\{-65\}$$

## Video 10: Solving a Linear Equation by First Clearing Decimals (3:24 min)

▶ Play the video and work along.

Solve the equation.

$$1.6x + 2 = 0.05x - 10.4$$

Solve the equation by the second method.

$$1.6x + 2 = 0.05x - 10.4$$

## Video 11: Solving a Linear Equation Containing Decimals and Parentheses (2:05 min)

▶ Play the video and work along.

Solve the equation.   $0.05x(x+2) + 0.12(x+6) = 2.52$

| Review and complete | You try |
|---|---|
| 20.  Solve the equation. | 21.  Solve the equation. |

20.  Solve the equation.

$$0.4(n+10) + 0.6n = 2$$

$\quad 0.4n + 4 + 0.6n = 2$    Clear parentheses.

$10 \cdot (0.4n + 4 + 0.6n) = (2) \cdot 10$

$\quad\quad 4n + 40 + 6n = 20$    Multiply both sides by 10.

$\quad\quad\quad\quad 10n = -20$    Combine like terms.

21.  Solve the equation.

$$0.15(w+50) - 0.05(w-20) = 13.5$$

| $n = -2$ |  |
|:---:|:---|
| $\{-2\}$ |  |

**Concept 4:** Conditional Equations, Contradictions, and Identities

---

**Video 12: Conditional Equations, Contradictions, and Identities (2:32 min)**

▶ Play the video and work along.

**Conditional Equation**
An equation that is true for some values of the variables but false for other values of the variable.

Ex:

**Contradiction**
An equation that has no solution.

Ex:

**Identity**
An equation in which the left- and right-hand sides are true for all real numbers.

Ex:

---

**Video 13: Identifying Conditional Equations, Contradictions, and Identities (3:45 min)**

▶ Play the video and work along.

Solve the equation, if possible.

$2x + 4 = 11$                              Check: $2x + 4 = 11$

⏸ Pause the video and try these yourself.

---

$$6x - 5(x-1) = x + 3 \qquad\qquad 4(x-2) = x - 8 + 3x$$

▶ Play the video and check your answers.

| Review and complete | You try |
|---|---|
| 22.  Solve the equation, if possible. | 23.  Solve the equation, if possible. |

22.  Solve the equation, if possible.

$$3x - 2(x-3) = x + 5$$

$3x - 2x + 6 = x + 5$    Simplify the left hand side.

$x + 6 = x + 5$    The left and right sides will never be equal.

$6 = 5$    This is a contradiction.

$\{\ \}$    No solution

23.  Solve the equation, if possible.

$$3(x-1) = x - 3 + 2x$$

## Concept 1: Introduction to Problem Solving

### Video 1: Introduction to Problem-Solving (3:18 min)

▶ Play the video and work along.

The difference of two numbers is 24. The larger number is 9 more than twice the smaller. Find the two numbers.

| | |
|---|---|
| Step 1 | Read the problem carefully. |
| Step 2 | Assign labels to unknowns. |
| Step 3 | Develop a verbal model. |
| Step 4 | Write an equation. |
| Step 5 | Solve the equation. |
| Step 6 | Interpret the results. |

Check:

| Review and complete | You try |
|---|---|
| 1. One number is 3 less than another number. Their sum is 15. Find the numbers. | 2. The difference of two numbers is 25. The larger number is 9 less than twice the smaller. Find the two numbers. |

**Review and complete (continued):**

1. One number is 3 less than another number. Their sum is 15. Find the numbers.

Step 1: Read the problem carefully.

Step 2: Assign labels to unknowns.

Let $x$ be the larger number.
Let $x-3$ be the smaller number.

Step 3: Develop a verbal model.

$$\text{Smaller number} + \text{Larger number} = 15$$

Step 4: Write an equation.

$$x-3+x=15$$

Step 5: Solve the equation.

$$x=9$$

Step 6: Interpret the results.

The larger number is 9.
The smaller number is 6.

## Concept 2:  Applications Involving Consecutive Integers

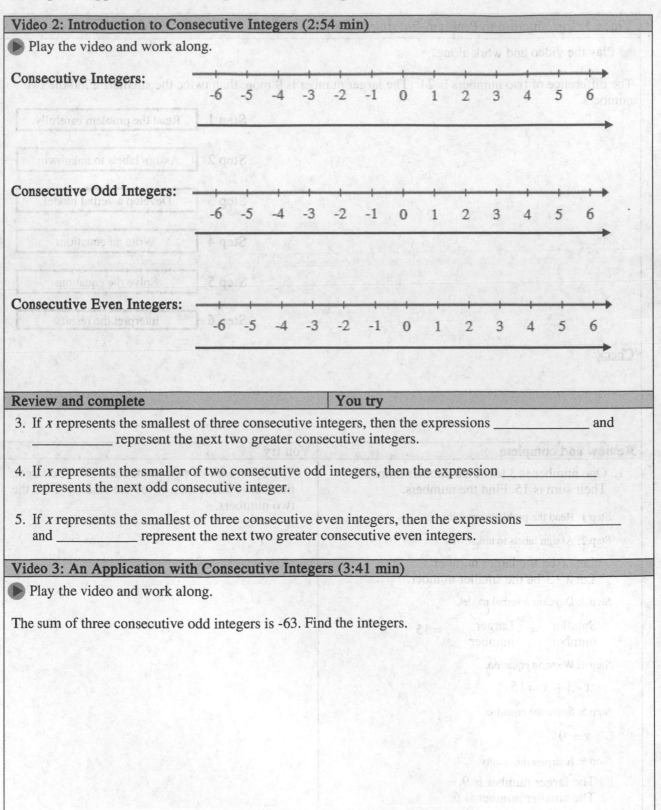

**Video 2: Introduction to Consecutive Integers (2:54 min)**

▶ Play the video and work along.

**Consecutive Integers:**

**Consecutive Odd Integers:**

**Consecutive Even Integers:**

| Review and complete | You try |
| --- | --- |

3. If *x* represents the smallest of three consecutive integers, then the expressions _____ and _____ represent the next two greater consecutive integers.

4. If *x* represents the smaller of two consecutive odd integers, then the expression _____ represents the next odd consecutive integer.

5. If *x* represents the smallest of three consecutive even integers, then the expressions _____ and _____ represent the next two greater consecutive even integers.

**Video 3: An Application with Consecutive Integers (3:41 min)**

▶ Play the video and work along.

The sum of three consecutive odd integers is -63. Find the integers.

## Video 4: An Application with Consecutive Integers (2:51 min)

 Before starting the video, try this yourself.

Seven times the smallest of three consecutive integers is 144 more than twice the sum of the other two integers. Find the integers.

▶ Play the video and check your answer.

| Review and complete | You try |
|---|---|
| 6. The sum of two consecutive even integers is 126. Find the integers.<br><br>Let $x$ be the smaller even integer.<br>Let $x+2$ be the consecutive even integer.<br><br>$\dfrac{\text{the first}}{\text{even}}_{\text{integer}} + \dfrac{\text{the second}}{\text{even integer}} = 126$<br><br>$(x)+(x+2)=126$<br>$2x+2=126$<br>$x=62$<br><br>**The smaller even integer is 62.**<br>**The next even integer is 64.**<br><br>Check: $62+64=126$ ✓ | 7. The sum of three consecutive integers is 186. Find the integers. |

## Concept 3: Applications Involving Percents and Rates

### Video 5: An Application Involving Simple Interest (3:48 min)

▶ Play the video and work along.

$$\begin{pmatrix} \text{Simple} \\ \text{interest} \end{pmatrix} = \begin{pmatrix} \text{principal} \\ \text{invested or borrowed} \end{pmatrix} \begin{pmatrix} \text{annual} \\ \text{interest rate} \end{pmatrix} \begin{pmatrix} \text{time in} \\ \text{years} \end{pmatrix}$$

- Determine the amount of simple interest owed on a loan of $5000 at 6% interest for 2 years.

- Andrea plans to keep her current car for four more years. She wants to save money for a new car and estimates that she'll need $21,600 four years from now. How much should she invest now if she can get 5% simple interest over the next four years?

### Video 6: An Application Involving Discount (2:23 min)

▶ Play the video and work along.

A set of patio furniture went on sale at the end of the summer for 15% off the original price. If the sale price is $680, determine the original price.

Check:

| Review and complete | You try |
|---|---|
| 8. The price of a used textbook after a 35% markdown is $29.25. What was the original price?<br><br>Let $x$ be the original price of the textbook.<br><br>Sale price = original price - discount<br><br>$29.25 = x - 0.35x$<br>$29.25 = 0.65x$    Divide both sides by 0.65.<br>$45 = x$<br><br>The original price was $45. | 9. The price of a used textbook after a 25% markdown is $35.91. What was the original price? |

## Concept 4: Applications Involving Principal and Interest

### Video 7: An Application Involving Principal and Interest (2:59 min)

▶ Play the video and work along.

Zack borrowed a total of $8000 from two different banks. One bank charged 8% simple interest, and a credit union charged 6% simple interest. If the total interest owed after one year was $580, how much money did Zack borrow from each bank?

|  | 8% account | 6% account | Total |
|---|---|---|---|
| Amount of principal |  |  |  |
| Amount of interest |  |  |  |

## Video 8: An Application Involving Principal and Interest (2:28 min)

■ Before starting the video, try this yourself.

Zack borrowed a total of $8000 from two different banks. One bank charged 8% simple interest, and a credit union charged 6% simple interest. If the total interest owed after one year was $580, how much money did Zack borrow from each bank?

|  | 8% account | 6% account | Total |
|---|---|---|---|
| Amount of principal | $x$ | $8000 - x$ | $8000 |
| Amount of interest | $0.08x$ | $.06(8000 - x)$ | $580 |

$$580 = 0.08x + 0.06x(8000 - x)$$

▶ Play the video and check your answer.

| Review and complete | You try |
|---|---|

**Review and complete**

10. Amanda borrowed a total of $6000 from two different banks. One bank charged 3% simple interest, and a credit union charged 8% simple interest. If the total interest owed after one year was $255, how much money did she borrow from each bank?

|  | 3 % account | 8 % account | Total |
|---|---|---|---|
| Amount of principal | $x$ | $6000 - x$ | $6000 |
| Amount of interest | $0.03x$ | $.08(6000 - x)$ | $255 |

$$255 = 0.03x + 0.08(6000 - x)$$
$$255 = 0.03x + 480 - 0.08x$$
$$255 = -0.05x + 480$$
$$-225 = -0.05x \qquad \text{Divide both sides by -0.05.}$$
$$x = 4500$$

$4500 was borrowed at 3% interest.
$1500 was borrowed at 8% interest.

**You try**

11. Mr. Hall had a total of $9000 in two different banks. One bank was earning 4% simple interest, and at the credit union, he was earning 6% simple interest. If the total interest earned after one year was $400, how much money did he have in each bank?

## Concept 5: Applications Involving Mixtures

### Video 9: An Application Involving Mixtures (3:31 min)

▶ Play the video and work along.

How many liters of a 10% acid solution must be added to 80 liters of a 40% acid solution to make a 30% acid solution?

**40% Solution**

**10% Solution**

**30% Solution**

|  | 40% acid | 10% acid | 30% acid |
|---|---|---|---|
| Amount of solution |  |  |  |
| Amount of pure acid |  |  |  |

### Video 10: Solving an Application Involving Mixtures (2:25 min)

■ Before starting the video, try this yourself.

How many liters of a 10% acid solution must be added to 80 liters of a 40% acid solution to make a 30% acid solution?

$$0.40(80) + 0.10x = 0.30(x + 80)$$

|  | 40% acid | 10% acid | 30% acid |
|---|---|---|---|
| Amount of solution | 80 | $x$ | $x + 80$ |
| Amount of pure acid | $0.40(80)$ | $0.10x$ | $.30(x + 80)$ |

▶ Play the video and check your answer.

| Review and complete | You try |
|---|---|

**Review and complete**

12. How many gal of a 65% acid solution should be mixed with 40 gal a 20% acid solution to produce a 50% acid solution?

| | 65% acid | 20% acid | 50% acid |
|---|---|---|---|
| Amount of solution | $x$ | 40 | $x+40$ |
| Amount of pure acid | $0.65x$ | $0.20(40)$ | $0.50(x+40)$ |

$0.65x + 0.20(40) = 0.50(x+40)$

$0.65x + 8 = .50x + 20$    Subtract $0.50x$ from both sides.

$0.15x + 8 = 20$    Subtract 8 from both sides.

$0.15x = 12$

$x = 80$

**80 gal of the 65% acid solution should be mixed.**

**You try**

13. How much of a 40% saline solution should Bob mix with 10 gal of a 70% saline solution to produce a 45% saline solution?

## Concept 6: Applications Involving Distance, Rate, and Time

### Video 11: An Application of Uniform Motion Distance = (Rate)(Time) (3:06 min)

▶ Play the video and work along.

A car and a bus both leave a rest area on I-95 traveling north. The car travels 6 mph faster than the bus. The car takes 4 hr to get from the rest area to the Jacksonville Airport, and the bus takes 4.4 hr. Find the speed of each vehicle.

| | Distance | Rate | Time |
|---|---|---|---|
| Car | | | |
| Bus | | | |

Airport

Rest area

## Video 12: An Application of Uniform Motion (3:30 min)
### Distance = (Rate)(Time)

▶ Play the video and work along.

Two trains that are 156 mi apart approach a station from opposite directions. One train travels 8 mph slower than the other. If they meet at the station 1.5 hr later,

a. Determine the average speed for each train.
b. Determine the distance that each train traveled.

**156 mi**

|  | Distance | Rate | Time |
|---|---|---|---|
| Faster train |  |  |  |
| Slower train |  |  |  |

|  | Distance | Rate | Time |
|---|---|---|---|
| Faster train |  |  |  |
| Slower train |  |  |  |

**156 mi**

| Review and complete | You try |
|---|---|
| **14.** Two trains that are 190 mi apart approach a station from opposite directions. One train travels 5 mph slower than the other. If they meet at the station 2 hr later, | **15.** Two cars that are 378 mi apart approach each other from opposite directions. One car travels 6 mph slower than the other. If they meet 3 hr later, |

**14.** (continued)

a. Determine the average speed for each train.

Let $x$ = the speed of the faster train.

$d = rt$

|  | Distance | Rate | Time |
|---|---|---|---|
| Faster train | $2x$ | $x$ | 2 |
| Slower train | $2(x-5)$ | $x-5$ | 2 |

$$2x+2(x-5)=190$$

$$2x+2x-10=190$$

$$4x=200$$

$$x=50$$

**The faster train was going at 50 mph.**
**The slower train was going at 45 mph.**

b. Determine the distance that each train traveled.

|  | Distance | Rate | Time |
|---|---|---|---|
| Faster train |  | 50 mph | 2 |
| Slower train |  | 45 mph | 2 |

Since distance = (rate) (time), multiply across to fill in the

The faster train traveled _____.

The slower train traveled _____.

**15.** (continued)

a. Determine the average speed for each car.
b. Determine the distance that each car traveled.

**Concept 1**: Applications Involving Geometry

| Video 1: Solving an Application Involving Geometry (Rectangle) (3:07 min) |
| --- |

▶ Play the video and work along.

The perimeter of a rectangle is 52 ft. If the length is 4 ft less than twice the width, find the length and width.

| Review and complete | You try |
| --- | --- |
| 1. The length of a rectangle is twice as long as it is wide. The perimeter is 114 in. Find the length and width. | 2. The length of a rectangle is 6 in. less than twice the width. The perimeter is 114 in. Find the length and width. |

$P = 2l + 2w$

$114 = 2(2x) + 2x$

$114 = 4x + 2x$

$114 = 6x$

$19 = x$

width
$x$

length
$2x$

Length $= 2x = 2(19) = 38$

The width is 19 in.
The length is 38 in.

| Video 2: Complementary and Supplementary Angles and the Sum of the Angles in a Triangle (2:07 min) |
| --- |

▶ Play the video and work along.

- Determine the complement of a 62° angle.

- Determine the supplement of a 107° angle.

- Determine the measure of the missing angle.

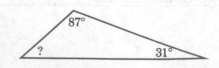

**Video 3: An Application of Complementary and Supplementary Angles (3:48 min)**

▶ Play the video and work along.

a.  Two supplementary angles are drawn such that one angle is 8° more than three times the other angle. Find the measure of each angle.

b.  How would the set up to this problem have changed if the angles were complementary?

**Review and complete**

3.  Two angles are complementary if the sum of their measures is ____.

4.  Two angles are supplementary if the sum of their measures is ____.

5.  The sum of the measures of the angles within a triangle is ____.

**You try**

6.  Two angles are complementary. One angle measures 14° less than 3 times the other. Find the measure of each angle.

7.  Two angles are complementary. If one angle measures 18° more than the other, find the measure of each angle.

$$\underset{\text{angle}}{\text{1st}} + \underset{\text{angle}}{\text{2nd}} = 90°$$

$$x + 3x - 14 = 90$$

$$4x - 14 = 90$$

$$4x = 104$$

$$x = 26$$

The 1st angle, $x$, is 26°.
The 2nd angle, $3x - 14$, is $3(26) - 14 = 64°$.

## Video 4: An Application Involving the Angles within a Triangle (3:01 min)

▶ Play the video and work along.

The measurement of one angle in a triangle is three times the measure of the smallest angle. The third angle is 2° less than ten times the measure of the smallest angle. Find the measure of each angle.

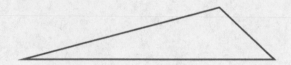

**Review and complete**                    **You try**

8. The smallest angle in a triangle is one-half the size of the largest. The middle angle measures 25° less than the largest. Find the measures of the three angles.

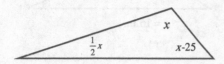

$$\underset{\text{angle}}{\text{smallest}} + \underset{\text{angle}}{\text{middle}} + \underset{\text{angle}}{\text{largest}} = 180°$$

$$\frac{1}{2}x \; + \; x - 25 \; + \; x = 180$$

$$\frac{5}{2}x = 205$$

$$\left(\frac{2}{5}\right) \cdot \frac{5}{2}x = \left(\frac{2}{5}\right) \cdot 205$$

$$x = 82$$

The smallest angle is $\dfrac{1}{2}x = \dfrac{1}{2}(82) = 41.$

The middle angle is
$$x - 25 = (82) - 25 = 57.$$

The 3 angles are 41°, 57°, and 82°.

9. The smallest angle in a triangle is one-third the size of the largest. The middle angle measures 9° less than the largest. Find the measures of the three angles.

## Concept 2: Literal Equations

### Video 5: Introduction to Formulas and Literal Equations (2:37 min)

▶ Play the video and work along.

Determine the perimeter of the triangle with the given lengths of the sides.

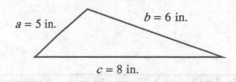

$a = 5$ in.  $b = 6$ in.

$c = 8$ in.

Suppose that the perimeter of a triangle is 32 in., and the lengths of two sides are given. Find the length of the third side.

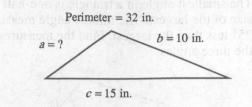

Perimeter = 32 in.

$b = 10$ in.

$a = ?$

$c = 15$ in.

| Review and complete | You try |

10. Write the formula for the perimeter of a triangle.  P = _____

11. Suppose that the perimeter of a triangle is 56 in., and the lengths of two sides are 12 in. and 27 in. Find the length of the third side.

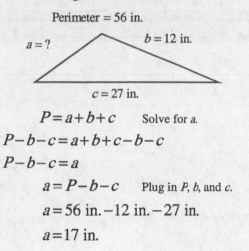

Perimeter = 56 in.

$a = ?$    $b = 12$ in.

$c = 27$ in.

$$P = a + b + c \qquad \text{Solve for } a.$$
$$P - b - c = a + b + c - b - c$$
$$P - b - c = a$$
$$a = P - b - c \qquad \text{Plug in } P, b, \text{ and } c.$$
$$a = 56 \text{ in.} - 12 \text{ in.} - 27 \text{ in.}$$
$$a = 17 \text{ in.}$$

12. Suppose that the perimeter of a triangle is 98 in., and the lengths of two of the sides are 23 in. and 35 in. Find the length of the third side.

Perimeter = 56 in.

$a = ?$

$c = 27$ in.

**Video 6: Solving a Formula for a Given Variable (1:13 min)**

▶ Play the video and work along.

Solve for $R$.    $V = IR$

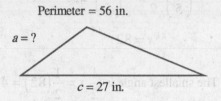

$a = ?$    $b = 12$ in.

$c = 27$ in.

**Video 7: Applying a Formula in an Application (2:10 min)**

▶ Play the video and work along.

The volume, $V$, of a pyramid is given by the formula $V = \dfrac{1}{3}Ah$, where $A$ is the area of the base and $h$ is the height.

a.  Solve the formula $V = \frac{1}{3}Ah$ for $h$.

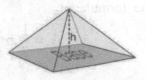

b.  If the volume of a pyramid is 100 in.$^3$ and the area of the base is 60 in.$^2$, determine the height.

| Review and complete | You try |
| --- | --- |

13. The volume, $V$, of a right circular cone is given by the formula

$$V = \frac{1}{3}\pi r^2 h$$

   a.  Solve the formula $V = \frac{1}{3}\pi r^2 h$ for $h$.

$$3 \cdot (V) = \frac{3}{1} \cdot \left( \frac{1}{3}\pi r^2 h \right)$$

$$3V = \pi r^2 h$$

$$\frac{3V}{\pi r^2} = \frac{\pi r^2 h}{\pi r^2}$$

$$\frac{3V}{\pi r^2} = h \quad \Rightarrow \quad h = \frac{3V}{\pi r^2}$$

   b.  If the volume of a right circular cone is $100\pi$ in.$^3$ and the radius is 5 in., what is the height?

$$h = \frac{3V}{\pi r^2}$$

$$h = \frac{3(100\pi)}{\pi(5)^2} \qquad \text{Plug in } V \text{ and } r.$$

$$h = \frac{300}{25}$$

$$h = 12 \text{ in.}$$

14. The area, $A$, of a triangle is given by the formula

$$A = \frac{1}{2}bh$$

   a.  Solve the formula $A = \frac{1}{2}bh$ for $b$.

   b.  If the area of a triangle is 100 in.$^2$ and the height is 50 in., determine the base.

**Video 8: Solving an Equation for a Given Variable (3:29 min)**

▶ Play the video and work along.

Solve for $P$.

$$D = \frac{5}{4}(P - R)$$

$$D = \frac{5}{4}(P - R)$$

| Review and complete | You try |
|---|---|
| 15. Solve for $F$. | 16. Solve for $R$. |
| | $$W = \frac{2}{3}(R - 26)$$ |

$$C = \frac{5}{9}(F - 32)$$

$$\frac{9}{5} \cdot (C) = \frac{9}{5} \cdot \frac{5}{9}(F - 32)$$

Multiply both sides by the reciprocal of 5/9.

$$\frac{9}{5} \cdot (C) = \frac{\cancel{9}^{1}}{\cancel{5}} \cdot \frac{\cancel{5}^{1}}{\cancel{9}}(F - 32)$$

$$\frac{9}{5}C = F - 32$$

$$\frac{9}{5}C + 32 = F$$

$$F = \frac{9}{5}C + 32$$

---

**Video 9: Writing an Equation $Ax + By = C$ in the Form $y = mx + b$ (2:54 min)**

■ Before starting the video, try this yourself.

Solve for $y$.    $3x + 4y = 8$

▶ Play the video and check your answer.

| Review and complete | You try |
|---|---|
| 17. Solve for $y$. | 18. Solve for $y$. |
| | $$-4x - 5y = 25$$ |
| $$5x - 7y = 15$$ | |
| $$-7y = -5x + 15 \quad \text{Subtract 5x from both sides.}$$ | |
| $$\frac{-7y}{-7} = -\frac{5x}{-7} + \frac{15}{-7} \quad \text{Divide every term by -7.}$$ | |
| $$y = \frac{5}{7}x - \frac{15}{7}$$ | |

### Video 10: Solving an Equation for a Given Variable where Factoring is Necessary (4:07 min)

▶ Play the video and work along.

Solve for $x$.

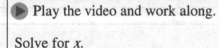

$$mx - 5 = ax + t$$

$$mx - 5 = ax + t$$

| Review and complete | You try |
|---|---|
| 19. Solve for $x$. | 20. Solve for $a$. |

**Review and complete**

19. Solve for $x$.

$$cx - 4 = dx + 9$$

$$cx - dx = 13$$

Get all the $x$ terms on one side and the non-x terms on the other side.

$$x(c - d) = 13$$

Factor out the GCF of $x$.

**KEY STEP**

$$\frac{x(c-d)}{(c-d)} = \frac{13}{(c-d)}$$

$$x = \frac{13}{c-d}$$

**You try**

20. Solve for $a$.

$$4a = 5 - ca$$

---

**Video 11: Recognizing the Equivalent Forms of Algebraic Expressions (2:33)**

▶ Play the video and work along.

Show that $\dfrac{-6}{y-x} = -\dfrac{6}{y-x} = \dfrac{6}{x-y}$

| Review and complete | You try |
|---|---|
| 21. Write two equivalent expressions for<br><br>$$\dfrac{a-b}{7}$$ | |

**Concept 1**: Solving Linear Inequalities

---

### Video 1: Introduction to Linear Inequalities in One Variable (1:47 min)

▶ Play the video and work along.

---

**Definition of a Linear Inequality in One Variable**

Let $a$, $b$, and $c$ be real numbers such that $a \neq 0$. A **linear inequality in one variable** is an inequality that can be written in one of the following forms.

$$ax + b < c \qquad\qquad ax + b \leq c \qquad\qquad ax + b > c \qquad\qquad ax + b \geq c$$

---

Linear equation:

$$x - 3 = 0$$

Linear inequality:

$$x - 3 < 0$$

---

### Video 2: Set-Builder Notation (2:10 min)

▶ Play the video and work along.

Write the solution set in set-builder notation and in interval notation.

$x \geq -1$

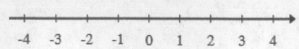

set-builder notation:

interval notation:

$x < 2$

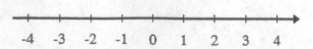

set-builder notation:

interval notation:

$-3 \leq x < 1$

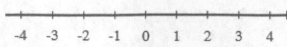

set-builder notation:

interval notation:

---

Video and Study Guide to accompany Intermediate Algebra, 5$^{th}$ ed., Miller/O'Neill/Hyde.
Copyright © 2018 McGraw-Hill Education

**Video 3: Interval Notation (3:56 min)**

▶ Play the video and work along.

Write the solution set in set-builder notation and in interval notation.

$x \geq -1$

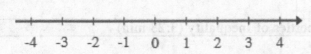

set-builder notation: $\left\{ x \mid x \geq -1 \right\}$

interval notation:

$x < 2$

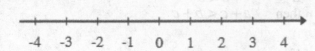

set-builder notation: $\left\{ x \mid x < 2 \right\}$

interval notation:

$-3 \leq x < 1$

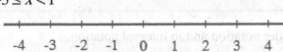

set-builder notation: $\left\{ x \mid -3 \leq x < 1 \right\}$

interval notation:

| Review and complete | You try |
|---|---|

1. Using Interval Notation

- The endpoints in interval notation are always written from left to right. In other words, the smaller number is written first, followed by a comma, followed by the larger number.

- Parentheses are always used with ∞ or −∞.

- Parentheses ) or ( indicate that an endpoint is (**choose one:** included/ not included) in the set.

- Square brackets ] or [ indicate that an endpoint is (**choose one:** included/not included) in the set.

2. Graph the set $x \leq 1$ on the number line and then express it in interval notation.

Set-builder notation: $\{x \mid x \leq 1\}$

Graph: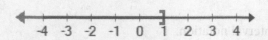

Interval Notation: $\left(-\infty, 1\right]$

3. Graph the set $x > -3$ on the number line and then express it in interval notation.

Set-builder notation:

Graph: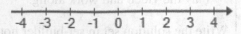

Interval Notation:

## Video 4: Using the Addition and Subtraction Properties of Inequality (3:23 min)

▶ Play the video and work along.

### Addition and Subtraction Properties of Inequality

Let $a$, $b$, and $c$ represent real numbers.

*Addition property of inequality:        If        $a < b$
                                                            then      $a + c < b + c$

*Subtraction property of inequality:     If        $a < b$
                                                            then      $a - c < b - c$

*These properties may also be stated for $a \leq b$, $a > b$, and $a \geq b$.

Solve the inequality. Express the solution set in set-builder notation and in interval notation.

$$6 + 3x - 8 < 2(x + 11)$$

Check:

$$6 + 3x - 8 < 2(x + 11)$$

## Video 5: Multiplication and Division Properties of Inequality (2:57 min)

 Play the video and work along.

### Multiplication and Division Properties of Inequality

Let $a$, $b$, and $c$ represent real numbers.

* If $c$ is positive and $a < b$, then $ac < bc$ and $\dfrac{a}{c} < \dfrac{b}{c}$ $\left( c \neq 0 \right)$.

* If $c$ is negative and $a < b$, then $ac > bc$ and $\dfrac{a}{c} > \dfrac{b}{c}$ $\left( c \neq 0 \right)$.

*These properties may also be stated for $a \leq b$, $a > b$, and $a \geq b$.

Solve the inequality. Express the solution set in set-builder notation and in interval notation.

$4 - 3x \geq 10$

| Review and complete | You try |
|---|---|

**4.** If you multiply or divide both sides of an inequality by a negative number, you (**choose one:** do or do not) reverse the inequality sign.

**5.** If you add or subtract a negative number from both sides of an inequality, you (**choose one:** do or do not) reverse the inequality sign.

| | |
|---|---|
| **6.** Solve the inequality. Express the solution set in set-builder notation and in interval notation. | **7.** Solve the inequality. Express the solution set in set-builder notation and in interval notation. |

**6.** (left column)

$-3w - 6 > 9$    Add 6 to both sides.

$-3w > 15$    Divide both sides by $-3$.

$\dfrac{-3w}{-3} < \dfrac{15}{-3}$    Reverse the inequality sign because we divided by a negative number.

$w < -5$

Set-builder notation: $\{w \mid w < -5\}$

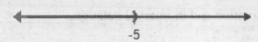

$-5$

Interval notation: _____

**7.** (right column)

$-2h - 4 \geq 6$

---

### Video 6: Solving a Linear Inequality Containing Decimals (3:12 min)

▶ Play the video and work along.

Solve the inequality. Express the solution set in set-builder notation and in interval notation.

$-0.03(x - 8) \leq 0.05x + 0.12$

Check:

Test:

Test:

Video and Study Guide to accompany Intermediate Algebra, 5th ed., Miller/O'Neill/Hyde
Copyright © 2018 McGraw-Hill Education

| Review and complete | You try |
|---|---|
| 8. Solve the inequality. Express the solution set in set-builder notation and in interval notation. | 9. Solve the inequality. Express the solution set in set-builder notation and in interval notation. |

**Review and complete**

8. Solve the inequality. Express the solution set in set-builder notation and in interval notation.

$$6h - 2.92 \leq 16.58$$
$$6h < 19.50 \qquad \text{Add 2.92 to both sides.}$$
$$h \leq 3.25 \qquad \text{Divide both sides by 6.}$$

Set-builder notation: $\{h \mid h \leq 3.25\}$

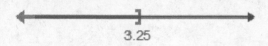

3.25

Interval notation: _____

**You try**

9. Solve the inequality. Express the solution set in set-builder notation and in interval notation.

$$-0.4x + 1.2 < -2$$

---

**Video 7: Solving a Linear Inequality Containing Fractions (2:31 min)**

▶ Play the video and work along.

Solve the inequality. Express the solution set in set-builder notation and in interval notation.

$$\frac{1}{3}x - \frac{5}{4} > \frac{1}{2}x + 1$$

**Concept 2:** Applications of Inequalities

### Video 8: An Application of a Linear Inequality in One Variable (2:20 min)

▶ Play the video and work along.

The average monthly snowfall for December, January, and February for Clarkesville is 22.5 in. Suppose Clarkesville received 15 in. in December and 19 in. in January. How many inches would be necessary in February to exceed the average monthly snowfall for these three months?

| Review and complete | You try |
|---|---|
| 10. Susan received grades of 78, 84, and 45 on her first three algebra tests. To earn at least a C in the course, she needs an average of 70 or more. What scores can she receive on the fourth test to earn at least a C?<br><br>Let x represent the score on the fourth test.<br><br>The average of the four tests $\geq 70$. | 11. Jessica received grades of 90, 94, and 75 on her first three algebra tests. To earn at least a B in the course, she needs an average of 80 or more. What scores can she receive on the fourth test to earn at least a B? |

$$\frac{78+84+45+x}{4} \geq 70$$

$$\cancel{4} \cdot \left(\frac{78+84+45+x}{\cancel{4}}\right) \geq 4 \cdot (70)$$

$$78+84+45+x \geq 280$$

$$x \geq 73$$

To earn at least a C, Susan would have to score at least 73 on her fourth test.

---

## Video 9: An Application Involving Linear Inequalities (1:25 min)

▶ Play the video and work along.

The average height for boys can be approximated based on age. The formula $h=2.5a+31$ represents the average height $h$ (in inches) based on age, $a$ (in years). Determine the age at which the average height for boys will be at least 56 in.

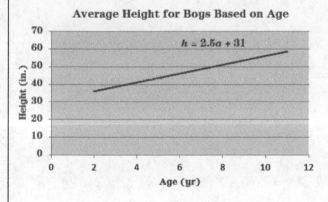

**Average Height for Boys Based on Age**

$h = 2.5a + 31$

| **Review and complete** | **You try** |
| --- | --- |

---

12. The monthly revenue $R$ for selling $x$ custom hand-made journals is $R = 250x$. The cost to produce $x$ journals is $C = 19x + 12,474$. Find the number of journals that Joe needs to sell each month to produce a profit. (*Hint*: A profit occurs when revenue exceeds cost.)

$$\text{Revenue} > \text{Cost}$$

$$250x > 19x + 12,474$$

$$231x > 12,474$$

$$x > 54$$

Joe must sell at least 55 journals a month.

13. The revenue $R$ for selling $x$ mountain bikes is $R = 249.95x$. The cost to produce $x$ bikes is $C = 140x + 56,000$. Find the number of bikes that the company needs to sell to produce a profit.

**Concept 1**: Union and Intersection of Sets

| Video 1: Introduction to Compound Inequalities (2:39 min) |
| --- |

▶ Play the video and work along.

| Definition of A Union B | Definition of A Intersection B |
| --- | --- |
| The **union** of the sets A and B, denoted $A \cup B$, is the set of elements that belong to set A or set B or to both sets A and B. | The **intersection** of the sets A and B, denoted $A \cap B$ is the set of elements common to both A and B. |

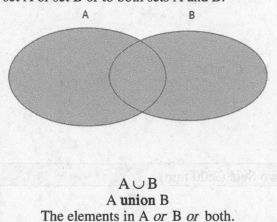

$A \cup B$
A union B
The elements in A *or* B *or* both.

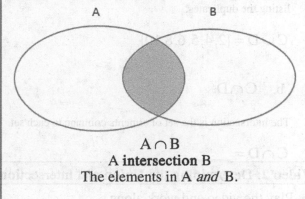

$A \cap B$
A intersection B
The elements in A *and* B.

Given the sets A, B, and C, determine the union or intersection as indicated.

$A = \{2, 4, 6, 8, 10\}$        $B = \{3, 6, 9, 12\}$        $C = \{5, 10, 15\}$

$A \cup B$                    $A \cap B$

⏸

Pause the video and try this yourself.

$B \cap C$

▶

Play the video and check your answer.

| Review and complete | You try |
|---|---|
| 1. Given $C = \{2,4,6,8,10\}$ and $D = \{2,5,6\}$, determine the union or intersection as indicated. <br><br>     a.   $C \cup D$ <br><br> The union includes all of the elments from each set without listing the duplicates. <br><br> $C \cup D = \{2,4,5,6,8,10\}$ <br><br><br>     b.   $C \cap D$ <br><br> The intersection is the set of elments common to each set. <br><br> $C \cap D = \underline{\hphantom{xxxx}}$ | 2. Given $C = \{1,3,5,7,9\}$ and $D = \{1,2,3\}$, find the union or intersection as indicated. <br><br>     a.   $C \cup D$ <br><br><br><br><br>     b.   $C \cap D$ |

## Video 2: Determining the Union and Intersection of Two Sets (2:00 min)

▶ Play the video and work along.

Given the sets B and C, determine the union or intersection as indicated.

$$B = \left\{ x \mid x \leq 4 \right\} \qquad\qquad C = \left\{ x \mid x < 0 \right\}$$

$\longrightarrow$ set B

B∩C

$\longrightarrow$ set C

The intersection of B and C   $\longrightarrow$

⏸ Pause the video and try this yourself.

$\longrightarrow$ set B

B∪C

$\longrightarrow$ set C

The union of B and C   $\longrightarrow$

▶ Play the video and check your answer.

| Review and complete | You try |
|---|---|
| 3. Find $C \cap D$ for the following sets. | 4. Find $C \cup D$ for the following sets. |

3. $C = \{x \mid x \le 3\}$ and $D = \{x \mid x < 1\}$

4. $C = \{x \mid x \le 3\}$ and $D = \{x \mid x < 1\}$

The intersection is the set of elements common to each set.

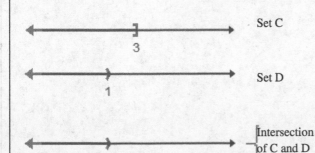

## Video 3: Determining Union and Intersection of Sets Defined in Interval Notation (1:54 min)

▶ Play the video and work along.

Given the sets of real numbers defined by the intervals, find the union or intersection as indicated.

$(-\infty, 2] \cup (-1, \infty)$

⏸ Pause the video and try this yourself.

$(-5, 2] \cap (-1, \infty)$

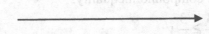

▶ Play the video and check your answer.

| Review and complete | You try |
|---|---|
| 5. Find the intersection of the given sets. Write your answer in interval notation. | 6. Find the intersection of the given sets. Write your answer in interval notation. |

5. $(-6,4)\cap[-1,\infty)$

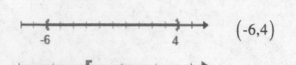 $(-6,4)$

 $[-1,\infty)$

The intersection is the set of elements common to each set.

6. $(-3,4]\cap(2,\infty)$

---

**Concept 2:** Solving Compound Inequalities: And

**Video 4: Solving a Compound Inequality Joined by "And" (2:36 min)**

▶ Play the video and work along.

Solve the compound inequality.

$-4x > 8$ and $-3 + x > -10$

| Review and complete | You try |
|---|---|

7. The solution to two inequalities joined by the word *and* is the (choose one: intersection or union) of their solution sets.
8. The solution to two inequalities joined by the word *or* is the (choose one: intersection or union) of their solution sets.

**Video 5: Solving a Compound Inequality Joined by "And" (3:06 min)**

▶ Play the video and work along.

Solve the compound inequality.

$\frac{1}{2}y - \frac{3}{4} \geq \frac{3}{8}y$

and $14.4 + y < -1.4y$

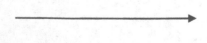

## Video 6: Solving a Compound Inequality Joined by "And" (2:02 min)

■ Before starting the video, try this yourself.

Solve the compound inequality.

$$-\frac{1}{4}d > 1 \qquad \text{and} \qquad \frac{3}{5}d < 3$$

▶ Play the video and check your answer.

| Review and complete | You try |
|---|---|
| 9. Solve the compound inequality. Write the answer in interval notation. | 10. Solve the compound inequality. Write the answer in interval notation. |

**Review and complete (9):**

$$-\frac{1}{6}h > 1 \qquad \text{and} \qquad \frac{2}{3}h \le 2$$

$$-6 \cdot \left(-\frac{1}{6}\right)h < -6 \cdot (1) \qquad \frac{3}{2} \cdot \left(\frac{2}{3}h\right) \le \frac{3}{2} \cdot (2)$$

$$h < -6 \qquad\qquad h \le 3$$

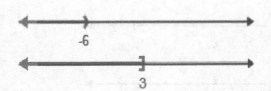

The intersection of $-\frac{1}{6}h > 1$ and $\frac{2}{3}h \le 2$ is

**You try (10):**

$$-\frac{1}{7}w > 1 \qquad \text{and} \qquad \frac{3}{4}w < 3$$

In interval notation, _____

Concept 3: Solving Inequalities of the Form $a < x < b$

### Video 7: Solving an Inequality of the Form $a < x < b$ (3:37 min)

▶ Play the video and work along.

Solve the inequality.

Method 1:    $-2 < 2x - 4 < 10$

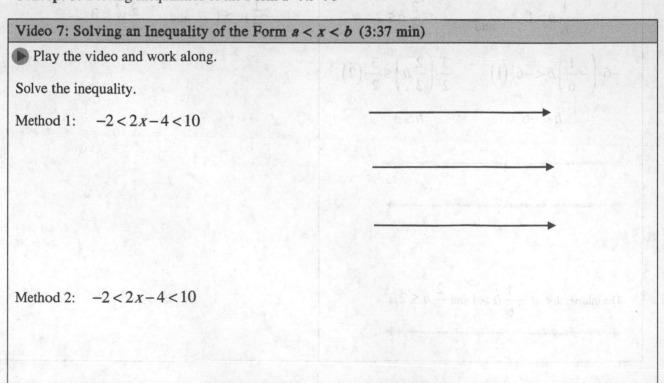

Method 2:    $-2 < 2x - 4 < 10$

---

**Video 8: Solving an Inequality of the Form $a < x < b$** (1:47 min)

■ Before starting the video, try this yourself.

Solve the inequality.    $3 \geq \dfrac{y-4}{-2} > 1$

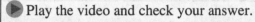

► Play the video and check your answer.

| Review and complete | You try |
|---|---|
| 11. An inequality of the form $a < x < b$ is equivalent to the compound inequality _____ and _____. | |

12. Solve the compound inequality. Write the answer in interval and set builder notation.

$$-1 < \frac{3-x}{2} \le 0$$

$$2 \cdot (-1) < 2 \cdot \left(\frac{3-x}{2}\right) \le 2 \cdot (0)$$

$$2 < 3 - x \le 0$$

$$-5 < -x \le -3$$

$$5 > x \ge 3$$

The intersection of $x \ge 3$ and $x < 5$

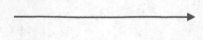

Interval notation: $[3, 5)$

Set builder notation: _____

13. Solve the compound inequality. Write the answer in interval and set builder notation.

$$-1 < \frac{2-x}{3} \le 2$$

**Concept 4:** Solving Compound Inequalities: Or

### Video 9: Solving a Compound Inequality Joined by "Or" (2:11 min)

▶ Play the video and work along.

Solve the compound inequality.

$$-5t + 6 > 26 \qquad \text{or} \qquad 4 < -3 + t$$

### Video 10: Solving a Compound Inequality Joined by "Or" (1:47 min)

■ Before starting the video, try this yourself.

Solve the compound inequality.

$$\frac{1}{2}p > -3 \qquad \text{or} \qquad 0.02p + 0.03 \le 0.05$$

▶ Play the video and check your answer.

**Concept 5:** Applications of Compound Inequalities

| Video 11: Writing an English Phrase as a Compound Inequality (1:18 min) |
| --- |

▶ Play the video and work along.

"Fasting blood sugar" is a lab test usually done six to eight hours after eating. The normal range for this test is 70 to 100 mg/dL (milligrams per deciliter), inclusive.

- Write a compound inequality representing the normal range for a fasting blood sugar test.


- Write a compound inequality representing abnormal levels.


| Video 12: Writing an English Phrase as a Compound Inequality and Solving the Inequality (2:11 min) |
| --- |

▶ Play the video and work along.

- Four more than three times a number is strictly between -5 and 16. Find all such numbers.


⏸ Pause the video and try this yourself.

- One-half of a number is either less than 1 or greater than or equal to 6. Find all such numbers.


▶ Play the video and check your answer.

| Review and complete | You try |
| --- | --- |
| 14. Three less than twice a number is either greater than 15 **or** less than 5. Find all such numbers. | 15. Two more three times a number is either greater than 11 **or** less than -1. Find all such numbers. |
| Let $x$ represent the unknown numbers. $$2x-3>15 \quad \text{or} \quad 2x-3<5$$ $$2x>18 \qquad\qquad 2x<8$$ $$x>9 \qquad\qquad x<4$$ All real numbers less than 4 or greater than 9 | |

Video and Study Guide to accompany Intermediate Algebra, 5ᵗʰ ed., Miller/O'Neill/Hyde

**Concept 1**: Solving Absolute Value Equations

| Video 1: Introduction to Absolute Value Equations (3:18 min) |
| --- |

▶ Play the video and work along.

Solve the equation.

$$|x| = 4$$

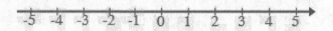

---

**Definition of Absolute Value**

$$|x| = \begin{cases} x & \text{if } x \geq 0 \\ -x & \text{if } x < 0 \end{cases}$$

This means that if $x$ is zero or a positive number, then the absolute value is $x$ itself.

This means that if $x$ is a negative number, then the absolute value is the opposite of $x$.

---

**Procedure for Solutions to Absolute Value Equations**

If $a$ is a nonnegative real number, then the equation $|x| = a$ is equivalent to $x = a$ or $x = -a$.

---

Solve the equation.

$$|x| = 2 \qquad\qquad |x| = 0 \qquad\qquad |x| = -5$$

---

| Review and complete | You try |
| --- | --- |
| 1. If $|x| = 9$, then the pair of equivalent equations are $x =$ _____ or $x =$ _____. | |

| Video 2: Solving an Absolute Value Equation (2:14 min) |
| --- |

▶ Play the video and work along.

Solve.  $|3w + 2| - 5 = 12$

### Video 3: Solving an Absolute Value Equation (3:06 min)

■ Before starting the video, try this yourself.

Solve the equation.  $-3 = -7 + 2|x-9|$

Check: $x = $ _____                               Check: $x = $ _____

$-3 = -7 + 2|x-9|$                               $-3 = -7 + 2|x-9|$

► Play the video and check your answer.

### Video 4: Solving an Absolute Value Equation (1:44 min)

► Play the video and work along.

Solve the equation.  $\dfrac{3}{4} + \left|\dfrac{t}{2} + \dfrac{t}{8}\right| = \dfrac{3}{4}$

### Video 5: Solving an Absolute Value Equation (the Solution Set is the Empty Set) (1:13 min)

■ Before starting the video, try this yourself.

Solve the equation.  $7.3 = 1.2 - |2.3x - 5|$

► Play the video and check your answer.

| Review and complete | You try |
|---|---|
| 2. Solve the equation.  $|w+2| - 3 = 4$ | 3. Solve the equation.  $|x-3| + 5 = 7$ |

| | |
|---|---|
| $\|w+2\|=7$    Isolate the absolute value. <br><br> $w+2=7$ or $w+2=-7$    Rewrite as two equations. <br> $w=5$      $w=-9$    Solve each equation. <br><br> The solution set is $\{-9,5\}$. | |
| **4.** Solve the equation. $\|x+7\|+8=3$ <br><br>     $\|x+7\|=-5$    Isolate the absolute value. <br><br>   An absolute value cannot be negative. <br><br>       No solution, $\{\ \ \}$ | **5.** Solve the equation. $\|m+1\|+3=2$ |

**Concept 2:** Solving Equations Containing Two Absolute Values

---

**Video 6: Solving an Equation Containing Two Absolute Values (3:09 min)**

▶ Play the video and work along.

> **Procedure for Solving an Equation Involving Two Absolute Values**
>
> The equation $\|x\|=\|y\|$ is equivalent to the pair of equations $x=y$ or $x=-y$.

Solve the equation. $\|x+1\|=\|2x-3\|$

---

**Video 7: Solving an Equation Containing Two Absolute Values (1:46 min)**

▶ Play the video and work along.

Solve the equation. $\|x-4\|=\|x+1\|$

---

## Video 8: Solving an Equation Containing Two Absolute Values (2:26 min)

■ Before starting the video, try this yourself.

Solve the equation. $|2y - 4| = |4 - 2y|$

▶ Play the video and check your answer.

| Review and complete | You try |
|---|---|
| 6. Solve the equation. $\|3y + 1\| = \|2y - 7\|$ | 7. Solve the equation. $\|3x + 1\| = \|x - 4\|$ |

Rewrite as two equations.

$$3y + 1 = 2y - 7 \qquad 3y + 1 = -(2y - 7)$$

Solve each equation.

$$y = -8 \qquad\qquad 3y + 1 = -2y + 7$$

$$y = \frac{6}{5}$$

The solution set is $\left\{ -8, \dfrac{6}{5} \right\}$.

**Concept 1**: Solving Absolute Value Inequalities by Definition

**Video 1: Introduction to Absolute Value Inequalities (3:09 min)**

 Play the video and work along.

Write the solution set.

$|x| = 2$

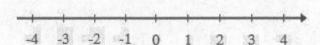

$|x| < 2$

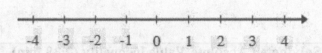

$|x| > 2$

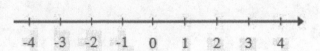

---

**Definition of Absolute Value Inequality**

Suppose that *a* represents a nonnegative real number. Then,

*1. The inequality $|x| < a$ is equivalent to the compound inequality: $-a < x < a$

*2. The inequality $|x| > a$ is equivalent to the compound inequality: $x < -a$ or $x > a$

*Both cases can also be stated using the inequality symbols $\leq$ or $\geq$.

---

Video and Study Guide to accompany Intermediate Algebra, 5ᵗʰ ed., Miller/O'Neill/Hyde
Copyright © 2018 McGraw-Hill Education

| Video 2:   Solving an Absolute Value Inequality (1:39 min) |
| --- |

▶ Play the video and work along.

Write the solution set.   $|t-4|+1<6$   ⟶

| Video 3: Solving an Absolute Value Inequality (2:38 min) |
| --- |

▶ Play the video and work along.

Write the solution set.      $13 \le 10 + \left|\dfrac{x}{2}\right|$   ⟶

| Review and complete | You try |
| --- | --- |
| 1.   What is the first step to solve the absolute value inequality $|x|+3>9$? _____ _____ | |
| 2. The solution set to the equation $|x+3|=-1$ is _____. | |
| 3. Write the solution set for $|2m-7|+4<5$ in interval notation.<br><br>   $|2m-7|<1$   Isolate the absolute value first.<br><br>   $-1<2m-7<1$   Rewrite in the equivalent form $-a<x<a$.<br>   $6<\ 2m\ <8$<br>   $3<\ \ m\ \ <4$   Solve for $m$.<br><br>The solution set is $(3,4)$. | 4. Write the solution set for $|x-3|+5<7$ in interval notation. |

5. Write the solution set for $|x-2|+3>9$ in interval notation.

$$|x-2|+3-3>9-3 \qquad \text{Isolate the absolute}$$
$$|x-2|>6 \qquad \text{value.}$$

$$x-2<-6 \text{ or } x-2>6 \qquad \text{Rewrite in the}$$
$$x<-4 \text{ or } \qquad x>8 \qquad \text{equivalent form.}$$

The solution set is $(-\infty,-4)\cup(8,\infty)$.

6. Write the solution set for $|2x-1|-4\geq-2$ in interval notation.

## Video 4: Solving Absolute Value Inequalities with "Special Case" Solution Sets (1:51 min)

▶ Play the video and work along.

Write the solution set.

$$|4x+7|+10>9 \qquad\qquad -2|3.5-x|\geq8.8$$

| Review and complete | You try |
|---|---|

7. The solution set to the equation $|x+3|<-1$ is _____.

8. The solution set to the equation $|x+3|>-1$ is _____.

## Video 5: Solving Absolute Value Inequalities with "Special Case" Solution Sets (2:45 min)

▶ Play the video and work along.

Write the solution set.

$$|x-5|\leq0 \qquad\qquad\qquad |x+1|>0$$

| Review and complete | You try |
|---|---|
| 9.  Write the solution set for $|x-4| > 0$ in interval and set builder notation. | 10. Write the solution set for $|x+8| > 0$ in interval and set builder notation. |

**9.** Write the solution set for $|x-4| > 0$ in interval and set builder notation.

$$|x-4| = 0$$
$$x-4 = 0$$
$$x = 4$$

An absolute value will be greater than zero at all points except where it is equal to zero. That is, the point(s) for which $|x-4| = 0$ must be excluded from the solution set.

Therefore, exclude $x = 4$ from the solution set.

Solution set: $\{x \mid x \neq 4\}$ or $(-\infty, 4) \cup (4, \infty)$

**Concept 2:** Solving Absolute Value Inequalities by the Test Point Method

| Video 6:  Solving an Absolute Value Inequality by Using the Test Point Method (4:50 min) |
|---|

▶ Play the video and work along.

Write the solution set.  $|x+3| - 4 \geq 1$

| | |
|---|---|
| **Step 1:** | Isolate the absolute value. |
| **Step 2:** | Solve the related equation. |
| **Step 3:** | Place the boundary points on the number line. |
| **Step 4:** | Substitute test points from each interval into the original inequality. Intervals that make the inequality true are part of the solution set. |
| **Step 5:** | Write the solution set in interval notation and in set-builder notation. |

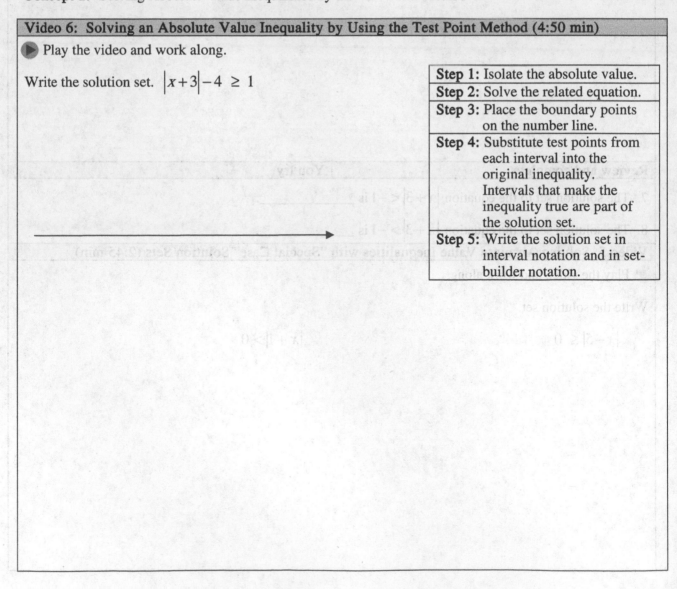

## Video 7: Solving an Absolute Value Inequality by Using the Test Point Method (2:03 min)

■ Before starting the video, try this yourself.

Write the solution set.  $6 > 2|4 + x|$

⟶

Test: $x =$                Test: $x =$                Test: $x =$

Play the video and check your answer.

| Review and complete | You try |
|---|---|
| 11. Write the solution set for $7\|y+1\|-3\geq11$ in interval and set builder notation. | 12. Write the solution set for $3\|x-1\|-2\geq10$ in interval notation. |

$\|y+1\|\geq2$     Isolate the absolute value.

$\|y+1\|=2$     Solve the related equation.

$y+1=2$ or $y+1=-2$    Write as two equations.

$y=1$        $y=-3$

Place the boundary points on the number line and choose a test point from each interval.

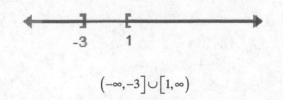

Interval I    Interval II    Interval III

-4   -3     0   1   2

Test: $y=-4$      Test: $y=0$      Test: $y=2$

$7\|(-4)+1\|-3\overset{?}{\geq}11$   $7\|(0)+1\|-3\overset{?}{\geq}11$   $7\|(2)+1\|-3\overset{?}{\geq}11$

$7\|-3\|-3\geq11$    $7\|1\|-3\geq11$    $7\|3\|-3\geq11$

$7(3)-3\geq11$    $7(1)-3\geq11$    $7(3)-3\geq11$

$21-3\geq11$     $7-3\geq11$     $21-3\geq11$

$18\geq11$      $4\geq11$      $18\geq11$

true         false        true

-3     1

$(-\infty,-3]\cup[1,\infty)$

**Concept 3:** Translating to an Absolute Value Expression

---

**Video 8: Using Absolute Value to Express Distance on a Number Line (3:03 min)**

▶ Play the video and work along.

Determine the distance between 3 and -1 on the number line.

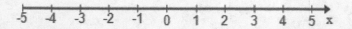

Write an absolute value expression to represent the following.

All real numbers, $x$, whose distance from zero is less than 4 units.

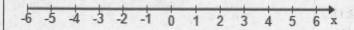

⏸ Pause the video and try this yourself.

All real numbers, $x$, whose distance from -4 is at least 2 units.

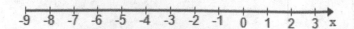

## Video 9: Using Absolute Value to Represent Measurement Error (1:27 min)

▶ Play the video and work along.

A chemist measures 40 grams of a compound on a scale. The scale is accurate to $\pm 0.2$ grams. Write an abolute value inequality to express an interval for the true mass of the compound.

Let $x$ represent the true mass of the compound.

| Review and complete | You try |
|---|---|
| 13. The length of a board is measured to be 32.3 in. where the maximum measurement error is $\pm 0.2$ in. Write an absolute value inequality that represents the range for the length of the board, $x$.<br><br>$$-0.2 \leq x - 32.3 \leq 0.2$$<br><br>$$\lvert x - 32.3 \rvert \leq 0.2$$ | 14. A 16-oz jug of orange juice may not contain exactly 16 oz of juice. The possibility of measurement error exists when the jug is filled in the factory. If the maximum measurement error is $\pm 0.025$ oz, write an absolute value inequality representing the range of volumes, $x$, in which the orange juice jug may be filled. |

**Answers:** Section 1.1

1. $ax + by = c$    3a.  nonlinear    3b. linear    4. $a + c = b + c$    5. $a - c = b - c$

7. $\{30\}$    8. $a \cdot c = b \cdot c$    9. $\dfrac{a}{c} = \dfrac{b}{c}$    11. $\left\{-\dfrac{10}{11}\right\}$    13. $\{5\}$

15. $\{17\}$    17. $\{6\}$    19. $\{-14\}$    21. $\{50\}$    23. **All Reals**

**Answers:** Section 1.2

2. 34 and 59    3. $x + 1, x + 2$    4. $x + 2$    5. $x + 2, x + 4$    7. 61, 62, 63

9. $47.88    11. $7000 at 4% and $2000 at 6%    13. 50 gal

14b. The faster train traveled 100 mi, and the slower train traveled 90 mi.

15a. The faster car traveled 66 mph, and the slower car traveled 60 mph.
15b. The faster car traveled 198 mi, and the slower car traveled 180 mi.

**Answers:** Section 1.3

2. width = 21 in., length = 36 in.    3. 90°    4. 180°    5. 180°

7. The angles are 36° and 54°.    9. The angles are 27°, 72°, and 81°.    10. $P = a + b + c$

12. 40 in.    14a. $b = \dfrac{2A}{h}$    14b. 4 in.    16. $R = \dfrac{3}{2}W + 26$

18. $y = -\dfrac{4}{5}x - 5$    20. $a = \dfrac{5}{4 + c}$    21. $\dfrac{a - b}{7} = -\dfrac{b - a}{7} = \dfrac{b - a}{-7}$

**Answers:** Section 1.4

1. not included; included

3. Graph: ; Set-builder notation: $\{x \mid x > -3\}$, Interval Notation: $[-3, \infty)$

4. do    5. do not    6. $(-\infty, -5)$    7. Set-builder notation: $\{h \mid h \le -5\}$, Interval Notation: $(-\infty, -5]$

Video and Study Guide to accompany Intermediate Algebra, 5th ed., Miller/O'Neill/Hyde
Copyright © 2018 McGraw-Hill Education

8. $\left(-\infty, 3.25\right]$     9. Set-builder notation: $\left\{x \mid x > 8\right\}$, Interval Notation: $(8, \infty)$

11. To earn at least a B, Jessica would have to score at least a 61 on her fourth test.

13. The company must sell at least 509 bikes to produce a profit.

**Answers**: Section 1.5 1b. $\{2,6\}$    2a. $C \cup D = \{1,2,3,5,7,9\}$    2b. $C \cap D = \{1,3\}$

4.

5. $\left[-1,4\right)$    6. $\left(2,4\right]$    7. intersection    8. union    9. $(-\infty,-6)$    10. $w < -7$, $w < 4$, $(-\infty,-7)$

11. $a < x;\ x < b$    12. $\left\{x \mid 3 \le x < 5\right\}$

13. $5 > x \ge -4$ ; Interval notation: $\left[-4,5\right)$ ; Set builder notation: $\left\{x \mid -4 \le x < 5\right\}$

15. Any real number less than -1 or greater than 3.

**Answers**: Section 1.6

1. 9,-9    3. $\{1,5\}$    5. $\{\ \ \}$    7. $\left\{\dfrac{3}{4}, -\dfrac{5}{2}\right\}$

**Answers**: Section 1.7

1. Subtract 3 from both sides.    2. $\{\ \ \}$    4. $(1,5)$    6. $\left(-\infty, -\dfrac{1}{2}\right] \cup \left[\dfrac{3}{2}, \infty\right)$    7. $\{\ \ \}$

8. $(-\infty, \infty)$    10. $(-\infty,-8) \cup (-8,\infty)$    12. $(-\infty,-3] \cup [5,\infty)$    14. $\left| x - 16 \right| \le 0.025$

**Concept 1**: The Rectangular Coordinate System

| Video 1: Representing Data Graphically (1:24 min) |
| --- |

▶ Play the video and work along.

Pam records the height of a bean plant after germination. The height (in cm) recorded at noon on day *x* is given in the table.

| Day | Height (cm) |
| --- | --- |
| 1 | 2 |
| 4 | 12.5 |
| 7 | 21 |
| 10 | 29.5 |
| 13 | 36.5 |
| 16 | 41 |
| 19 | 44 |
| 22 | 46 |

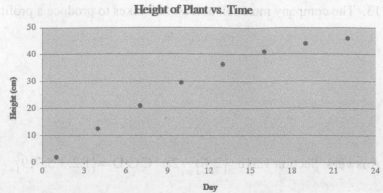

- Approximate the height of the plant on day 17.

- When did the height of the plant reach 21 cm?

| Review and complete | You try |
| --- | --- |
| 1. Use the table and graph from Video 1.<br><br>  a.  Approximate the height of the plant on day 6.<br><br>    On day 4 the height is 12.5 cm, and on day 7 the height is 21 cm. This is an 8.5 cm change in 3 days or about a 2.8 change in 1 day. On day 6, the height is approximately 18 cm. The following chart supports our answer.<br><br>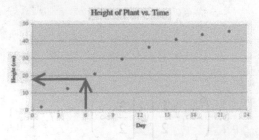<br><br>  b.  When did the height of the plant reach 41 cm?<br><br>    The height the plant reaches can be determined by using the table or graph. On day 16, the plant reached 41 cm. | 2. Using the table and graph from Video 1.<br><br>  a.  Approximate the height of the plant on day 12.<br><br><br><br>  b.  When did the height of the plant reach 44 cm? |

## Video 2: Introduction to a Rectangular Coordinate System (1:17 min)

▶ Play the video and work along.

Rectangular Coordinate System

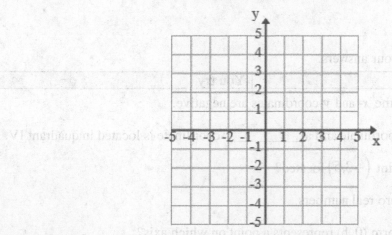

| Review and complete | You try |
|---|---|

3. In a rectangular coordinate system, two number lines are drawn at right angles to each other. The

horizontal line is called the _____-axis, and the vertical line is called the _____.

4. The x- and y-axes divide the coordinate plane into four regions called _____.

## Video 3: Plotting Points (2:32 min)

▶ Play the video and work along.

Plot the points.

$A(2,-4)$     $B(-4,2)$

$C\left(-\dfrac{9}{4},-3.1\right)$

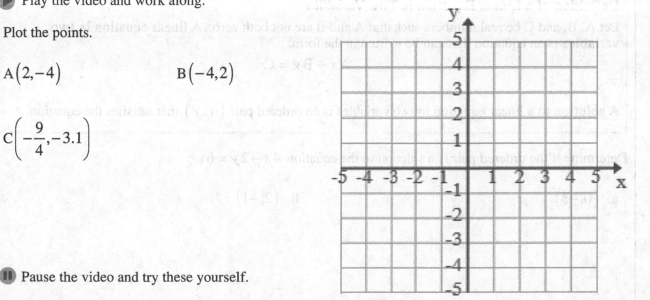

⏸ Pause the video and try these yourself.

$D\left(\sqrt{11},1\right)$

$E(0,5)$                               $F(-3,0)$

▶ Play the video and check your answers.

| Review and complete | You try |
|---|---|
| 5. In quadrant _____, both the *x*- and *y*-coordinates are negative. | |
| 6. A point with a positive *x*-coordinate and a _____ *y*-coordinate is located in quadrant IV. | |
| 7. In which quadrant is the point $(-3,5)$ located? | |
| 8. Let *a* and *b* represent nonzero real numbers. | |
|    a.  An ordered pair of the form (0, b) represents a point on which axis? _____ | |
|    b.  An ordered pair of the form (a, 0) represents a point on which axis? _____ | |

**Concept 2:**  Linear Equations in Two Variables

**Video 4: Solutions to Linear Equations in Two Variables (2:46 min)**

▶ Play the video and work along.

> **Definition of a Linear Equation in Two Variables**
>
> Let A, B, and C be real numbers such that A and B are not both zero.  A **linear equation in two variables** is an equation that can be written in the form:
>
> $$Ax + By = C$$
>
> A **solution** to a linear equation in two variables is an ordered pair $(x, y)$ that satisfies the equation.

Determine if the ordered pair is a solution to the equation $4x - 2y = 6$.

a.  $(0,-3)$                          b.  $(2,-1)$

⏸ Pause the video and try this yourself.

c. $\left(\dfrac{1}{2}, -2\right)$

▶ Play the video and check your answer.

| Review and complete | You try |
|---|---|
| 9. Determine if the ordered pair $(0, -4)$ is a solution to the linear equation | 10. Determine if the ordered pair $(0, 3)$ is a solution to the linear equation |
| $$y = -\dfrac{3}{2}x - 4$$ | $$-5x - 2y = 6$$ |
| $-4 = -\dfrac{3}{2}(0) - 4$   Substitute 0 for $x$ and -4 for $y$. | |
| $-4 = -4$ ✓ true | |
| So, $(0, -4)$ is a solution to $y = -\dfrac{3}{2}x - 4$. | |

## Video 5: Solutions to Linear Equations in Two Variables (2:17 min)

▶ Play the video and work along.

Find several solutions to the equation $2x + y = 4$.
Then graph the points.

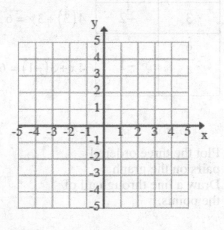

**Concept 3:** Graphing Linear Equations in Two Variables

## Video 6: Graphing a Linear Equation by Using a Table of Points (3:40 min)

Play the video and work along.

Graph the equation $-4x + 2y = 6$.

| x | y |
|---|---|
|   |   |
|   |   |
|   |   |

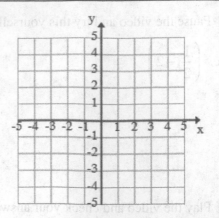

| Review and complete | You try |
|---|---|

11. Complete the table and then graph the equation $4x + 3y = 6$.

| x | y |
|---|---|
|   | 2 |
| 3 |   |
|   | -1 |

Substitute the given value into the equation to find the missing value of the ordered pair.

| x | y |  |
|---|---|---|
| 0 | 2 | $4x + 3(2) = 6 \Rightarrow x = 0$ |
| 3 | -2 | $4(3) + 3y = 6 \Rightarrow y = -2$ |
| $\frac{9}{4}$ | -1 | $4x + 3(-1) = 6 \Rightarrow x = \frac{9}{4}$ |

Plot the three ordered pairs on the graph. Draw a line through all of the points.

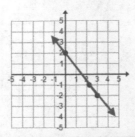

12. Complete the table and then graph the equation $2x - 3y = 12$.

| x | y |
|---|---|
| 0 |   |
| 3 |   |
|   | -6 |

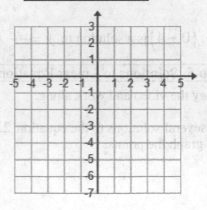

**Video 7: Graphing a Linear Equation by Using a Table of Points (2:25 min)**

■ Before starting the video, try this yourself.

Graph the equation $y = \dfrac{2}{5}x + 2$.

| x | y |
|---|---|
|   |   |
|   |   |

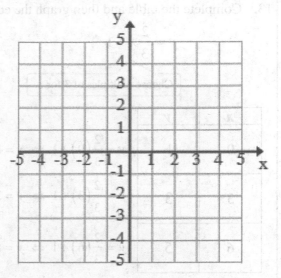

| **Review and complete** | **You try** |
|---|---|

Video and Study Guide to accompany Intermediate Algebra, 5ᵗʰ ed., Miller/O'Neill/Hyde
Copyright © 2018 McGraw-Hill Education

13. Complete the table and then graph the equation.

$$y = \frac{2}{3}x + 1$$

Choose multiples of 3 for $x$.

| x | y |
|---|---|
| 0 | 1 |
| 3 | 3 |
| 6 | 5 |

$y = \frac{2}{3}(0) + 1 \Rightarrow y = 1$

$y = \frac{2}{3}(3) + 1 \Rightarrow y = 3$

$y = \frac{2}{3}(6) + 1 \Rightarrow y = 5$

Plot the three ordered pairs on the graph.
Draw a line through all of the points.

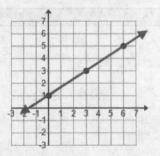

14. Complete the table and then graph the equation

$$y = \frac{1}{3}x - 4$$

| x | y |
|---|---|
| 0 | |
| 3 | |
| -3 | |

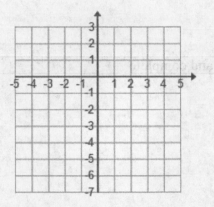

## Concept 4: x- and y-Intercepts

### Video 8: Introduction to x- and y-Intercepts (2:51 min)

▶ Play the video and work along.

Determine the x- and y-intercepts.

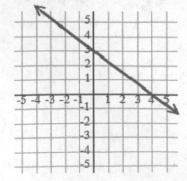

**Definition of x- and y-Intercepts**

An **x-intercept** is a point $(a, 0)$ where a graph intersects the x-axis.

A **y-intercept** is a point $(0, b)$ where a graph intersects the y-axis.

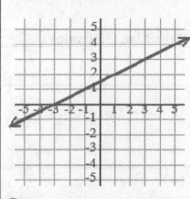

*x*-intercept:

*y*-intercept:

**Ⅱ** Pause the video and try these yourself.

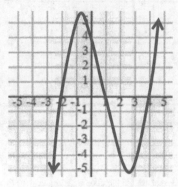

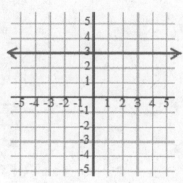

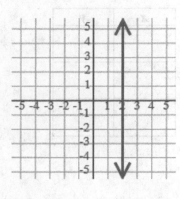

*x*-intercept:                     *x*-intercept:                     *x*-intercept:

*y*-intercept:                     *y*-intercept:                     *y*-intercept:

▶ Play the video and check your answers.

| Review and complete | You try |
|---|---|
| 15. Find the *x*-intercept and *y*-intercept of the graph. 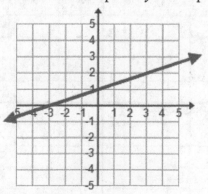 The *x*-intercept is (−3,0). | 16. Find the *x*-intercept and *y*-intercept of the graph. 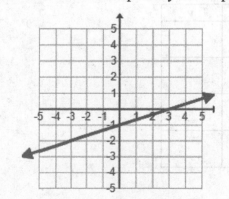 |

The *y*-intercept is _____.

## Video 9: Finding the *x*- and *y*-Intercepts of a Line Given an Equation of the Line (3:10 min)

▶ Play the video and work along.

### Procedure for Determining the *x*- and *y*-Intercepts from an Equation

To find the *x*-intercept(s), substitute 0 for *y* and solve for *x*.

To find the *y*-intercept(s), substitute 0 for *x* and solve for *y*.

Determine the *x*- and *y*-intercepts and graph the equation.   $2x - y = 5$

| *x* | *y* |
|---|---|
|  |  |
|  |  |
|  |  |

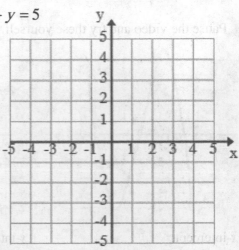

## Video 10: Determining the *x*- and *y*-Intercepts and Graphing a Line that Passes Through the Origin (2:23 min)

▶ Play the video and work along.

Determine the *x*- and *y*-intercepts and graph the equation.   $y = \dfrac{1}{4}x$

| *x* | *y* |
|---|---|
|  |  |
|  |  |
|  |  |

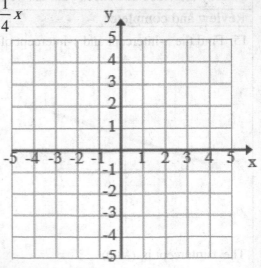

| Review and complete | You try |
|---|---|
| 17. The ordered pair $(5,0)$ is a (**choose one:** *x*-intercept or *y*-intercept). | |
| 18. The ordered pair $(0,4)$ is a (**choose one:** *x*-intercept or *y*-intercept). | |

| | |
|---|---|
| 19. Determine the *x*- and *y*-intercepts and graph the equation $$2x-5y=10$$ | 20. Determine the *x*- and *y*-intercepts and graph the equation $$3x-2y=6$$ |

**x-intercept:**

To find the *x*-intercept, substitute 0 for y and solve for x.

$$2x-5(0)=10$$
$$2x=10$$
$$x=5$$

The x-intercept is $(5,0)$.

**y-intercept:**

To find the *y*-intercept, substitute 0 for x and solve for y.

$$2(0)-5y=10$$
$$-5y=10$$
$$y=-2$$

The y-intercept is $(0,-2)$.

Plot both intercepts and draw a line through the points.

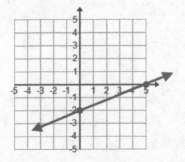

## Video 11: An Application Involving the *x*- and *y*-Intercept (1:15 min)

▶ Play the video and work along.

Joanne borrowed money from her father to buy a new computer for school. She promised to pay the money back in equal monthly payments. The graph shows the amount Joanne still owes, *x* months after borrowing the money.

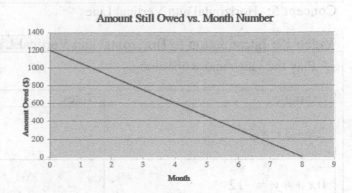

a.  Determine the *y*-intercept and interpret its value in the context of the problem.

b.  Determine the *x*-intercept and interpret its value in the context of the problem.

| Review and complete | You try |
|---|---|

21. Lizette borrowed money from the bank to buy a new car. She promised to pay the money back in equal monthly payments. The graph shows the amount she still owes, *x* months after borrowing the money.

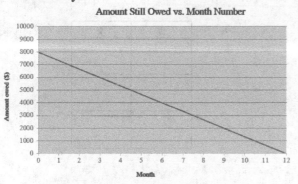

a.  Determine the *y*-intercept and interpret its value in the context of the problem.

The *y*-intercept $(0, 8000)$ indicates Lizette borrowed $8000 for the car.

b.  Determine the *x*-intercept and interpret its value in the context of the problem.

The *x*-intercept $(12, 0)$ indicates that when the car is 12 months old, Lizette will owe no more money to the bank.

22. Amber borrowed money from a friend to buy a new dress. She promised to pay the money back in equal monthly payments. The graph shows the amount she still owes, *x* months after borrowing the money.

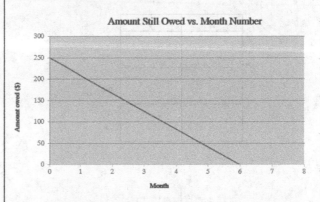

a.  Determine the *y*-intercept and interpret its value in the context of the problem.

b.  Determine the *x*-intercept and interpret its value in the context of the problem.

**Concept 5:** Horizontal and Vertical Lines

### Video 12: Introduction to Horizontal and Vertical Lines (2:04 min)

▶ Play the video and work along.

| Example | Notes |
|---|---|
| $2x + 3y = 6$ | |
| $0x + 4y = -12$ | |
| $2x + 0y = 8$ | |

### Video 13: Graphing Horizontal and Vertical Lines (3:11 min)

▶ Play the video and work along.

Graph the equations.

$y = -3$

| x | y |
|---|---|
| | |
| | |
| | |

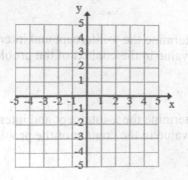

$x = 4$

| x | y |
|---|---|
| | |
| | |
| | |

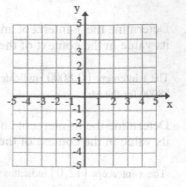

| Review and complete | You try |
|---|---|
| 23. Graph the equation $x = 2$. Identify the $x$- and $y$-intercepts. | 24. Graph the equation $y = 1$. Identify the $x$- and $y$-intercepts. |

**Review and complete**

23. Graph the equation $x = 2$. Identify the $x$- and $y$-intercepts.

Because the equation is in the form $x = k$, the line is **vertical and must cross the** $x$-axis at $x = 2$.

We can also construct a table of solutions to the equation $x = 2$. The choice for the $x$-coordinate must be 2, but $y$ can be any real number.

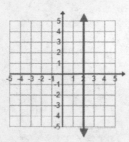

**$x$-intercept:** $(2, 0)$

**$y$-intercept:** none

**You try**

24. Graph the equation $y = 1$. Identify the $x$- and $y$-intercepts.

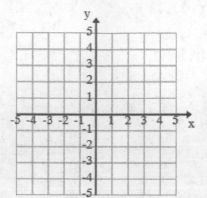

**Concept 1**: Introduction to the Slope of a Line

| Video 1: Introduction to Slope (2:36 min) |
|---|

▶ Play the video and work along.

Determine the slope of a wheelchair ramp.

$$\text{slope} = \frac{\text{vertical change}}{\text{horizontal change}}$$

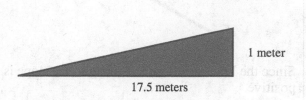

1 meter

17.5 meters

| Video 2: Orientation of a Line and the Sign of the Slope (3:53 min) |
|---|

▶ Play the video and work along.

positive slope

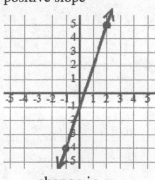

$$m = \frac{\text{change in } y}{\text{change in } x}$$

negative slope

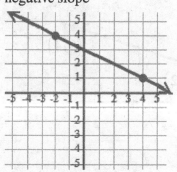

$$m = \frac{\text{change in } y}{\text{change in } x}$$

zero slope

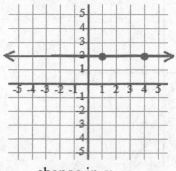

$$m = \frac{\text{change in } y}{\text{change in } x}$$

undefined slope

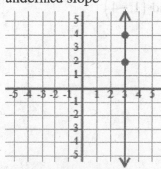

$$m = \frac{\text{change in } y}{\text{change in } x}$$

| Review and complete | You try |
|---|---|
| 1. Specify if the line has a positive, negative, zero, or undefined slope.<br><br><br><br>Since the line rises from left to right, the slope is positive. | 2. Specify if the line has a positive, negative, zero, or undefined slope.<br><br> |

**Concept 2:** The Slope Formula

### Video 3: Introduction to the Slope Formula (3:50 min)

▶ Play the video and work along.

Find the slope of the line containing the points $(3,2)$ and $(9,4)$.

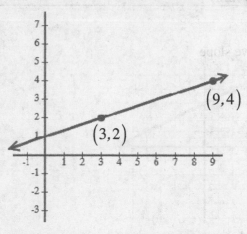

### Video 4: Using the Slope Formula (1:37 min)

■ Before starting the video, try this yourself.

Determine the slope of the line containing the points $(-1,0)$ and $(1,-4)$ .

$$m = \frac{y_2 - y_1}{x_2 - x_1}$$

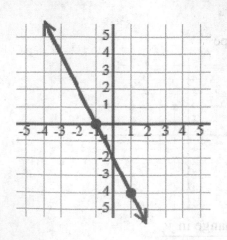

▶ Play the video and check your answer.

---

**Video 5: Applying the Slope Formula in which the Coordinates of the Given Points are Fractions**
  **(2:45 min)**

▶ Play the video and work along.

Determine the slope of the line containing the points $\left(\dfrac{3}{4}, -\dfrac{1}{4}\right)$ and $\left(\dfrac{5}{6}, \dfrac{5}{4}\right)$.

$$m = \dfrac{y_2 - y_1}{x_2 - x_1}$$

---

| **Review and complete** | **You try** |
|---|---|
| 3. Determine the slope of the line containing the points $(4, -2)$ and $(6, -8)$. | 4. Determine the slope of the line containing the points $(-5, -4)$ and $(1, -7)$. |

Using the formula for slope: $m = \dfrac{y_2 - y_1}{x_2 - x_1}$

$$m = \dfrac{-8 - (-2)}{6 - 4} = \dfrac{-8 + 2}{6 - 4} = \dfrac{-6}{2} = -3$$

The slope is $-3$.

---

**Video 6: Estimating the Slope of a Line from the Graph of the Line (1:43 min)**

▶ Play the video and work along.

Determine the slope of the line shown.

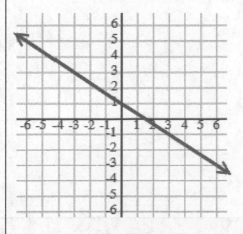

---

| Review and complete | You try |
|---|---|
| **5. Determine the slope of the line shown.** | **6. Determine the slope of the line shown.** |

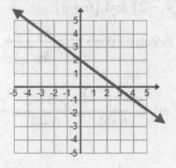

Choose 2 points on the graph.

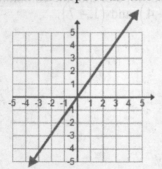

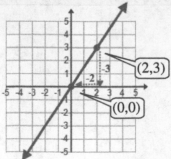

Moving from the point (2,3) to (0,0), there is a vertical change of -3 units and a horizontal change of -2 units.

$$m = \frac{-3}{-2} = \frac{3}{2}$$

**Video 7: Determining the Slope of a Vertical Line (1:18 min)**

▶

Play the video and work along.

Determine the slope of the line containing the points $(-5,4)$ and $(-5,2)$.

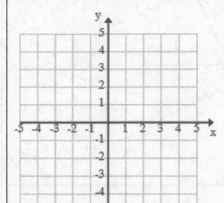

$$m = \frac{y_2 - y_1}{x_2 - x_1}$$

## Video 8: Determining the Slope of a Horizontal Line (1:42 min)

■ Before starting the video, try this yourself.

Determine the slope of the line containing the points $(4.2, 3.1)$ and $(-2.2, 3.1)$.

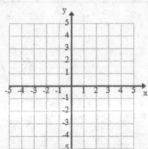

$$m = \frac{y_2 - y_1}{x_2 - x_1}$$

▶ Play the video and check your answer.

| Review and complete | You try |
|---|---|
| 7. Determine the slope of the line containing the points $(4, -2)$ and $(6, -2)$.  $$m = \frac{y_2 - y_1}{x_2 - x_1} = \frac{-2 - (-2)}{6 - 4} = \frac{-2 + 2}{6 - 4} = \frac{0}{2} = 0$$  The slope is 0. | 8. Determine the slope of the line containing the points $(2, -3)$ and $(2, -5)$. |

**Concept 3:** Parallel and Perpendicular Lines

## Video 9: Slopes of Parallel and Perpendicular Lines (2:59 min)

▶ Play the video and work along.

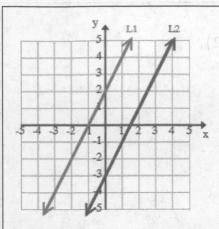
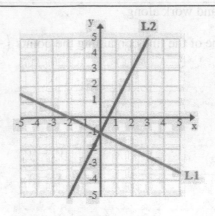

### Slopes of Parallel and Perpendicular Lines

Let $m_1$ and $m_2$ represent the slopes of two nonvertical lines.

- If the lines are parallel, then _____

- If the lines are perpendicular then _____

### Video 10: Identifying Slopes of Parallel and Perpendicular Lines (1:06 min)

▶Play the video and work along.

Suppose a given line has a slope of -2.

    a.   Find the slope of a line parallel to the given line.

    b.   Find the slope of a line perpendicular to the given line.

| Review and complete | You try |
|---|---|
| 9. Suppose a line is defined by the equation $x = 2$. The slope of a line perpendicular to this line is _____. | |
| 10. Suppose a line is defined by the equation $y = 3$. The slope of a line perpendicular to this line is _____. | |
| 11. A line has slope $m = 3$. The slope of a line perpendicular to this line is _____. | |

### Video 11: Determine if Two Lines are Parallel, Perpendicular, or Neither, Based on Two Points from Each Line (1:50 min)

▶ Play the video and work along.

In the following examples, information is given about two different lines. Use the slopes to determine if the lines are parallel, perpendicular, or neither.

- The slopes of the lines are $m_1 = 4$ and $m_2 = -\dfrac{1}{4}$, respectively.

- The slopes of the lines are $m_1 = \dfrac{1}{3}$ and $m_2 = \dfrac{2}{6}$, respectively.

- The slopes of the lines are $m_1 = \dfrac{3}{4}$ and $m_2 = \dfrac{4}{3}$, respectively.

---

**Video 12: Determine if Two Lines are Parallel, Perpendicular, or Neither, Based on Two Points from Each Line (2:53 min)**

▶ Play the video and work along.

Lines $L_1$ and $L_2$ pass through the points given. Determine if $L_1$ and $L_2$ are parallel, perpendicular, or neither without actually graphing the points or the lines.

$L_1$: $(1,4)$ and $(-1,-2)$          $L_2$: $(0,1)$ and $(3,2)$

Line 1:                                                 Line 2:

$m = \dfrac{y_2 - y_1}{x_2 - x_1}$          $m = \dfrac{y_2 - y_1}{x_2 - x_1}$

| Review and complete | You try |
|---|---|
| 12. Lines 1 and 2 pass through the points given. Determine if the lines are parallel, perpendicular, or neither without actually graphing the points or the lines.<br><br>$L_1$: $(3,5)$ and $(2,5)$    $L_2$: $(2,4)$ and $(0,4)$<br><br>Line 1:                Line 2:<br>$m = \dfrac{y_2 - y_1}{x_2 - x_1}$        $m = \dfrac{y_2 - y_1}{x_2 - x_1}$<br><br>$= \dfrac{5-5}{3-2} = \dfrac{0}{1} = 0$    $= \dfrac{4-4}{2-0} = \dfrac{0}{2} = 0$<br><br>The slope is 0.          The slope is 0.<br><br><br>The slopes are the same, therefore they are **parallel**. | 13. Lines 1 and 2 pass through the points given. Determine if the lines are parallel, perpendicular, or neither without actually graphing the points or the lines.<br><br>$L_1$: $(-3,-5)$ and $(-1,2)$    $L_2$: $(0,4)$ and $(7,2)$ |

**Concept 4:** Applications and Interpretation of Slope

## Video 13: An Application of Slope (:16 min)

▶ Play the video and work along.

The graph shown gives the price of one share of stock over a period of 10 days. For the first 3 days, the stock rose approximately linearly. Over the next 7 days, the stock showed a steady decline.

a. Determine the slope of the line over the first three days. Interpret the meaning of the slope in the context of the problem.

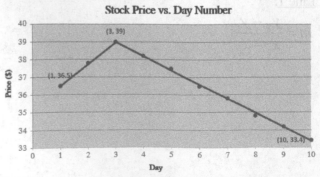

b. Determine the slope of the line from the third day to the tenth day. Interpret the meaning of the slope in the context of this problem.

| Review and complete | You try |
|---|---|
| 14. The graph shown gives the revenue for a company over a period of 10 months. For the first four months, the revenue fell linearly. Over the next six months, the revenue showed a steady incline. | 15. The graph shown gives the revenue for a company over a period of 10 months. For the first four months, the revenue fell linearly. Over the next six months, the revenue showed a steady incline. |
| 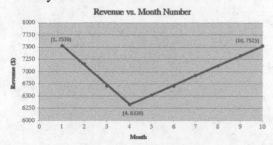 | 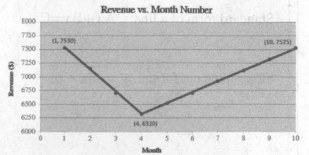 |
| Determine the slope of the line over the first four months. Interpret the meaning of the slope in the context of the problem. | Determine the slope of the line over the last six months. Interpret the meaning of the slope in the context of the problem. |
| Use the points $(1, 7530)$ and $(4, 6320)$ in the slope formula.<br><br>$$m = \frac{6320 - 7530}{4 - 1} \approx -403.33$$<br><br>The revenue decreased at a rate of \$403.33 per month during this time period. | |

**Concept 1**: Slope-Intercept Form

---

**Video 1: Introduction to Slope-Intercept Form (3:59 min)**

▶ Play the video and work along.

> **Definition of Forms of a Line**
>
> Standard form of a line:   $Ax + By = C$
>
> Slope-intercept form:   $y = mx + b$

$y = mx + b$           $(0, b)$ is the $y$-intercept           $m$ is the slope

$$m = \frac{y_2 - y_1}{x_2 - x_1}$$

---

**Video 2: Graphing a Line from Its Slope and $y$-Intercept (1:02 min)**

▶ Play the video and work along.

Graph the equation by using the slope and $y$-intercept.

$y = 3x + 2$

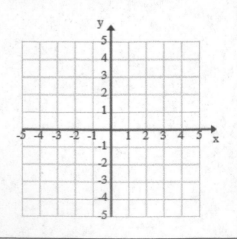

---

## Video 3: Graphing a Line from Its Slope and y-Intercept (1:16 min)

■ Before starting the video, try this yourself.

Graph the equation by using the slope and the y-intercept.

$$y = -\frac{3}{4}x - 1$$

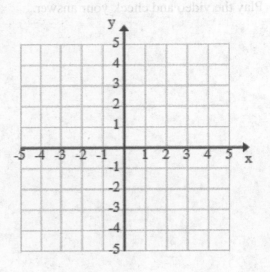

Play the video and check your answer.

---

| Review and complete | You try |
| --- | --- |

1. Graph the equation by using the slope and the
   $y$-intercept.

   slope

   $$y = -\frac{1}{2}x + 1$$

   $y$-intercept

   First plot the $y$-intercept $(0,1)$.

   The slope $m = -\dfrac{1}{2}$ can be written as $m = \dfrac{-1}{2}$.

   To find a second point on the line, start at the $y$-intercept
   and move down 1 unit and to the right 2 units. Draw the
   line through the two points.

   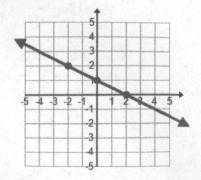

2. Graph the equation by using the slope and the
   $y$-intercept.

   $$y = -\frac{2}{3}x - 2$$

---

**Video 4: Determining if Two Lines are Parallel, Perpendicular, or Neither (3:38 min)**

Video and Study Guide to accompany Intermediate Algebra, 5th ed., Miller/O'Neill/Hyde

▶ Play the video and work along.

Determine if the lines are parallel, perpendicular, or neither.

- $L_1$: $y = \dfrac{2}{3}x - 4$

  $L_2$: $y = \dfrac{2}{3}x + 2$

- $L_1$: $4x + 2y = 2$

  $L_2$: $3x + 6y = 4$

- $L_1$: $y = 4$

  $L_2$: $2x = 6$

| Review and complete | You try |
|---|---|
| 3. Determine if the lines are parallel, perpendicular, or neither.<br><br> $L_1$: $3x + 2y = 6$   and  $L_2$: $y = \dfrac{2}{3}x$ | 4. Determine if the lines are parallel, perpendicular, or neither.<br><br> $L_1$: $3x - 4y = 12$   and  $L_2$: $4y = 3x + 5$ |

Video and Study Guide to accompany Intermediate Algebra, 5th ed., Miller/O'Neill/Hyde
Copyright © 2018 McGraw-Hill Education

We need to find the slope of each line.

Line 1:                          Line 2:

$3x + 2y = 6$

$y = -\dfrac{3}{2}x + 3$          $y = \dfrac{2}{3}x$

$m_1 = -\dfrac{3}{2}$            $m_2 = \dfrac{2}{3}$

They are **perpendicular** because the slopes are negative reciprocals.

---

## Video 5: Using Slope-Intercept Form to Determine an Equation of a Line Given a Point on the Line and the Slope (3:03 min)

▶ Play the video and work along.

Use the slope-intercept form to write an equation of a line passing through the point $(-4, 3)$ and having a slope of -2. Write the answer in slope-intercept form and in standard form.

$y = mx + b$

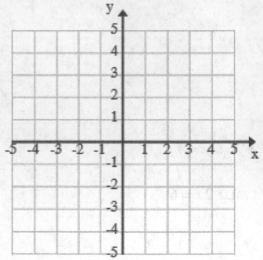

---

| Review and complete | You try |
| --- | --- |

5. Use the slope-intercept form to write an equation of a line passing through the point $(3,10)$ and having a slope of $-2$. Write the answer in slope-intercept form and in standard form.

Start with $y = mx + b$ and substitute $m = -2$.

$y = -2x + b$

$10 = -2(3) + b$    Plug in $(3,10)$ for $x$ and $y$.

$10 = -6 + b$

$16 = b$    Solve for $b$.

$y = -2x + 16$    Substitute $m$ and $b$.

To change to standard form, add $2x$ to both sides.

Standard form: _____

6. Use the slope-intercept form to write an equation of a line passing through the point $(-1,-6)$ and having a slope of 4. Write the answer in slope-intercept form and in standard form.

**Concept 2:** The Point-Slope Formula

**Video 6: Introduction to the Point-Slope Formula (3:35 min)**

▶ Play the video and work along.

**Definition of Point-Slope Formula**

Point-Slope Formula:  $y - y_1 = m(x - x_1)$
$m$ is the slope, $(x_1, y_1)$ is a known point on the line.

Write an equation of the line passing through the point $(-4,3)$ and having a slope of -2. Write the answer in slope-intercept form and in standard form.

$m = \dfrac{y_2 - y_1}{x_2 - x_1}$          $(x_1, y_1)$          $(x, y)$

| Review and complete | You try |
|---|---|
| 7. Write an equation of a line passing through the point $(3,10)$ and having a slope of -2. Write the answer in slope-intercept form and in standard form.<br><br>$\quad y-10=-2(x-3)$    Substitute $m, x_1$ and $y_1$ in $y=mx+b$.<br>$\quad y-10=-2x+6$<br>$\qquad y=-2x+16$    Solve for $y$.<br><br>To change to standard form, add 2x to both sides to get<br><br>Standard form:_____ | 8. Write an equation of a line passing through the point $(-1,-5)$ and having a slope of 3. Write the answer in slope-intercept form and in standard form. |

## Video 7: Writing an Equation of a Line Given Two Points on the Line (4:45 min)

▶ Play the video and work along.

Write an equation of the line passing through the points $(2,-3)$ and $(5,-2)$. Write the answer in slope-intercept form. (Use the point-slope form.)

Write an equation of the line passing through the points $(2,-3)$ and $(5,-2)$. Write the answer in slope-intercept form. (Use the slope-intercept form.)

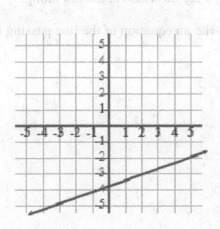

| Review and complete | You try |
|---|---|
| 9. Write an equation of the line passing through the points $(-8,-1)$ and $(-5,9)$. Write the answer in slope-intercept form. | 10. Write an equation of the line passing through the points $(-2,-1)$ and $(3,-4)$. Write the answer in slope-intercept form. |

First find the slope.

$$m = \frac{y_2 - y_1}{x_2 - x_1} = \frac{9-(-1)}{-5-(-8)} = \frac{9+1}{-5+8} = \frac{10}{3}$$

Next apply the point-slope formula.

$$y - y_1 = m(x - x_1)$$

$$y - (-1) = \frac{10}{3}\left(x - (-8)\right)$$    Substitute $m = \frac{10}{3}$ and

$$y + 1 = \frac{10}{3}(x + 8)$$    We will use $(-8,-1)$ for $(x_1, y_1)$.

$$y + 1 = \frac{10}{3}x + \frac{80}{3}$$    Clear parentheses.

$$y + \frac{3}{3} = \frac{10}{3}x + \frac{80}{3}$$

$$y = \frac{10}{3}x + \frac{77}{3}$$    Solve for y.

**Video 8: Writing an Equation of a Line Passing through a Given Point and Parallel to a Given Line (3:37 min)**

▶ Play the video and work along.

Write an equation of the line passing through the point $(-3,-1)$ and parallel to the line $-2x + y = 1$.

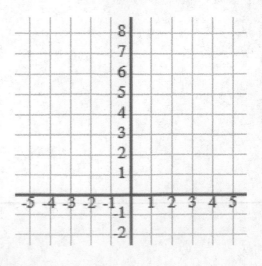

| Review and complete | You try |
|---|---|
| 11. Write an equation of the line passing through the point $(-6,-1)$ and parallel to the line $2x+3y=-12$. Write the answer in slope-intercept form. | 12. Write an equation of the line passing through the point $(6,1)$ and parallel to the line $2x+3y=-12$. Write the answer in slope-intercept form. |

Find the slope for the line $2x+3y=-12$.

$2x+3y=-12$

$3y=-2x-12$

$y=-\dfrac{2}{3}x-4$

$m=-\dfrac{2}{3}$

Substitute $m, x_1$ and $y_1$ in the point-slope formula.

$y-y_1=m(x-x_1)$

$y-(-1)=-\dfrac{2}{3}(x-(-6))$    Simplify.

$y+1=-\dfrac{2}{3}(x+6)$

$y+1=-\dfrac{2}{3}x-4$    Subtract 1 from both sides.

$y=-\dfrac{2}{3}x-5$

### Video 9: Determining an Equation of a Line Passing through a Given Point and Perpendicular to a Given Line (2:39 min)

▶ Play the video and work along.

Write an equation of a line passing through the point $(3,1)$ and perpendicular to the line $3x+2y=8$.

$y-y_1=m(x-x_1)$

Video and Study Guide to accompany Intermediate Algebra, 5th ed., Miller/O'Neill/Hyde

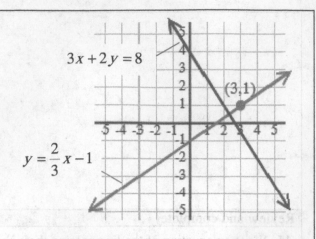

$3x + 2y = 8$

$(3,1)$

$y = \dfrac{2}{3}x - 1$

| Review and complete | You try |
|---|---|
| 13. Write an equation of the line passing through the point $(-1,-5)$ and perpendicular to the line $y = -4x$. Write the answer in slope-intercept form. | 14. Write an equation of the line passing through the point $(4,-2)$ and perpendicular to the line $4x + 3y = -6$. Write the answer in slope-intercept form. |

The slope of the line $y = -4x$ is the coefficient of $x$.
$m = -4$

The slope of a perpendicular line is $m = \dfrac{1}{4}$.

Substitute $m, x_1$ and $y_1$ in the point-slope formula.

$$y - y_1 = m(x - x_1)$$

$$y - (-5) = \frac{1}{4}(x - (-1))$$

$$y + 5 = \frac{1}{4}(x + 1)$$

$$y + 5 = \frac{1}{4}x + \frac{1}{4}$$

$$y = \frac{1}{4}x + \frac{1}{4} - 5$$

$$y = \frac{1}{4}x + \frac{1}{4} - \frac{20}{4}$$

$$y = \frac{1}{4}x - \frac{19}{4}$$

**Video 10:  Determining an Equation of a Horizontal or Vertical Line (1:57 min)**

▶ Play the video and work along.

- Determine an equation of the line passing through the point $(3,-2)$ and parallel to the line defined by $x = -4$.

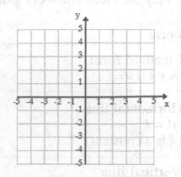

- Determine an equation of the line passing through the point $(1,4)$ and perpendicular to the $y$-axis.

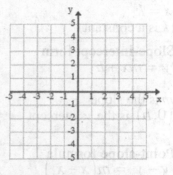

| Review and complete | You try |
|---|---|
| 15. Determine an equation of the line passing through the point $(3,-7)$ and perpendicular to the line defined by $x = -4$. <br><br> The equation $x = -4$ is a vertical line. A line perpendicular to this line is a horizontal line, an equation of the form $y = k$. <br><br> In this case, the line passes through a point with a y-coordinate of -7, so the equation is <br><br> $$y = -7$$ | 16. Determine an equation of the line passing through the point $(7,-2)$ and parallel to the line defined by $y = 3$. |

**Concept 3:** Equations of a Line: A Summary

**Video 11: Equations of Lines: A summary (2:39 min)**

▶ Play the video and work along.

| Form | Example | Comments |
|---|---|---|
| **Standard form**<br>$Ax + By = C$ | $3x - 4y = 7$ | The coefficients A and B must not both be 0. |
| **Horizontal line**<br>$y = k$<br>($k$ is a constant) | $y = 4$ | The slope is 0, and the $y$-intercept is $(0, k)$. |
| **Vertical line**<br>$x = k$<br>($k$ is a constant) | $x = -3$ | The slope is *undefined*, and the $x$-intercept is $(k, 0)$. |
| **Slope-intercept form**<br>$y = mx + b$<br><br>$m$ is the slope<br>$(0, b)$ is the $y$-intercept | $y = mx + b$<br>$y = -3x + 5$<br>Slope is -3<br>$y$-intercept is $(0, 5)$ | • Useful to determine the slope and $y$-intercept of a line.<br>• Useful to graph a line.<br>• Useful to write an equation of a line given the slope and a point on the line. |
| **Point-slope formula**<br>$y - y_1 = m(x - x_1)$<br><br>Slope is $m$ and $(x_1, y_1)$ is a point on the line. | Given $m = 2$ and $(x_1, y_1) = (1, 3)$,<br>$y - y_1 = m(x - x_1)$<br>$y - 3 = 2(x - 1)$<br>$y - 3 = 2x - 2$<br>$y = 2x + 1$ | • This formula is typically used to write an equation of a line when a point on the line and the slope are known. |

**Concept 1**: Writing a Linear Model

| Video 1: Determining an Equation of a Line in an Application Given a Rate and a Fixed Value (3:38 min) |
| --- |

▶ Play the video and work along.

A small business needs to rent additional office space. The rental company requires a nonrefundable deposit of $1000 plus rent for $600 per month.

a. Write a linear equation that represents the cost $y$ (in dollars) to rent the additional office space for $x$ months.

b. Graph the equation from part (a).

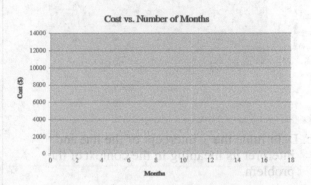

c. Determine the $y$-intercept of the line and interpret its meaning in the context of the problem.

d. Use the equation from part (a) to determine the cost to rent the additional office space for 1 year.

| Review and complete | You try |
|---|---|

**Review and complete**

1. The cost of a rental car is $110 per week plus an additional $0.20 per mile.

   a. Write a linear equation that represents the cost $y$ (in dollars) to rent a car for one week for $x$ m.

   $$y = 0.20x + 110$$

   b. Graph the equation from part (a).

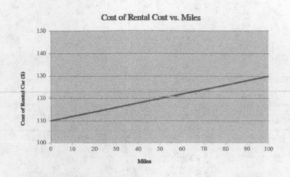

   c. Determine the $y$-intercept of the line and interpret its meaning in the context of the problem.

   The y-intercept is $(0, 110)$. It means that the cost to rent a car for one week without driving any miles is $110.

   d. Use the equation from part (a) to determine the cost to rent the car for one week with 125 mi.

   $$y = 0.20(125) + 110$$
   $$y = 135$$

   It will cost $135 to rent the car for one week with 125 mi.

**You try**

2. Brian earns a base salary of $1000 per month plus a 4% commission on his sales for the month.

   a. Write a linear equation that represents Brian's monthly salary $y$ (in dollars) if he has a total of $x$ sales (in dollars) for the month.

   b. Graph the equation from part (a).

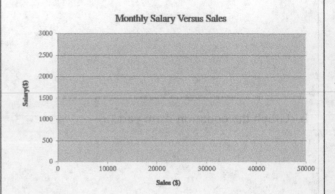

   c. Determine the $y$-intercept of the line and interpret its meaning in the context of the problem.

   d. How much will Brian make if his sales for a given month are $30,000?

## Concept 2: Interpreting a Linear Model

### Video 2: Interpreting a Linear Model (4:26 min)

 Play the video and work along.

A pool technician empties a pool so that the tile and concrete on the sides of the pool can be cleaned. The water level, $y$, (in ft) can be approximated by the equation $y = -0.8x + 6$. In this equation, $x$ represents the number of hours after the pumps have been turned on.

a. Determine the water level after 2 hr.

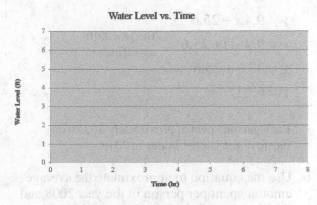

Water Level vs. Time

b. After how many hours will the water level reach 2 ft?

c. Interpret the slope of the line in the context of this problem.

d. Determine the $y$-intercept and interpret its meaning in the context of this problem.

e. Determine the $x$-intercept and interpret its meaning in the context of this problem.

| Review and complete | You try |
|---|---|
| 3. The equation $y = 9.4x + 25.6$ represents the average amount $y$ (in \$) spent on music per person in the United States where $x$ is the number of years since 2006. | 4. Let $y$ represent the average number of miles driven per year for passenger cars in the United States since 1980. Let $x$ represent the years since 1980. The average yearly mileage for passenger cars can be approximated by the equation $y = 142x + 9060$, where $x \geq 0$. |

a. Use the equation to approximate the average amount spent per person in the year 2010.

$y = 9.4x + 25.6$

$y = 9.4(4) + 25.6$    The year 2010 corresponds to $x = 4$.

$y = 37.6 + 25.6$

$y = 63.2$

Each person spent approximately \$63.20 on music in the United States in 2010.

a. Use the linear equation to approximate the average yearly mileage for passenger cars in the United States in the year 2005.

b. Use the equation to approximate the average amount spent per person in the year 2008 and compare it with the actual spent of \$45.80.

$y = 9.4x + 25.6$

$y = 9.4(2) + 25.6$    The year 2008 corresponds to $x = 2$.

$y = 18.8 + 25.6$

$y = 44.4$

The model estimates that each person spent about \$44.40 on music in 2008. The approximate value differs from the actual value by \$45.80 - \$44.40 = \$1.40.

b. Use the linear equation to approximate the average mileage for the year 1985 and compare it with the actual value of 9700 mi.

c. What is the slope of the line and what does it mean in the context of this problem?

The slope is $m = 9.4$. It means that the amount spent per person on music in the United States increased by an average of \$9.40 per year.

c. What is the slope of the line and what does it mean in the context of this problem?

d. Determine the $y$-intercept and interpret its meaning in the context of this problem.

The y-intercept is $(0, 25.60)$. It means that the average amount spent on music per person was \$25.60 in year 0, or 2006.

d. Determine the $y$-intercept and interpret its meaning in the context of this problem.

**Concept 3:** Finding a Linear Model from Observed Data Points

> **Video 3: Determining an Equation of a Line in an Application where Two Observed Points are Given (3:47 min)**

▶ Play the video and work along.

A coffee shop in Ormond Beach opened shortly before a dramatic fall in the U.S. economy. Unfortunately for the shop owners, many people stopped treating themselves to coffee and other café treats. The graph shows the monthly profit from Month 1 to Month 10. A negative profit indicates that monthly overhead such as rent and supplies cost more than revenue brought in by sales.

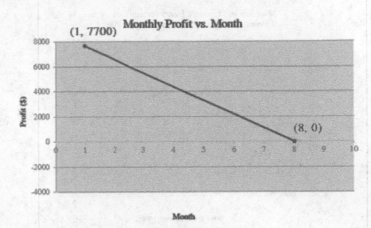

a.  Determine the slope and interpret its meaning in the context of this problem.

b.  Determine the *x*-intercept and interpret its meaning in the context of this problem.

c.  Write an equation of the line.

d.  Use the equation found in part (c) to predict the profit in month 12.

| Review and complete | You try |
|---|---|
| 5. The number of prisoners (in thousands) in federal or state correctional facilities from 1990 – 2014 is shown. | 6. During a drought, the average water level in a retention pond decreased linearly with time. |

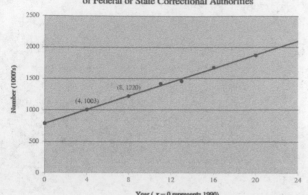

**Number of Prisoners Under Jursidiction of Federal or State Correctional Authorities**

Year ( $x=0$ represents 1990)

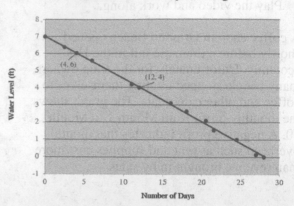

**Water Level Versus Days of Drought**

Number of Days

a. Use the given ordered pairs to find a linear equation to model the number of prisoners by year.

$$m = \frac{y_2 - y_1}{x_2 - x_1}$$

$$m = \frac{1220 - 1003}{8 - 4}$$   Apply the slope formula.

$$m = \frac{217}{4} = 54.25$$

$$y - y_1 = m(x - x_1)$$   Apply the point-slope formula.

$$y - 1003 = 54.25(x - 4)$$

$$y - 1003 = 54.25x - 217$$   Solve for $y$.

$$y = 54.25 + 786$$

a. Use the given ordered pairs (12, 4) and (4, 6) to find a linear equation to model the water by the number of days since the drought started.

b. Use the equation in part (a) to predict the number of prisoners in federal or state correctional facilities for the year 2015.

$$y = 54.25(25) + 786$$
$$y = 2142.25$$

There will be approximately 2,142.25 thousand prisoners in 2015.

b. Use the linear equation to approximate the water level after 15 days.

c. What is the slope of the line and what does it mean in the context of this problem?

The slope $m = 54.25$ means that the number of prisoners increased at a rate of 54.25 thousand per year during this time period.

c. What is the slope of the line and what does it mean in the context of this problem?

**Concept 1**: Definition of a Relation

▶ Play the video and work along.

---

**Definition of a Relation**

A **relation in $x$ and $y$** is a set of ordered pairs of the form $(x, y)$.

- The set of $x$ values is called the **domain** of the relation.

- The set of $y$ values is called the **range** of the relation.

---

The table shows a relation between the number of minutes played and the number of points scored for a college basketball player.

a. Write the relation given as a set of ordered pairs.

b. Determine the domain and range of the relation.

   Domain:

   Range:

| Minutes played $x$ | Points scored $y$ |
|---|---|
| 32 | 18 |
| 24 | 15 |
| 30 | 12 |
| 38 | 26 |
| 30 | 22 |

---

| Review and complete | You try |
|---|---|

1.  The (**choose one**: domain or range) of a relation is the set of first components in the ordered pairs.

2.  The (**choose one**: domain or range) of a relation is the set of second components in the ordered pairs.

3.  The table shows a relation.

| $x$ | 0 | 3 | -5 | 6 | 5 |
|---|---|---|---|---|---|
| $y$ | 3 | 1 | 5 | 1 | -2 |

a. Write the relation as a set of ordered pairs.

$$\{(0,3),(3,1),(-5,5),(6,1),(5,-2)\}$$

b. Determine the domain and range of the relation.

Domain: $\{0,3,-5,6,5\}$   Find the set of $x$-values.

Range: _____   Find the set of $y$-values without duplicates.

4.  The table shows a relation between a state and the year it became a state.

| State, $x$ | Year of Statehood, $y$ |
|---|---|
| Maine | 1820 |
| Nebraska | 1823 |
| Utah | 1847 |
| Hawaii | 1959 |
| Alaska | 1959 |

a. Write the relation given as a set of ordered pairs.

b. Determine the domain and range of the relation.

---

Video and Study Guide to accompany Intermediate Algebra, 5ᵗʰ ed., Miller/O'Neill/Hyde
Copyright © 2018 McGraw-Hill Education

**Concept 2:**  Domain and Range of a Relation

---

**Video 2: Determining the Domain and Range of a Relation Containing a Finite Number of Points (1:50 min)**

▶ Play the video and work along.

Determine the domain and range of each relation.

a.

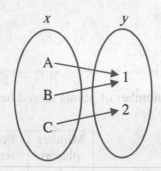

b.

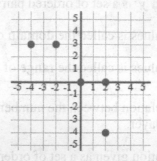

Domain:

Range:

Domain:

Range:

---

**Video 3: Determining the Domain and Range of a Relation Containing an Infinite Number of Points (3:35 min)**

▶ Play the video and work along.

Determine the domain and range of the relation.

a.

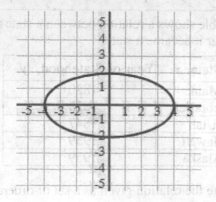

Domain:

Range:

⏸ Pause the video and try this yourself.

Determine the domain and range of the relation.

b.

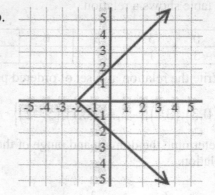

Domain:

Range:

▶ Play the video and check your answer.

---

**Video 4: Determining the Domain and Range of a Relation Given the Graph of the Relation (2:44 min)**

▶ Play the video and work along.

Determine the domain and range of each relation.

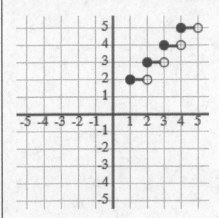

Domain:

Range:

_____ The point is included.


_____ The point is *not* included.


---

**Avoiding Mistakes**
Be sure to distinguish between set notation and interval notation.

- The set $\{2, 3, 4, 5\}$ consists of four elements.

- The set $[2, 5]$ consists of an infinite number of elements.

| Review and complete | You try |
|---|---|
| 5. Determine the domain and range of the relation. | 6. Determine the domain and range of the relation. |

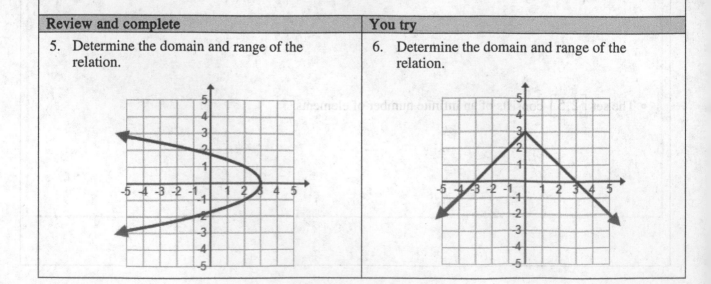

Domain: $\left(-\infty, 3\right]$  The x-values range from negative infinity to 3, included.

Range: $\left(-\infty, \infty\right)$  Only use curved parentheses.

**Concept 1**: Definition of a Function

### Video 1: Definition of a Function (1:56 min)

▶ Play the video and work along.

For each relation, determine if the relation defines $y$ as a function of $x$.

a. $\{(6,3),(4,5),(6,-1)\}$          b. $\{(3,4),(5,1),(2,-1)\}$

### Video 2: Determining if a Relation is a Function (2:08 min)

▶ Play the video and work along.

a.

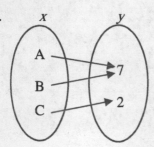

Yes or No?

⏸ Pause the video and try these yourself.

b.

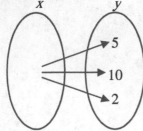

Yes or No?

c.

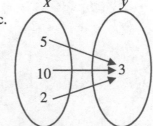

Yes or No?

▶ Play the video and check your answers.

Video and Study Guide to accompany Intermediate Algebra, 5th ed. Miller/O'Neill/Hyde
Copyright © 2018 McGraw-Hill Education

| Review and complete | You try |
|---|---|

**Review and complete**

1.  For each relation, determine if the relation defines $y$ as a function of $x$.

    a.  $\{(5,2),(7,1),(5,-1)\}$

    The first ordered pair and the last ordered pair have the same $x$-value, but different $y$-values. This contradicts the definition of a function. In a function, for every $x$-value in the domain, there can be only <u>one unique</u> $y$-value.

    Therefore, this relation is **not** a function.

    b.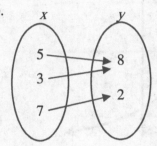

    The mapping corresponds to the ordered pairs

    $(5,8)\quad(3,8)\quad(7,2)$

    Since each value of $x$ is paired with only one $y$-value, this relation **is a function**.

    Remember, it doesn't matter if a $y$ value is repeated with different $x$ values.

**You try**

2.  For each relation, determine if the relation defines $y$ as a function of $x$.

    a.  $\{(3,2),(6,4),(3,-5)\}$

    b.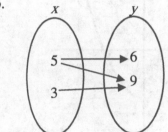

**Concept 2:** Vertical Line Test

| Video 3: Vertical Line Test (2:34 min) |
|---|

▶ Play the video and work along.

Determine if the relation is a function.

a.  $\{(3,4),(3,-1),(-2,4),(1,3)\}$

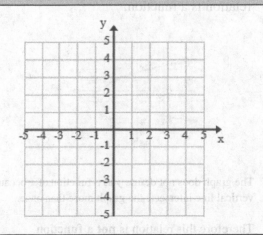

Yes or No?

### Procedure for a Vertical Line Test

Given a relation in $x$ and $y$, the graph defines $y$ as a function of $x$ if no vertical line intersects the graph in more than one point.

 Pause the video and try these yourself.

b.

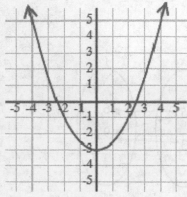

Yes or No?

c.

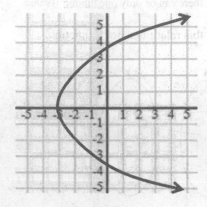

Yes or No?

▶ Play the video and check your answers.

| Review and complete | You try |
|---|---|
| 3. Use the vertical line test to determine if the relation is a function.<br><br>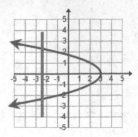<br><br>The graph does not define $y$ as a function of $x$ because a vertical line intersects the graph more than once.<br><br>Therefore this relation is **not a function**. | 4. Use the vertical line test to determine if the relation is a function.<br><br>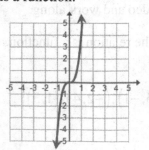 |

Video and Study Guide to accompany Intermediate Algebra, 5th ed., Miller/O'Neill/Hyde

**Concept 3:** Function Notation

### Video 4: Function Notation (3:13 min)

▶ Play the video and work along.

$$y = 2x$$

Given the function defined by $f(x) = 2x$, determine the function values.

a.  f(-2)                          b.  f(-1)

⏸ Pause the video and try these yourself.

c.  f(0)                           d.  f(1)

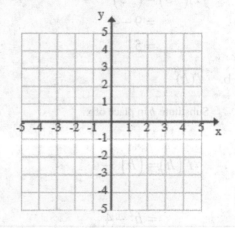

e.  f(2)

▶ Play the video and check your answers.

### Video 5: Evaluating Functions (2:42 min)

▶ Play the video and work along.

Evaluate the functions for the given values of $x$.

$$g(x) = x^2 + 4 \qquad\qquad h(x) = \frac{1}{5 + x}$$

a.  $g(t)$                         b.  $h(x+3)$

⏸ Pause the video and try these yourself.

c.  $g(-a)$              d.  $h(-5)$

▶ Play the video and check your answers.

| Review and complete | You try |
|---|---|
| 5.  Evaluate the functions for the given values. $f(x) = x^2 - 4$ | 6.  Evaluate the function for the given values. $f(x) = -x^2 - 3x + 1$ |

5.  Evaluate the functions for the given values.
$f(x) = x^2 - 4$

    a.  $f(-3)$

       Substitute $-3$ in place of $x$.

$$f(-3) = (-3)^2 - 4$$
$$= 9 - 4$$
$$= 5$$

    b.  $f(h)$

       Substitute $h$ in place of $x$.

$$f(h) = (h)^2 - 4$$

$$= h^2 - 4$$

6.  Evaluate the function for the given values.
$f(x) = -x^2 - 3x + 1$

    a.  $f(-5)$

    b.  $f(w)$

**Concept 4:** Finding Function Values from a Graph

**Video 6: Estimating Function Values from a Graph (4:31 min)**

▶ Play the video and work along.

The graph of $y = f(x)$ is shown here.

- Find $f(-2)$.

⏸ Pause the video and try these yourself.

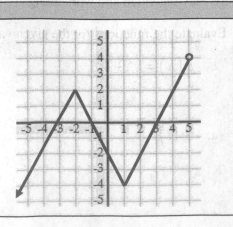

- Find $f(1)$.

- Find $f(5)$.

▶ Play the video and check your answers.

▶ Play the video and work along.

The graph of $y = f(x)$ is shown here.

- For what values of $x$ is $f(x) = 0$ ?

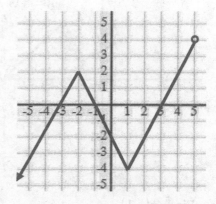

⏸ Pause the video and try this yourself.

- For what values of $x$ is $f(x) = 3$ ?

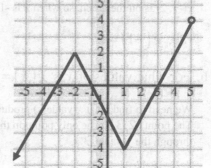

▶ Play the video and check your answer.

▶ Play the video and work along.

The graph of $y = f(x)$ is shown here.

- Write the domain of $f$.

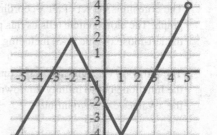

- Write the range of $f$.

| Review and complete | You try |
|---|---|

7.  The graph of $y = g(x)$ is shown here.

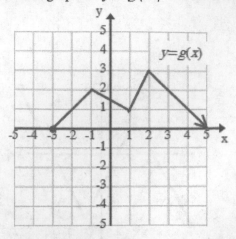

a.  Find $g(-1)$.

    We need to find the *y*-value when *x* is -1. The point (-1, 2) is on the graph, so when *x* is -1, *y* is 2.

    So, $g(-1) = 2$.

b.  For what values of *x* is $g(x) = 3$?

    We need to find what *x*-value(s) produce a *y*-value of 3. The point $(2, 3)$ is the only point on the graph that goes through the *y*-value, 3.

    So, the value of *x* that produces a *y*-value of 3 is **2**.

c.  Write the domain of *g*.

    The domain is the set of all *x*-coordinates for points on the graph. The closed circle tells us this point is included. So, the left-most *x*-value is -3, including -3.

    The arrow on the right side of the function indicates that the function goes to infinity far down and to the right. So, the *x*-values extend to infinity.

    The domain of *g* is $[-3, \infty)$.

8.  The graph of $y = g(x)$ is shown here.

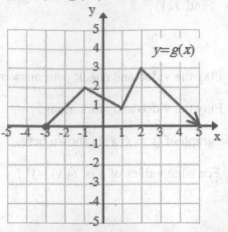

a.  Find $g(4)$.

b.  For what values of *x* is $g(x) = 0$?

c.  Write the range of *g*.

**Concept 5:** Domain of a Function

**Video 7: Determining the Domain of a Rational Function and a Square Root Function (3:45 min)**

▶ Play the video and work along.

Use algebraic methods to determine the domain of each function.

a. $f(x) = \dfrac{x+3}{x-2}$

b. $h(t) = \sqrt{t-3}$

---

### Video 8: Determining the Domain of a Function Analytically (3:14 min)

■ Before starting the video, try these yourself.

Use algebraic methods to determine the domain of each function.

a. $w(x) = \dfrac{x+2}{2x-5}$

b. $f(x) = \sqrt{12-x}$

▶ Play the video and check your answers.

▶ Play the video and work along.

c. $g(x) = \dfrac{x+1}{x^2+4}$

d. $p(x) = x^2 + 5x$

---

| Review and complete | You try |
|---|---|
| 9. The domain is the set of real numbers *excluding* those that<br><br> a. make a denominator equal to _____.<br> b. make the expression within a square root (or even-indexed root) _____. | |
| 10. Use algebraic methods to determine the domain of the function. Write the answer in interval notation.<br><br>$$m(x) = \dfrac{x-1}{x-4}$$<br><br>The domain is the set of real numbers *excluding* those that make the denominator equal to zero. Set the denominator equal to zero and solve for x.<br><br>$$x - 4 = 0 \implies x = 4$$<br><br>The domain is the set of all real numbers *excluding* 4.<br><br>In interval notation, _____. | 11. Use algebraic methods to determine the domain of the function. Write the answer in interval notation.<br><br>$$n(x) = \sqrt{12 - 6x}$$ |

## Video 9: An Application of a Function (2:17 min)

▶ Play the video and work along.

The temperature $T(x)$ (in °C) of an object placed in a freezer can be approximated by

$$T(x) = \frac{750}{x^2 + 3x + 10}$$

where $x$ is the time in hours after the object was placed in the freezer.

• Find $T(2)$ and interpret the meaning in the context of this problem.

⏸ Pause the video and try this yourself.

• Find $T(5)$ and interpret the meaning in the context of this problem.

▶ Play the video and check your answer.

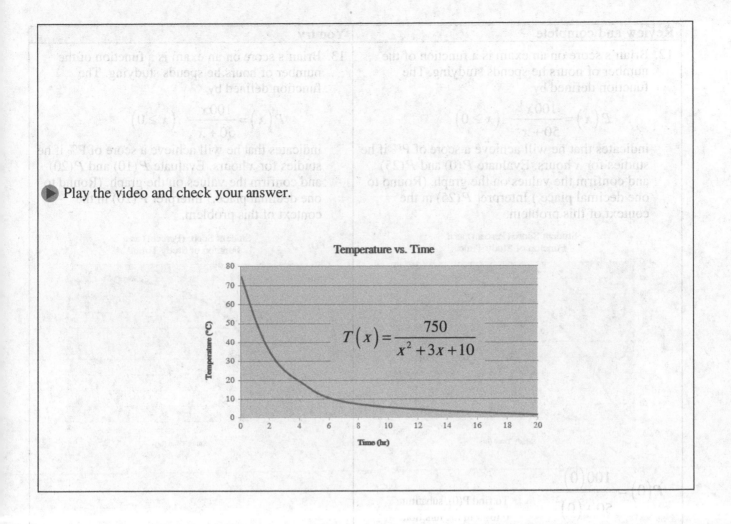

**Temperature vs. Time**

$$T(x) = \frac{750}{x^2 + 3x + 10}$$

| Review and complete | You try |
|---|---|

**12.** Brian's score on an exam is a function of the number of hours he spends studying. The function defined by

$$P(x) = \frac{100x^2}{50+x^2} \quad (x \geq 0)$$

indicates that he will achieve a score of P% if he studies for $x$ hours. Evaluate $P(0)$ and $P(25)$ and confirm the values on the graph. (Round to one decimal place.) Interpret $P(25)$ in the context of this problem.

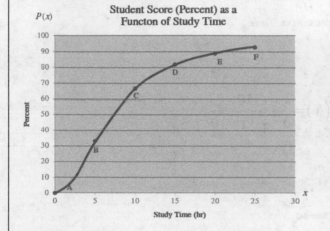

$$P(0) = \frac{100(0)^2}{50+(0)^2}$$

To find P(0), substitute 0 for $x$ in the function.

$$P(0) = \frac{0}{50} = 0$$

$$P(25) = \frac{100(25)^2}{50+(25)^2}$$

To find $P(25)$, substitute 25 for $x$ in the function.

$$P(25) = \frac{62{,}500}{675}$$

$$P(25) \approx 92.6$$

$P(25)$ means that if Brian studies for 25 hr, he will get a score of 92.6%.

**13.** Brian's score on an exam is a function of the number of hours he spends studying. The function defined by

$$P(x) = \frac{100x^2}{50+x^2} \quad (x \geq 0)$$

indicates that he will achieve a score of P% if he studies for $x$ hours. Evaluate $P(10)$ and $P(20)$ and confirm the values on the graph. (Round to one decimal place.) Interpret $P(20)$ in the context of this problem.

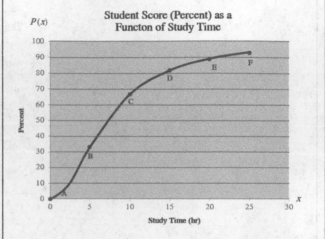

## Concept 1: Linear and Constant Functions

### Video 1: Definition of Constant and Linear Functions (1:51 min)

▶ Play the video and work along.

Let $b$ represent a real number. Then a function that can be written in the form $f(x) = b$ is called a **constant function**.

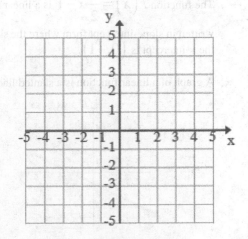

Let $m$ and $b$ represent real numbers and $m \neq 0$. Then a function that can be written in the form $f(x) = mx + b$ is called a **linear function**.

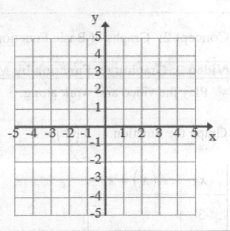

| Review and complete | You try |
| --- | --- |

1. A horizontal line (**choose one:** is or is not) a function.

2. A vertical line (**choose one:** is or is not) a function.

3. Graph the constant function $f(x) = -1$.

   A graph of a constant function is a horizontal line.

   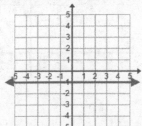

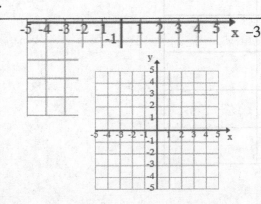

3. Graph the constant function $f(x) = -3$.

5.  Graph the linear function

$$f(x) = \frac{1}{2}x - 1$$

The function $f(x) = \frac{1}{2}x - 1$ is a linear function

written in slope-intercept form where the slope is ½ and the y-intercept is $(0, -1)$.

A graph of a linear function is a slanted line.

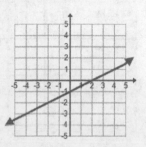

6.  Graph the linear function

$$f(x) = \frac{2}{5}x - 2$$

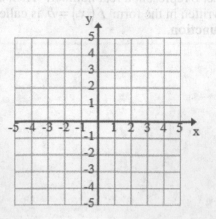

**Concept 2:** Graphs of Basic Functions

---

**Video 2: Graphing a Function by Making a Table of Points (2:06 min)**

▶ Play the video and work along.

Graph the function.    $f(x) = x^2$

| $x$ | $f(x) = x^2$ |
|-----|--------------|
| -3  |              |
| -2  |              |
| -1  |              |
| 0   |              |
| 1   |              |
| 2   |              |
| 3   |              |

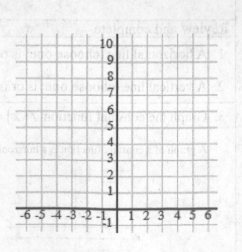

### Video 3: Graphing a Function by Creating a Table of Points (2:16 min)

■ Before starting the video, try this yourself.

Graph the function.    $f(x) = \dfrac{1}{x}$

| $x$ | $f(x) = \dfrac{1}{x}$ |
|-----|------------------------|
| -3  |                        |
| -2  |                        |
| -1  |                        |
| 0   | undefined              |
| 1   |                        |
| 2   |                        |
| 3   |                        |

| $x$ | $f(x) = \dfrac{1}{x}$ |
|-----|------------------------|
| $-\dfrac{1}{2}$ |            |
| $-\dfrac{1}{3}$ |            |
| $-\dfrac{1}{4}$ |            |
| $\dfrac{1}{2}$  |            |
| $\dfrac{1}{3}$  |            |
| $\dfrac{1}{4}$  |            |

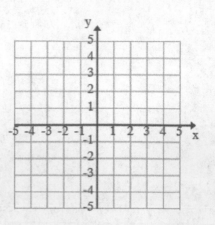

▶ Play the video and check your answer.

| Review and complete | You try |
|---|---|

7. Graph the function.    $f(x) = |x|$

Plug in each $x$ value into the function to find $f(x)$.

| $x$ | $f(x) = |x|$ | $f(x)$ |
|-----|--------------|--------|
| -3  | $f(-3) = |-3| = 3$ | 3 |
| -2  | $f(-2) = |-2| = 2$ | 2 |
| -1  | $f(-1) = |-1| = 1$ | 1 |
| 0   | $f(0) = |0| = 0$ | 0 |
| 1   | $f(1) = |1| = 1$ | 1 |
| 2   | $f(2) = |2| = 2$ | 2 |
| 3   | $f(3) = |3| = 3$ | 3 |

8. Graph the function.    $f(x) = x^3$

| $x$ | $f(x) = x^3$ |
|-----|--------------|
| -3  |              |
| -2  |              |
| -1  |              |
| 0   |              |
| 1   |              |
| 2   |              |
| 3   |              |

**Concept 3:** Definition of a Quadratic Function

| Video 4: Definition of a Quadratic Function (2:35 min) |
|---|

▶ Play the video and work along.

Let $a$, $b$, and $c$ represent real numbers such that $a \neq 0$. Then a function that can be written in the form $f(x) = ax^2 + bx + c$ is called a **quadratic function**.

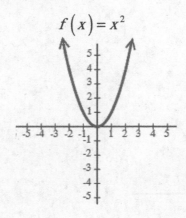
$f(x) = x^2$

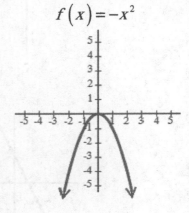
$f(x) = -x^2$

| Review and complete | You try |
|---|---|

9.  The function $f(x) = -3x^2 + 6x - 7$ is a quadratic function opening (**choose one:** upward or downward).

10. The function $f(x) = 4x^2 - 2x - 5$ is a quadratic function opening (**choose one:** upward or downward).

| Video 5: Identifying Constant, Linear, and Quadratic Functions (2:58 min) |
|---|

▶ Play the video and work along.

Identify the function as constant, linear, quadratic, or none of these.

$f(x) = 3 - 4x$                            $g(x) = -3x^2 - 4x$

$h(x) = -3$                                $k(x) = 3 - \dfrac{4}{x}$

Identify the function as constant, linear, quadratic, or none of these.

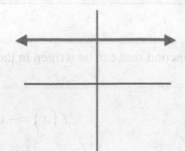

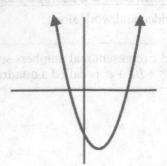

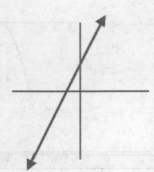

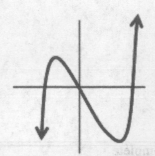

| Review and complete | You try |
| --- | --- |

11. The function $f(x) = 2x - 5$ is an example of a (**choose one**: constant, linear, or quadratic) function.

12. The function $g(x) = -5$ is an example of a (**choose one**: constant, linear, or quadratic) function.

13. The function $h(x) = 4x^2 - 2x - 5$ is an example of a (**choose one**: constant, linear, or quadratic) function.

| 14. Identify the function as constant, linear, or quadratic. | 15. Identify the function as constant, linear, or quadratic. |
| --- | --- |
| 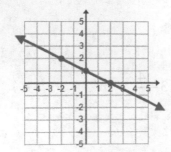 | 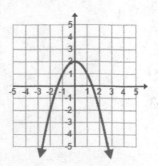 |
| This is a graph of a slanted line, therefore the function is **linear**. | |

**Concept 4:** Finding the *x*- and *y*-Intercepts of a Graph Defined by $y = f(x)$

▶ Play the video and work along.

Given the function defined by $f(x) = -3x + 4$,

  a. Determine the *y*-intercept.

                                    b. Determine the *x*-intercept.

  c. Graph the function.

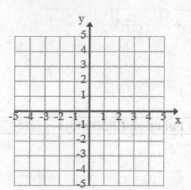

> Given a function defined by $y = f(x)$,
> - To find the *y*-intercept, evaluate $f(0)$.
> - To find the *x*-intercept, set $f(x) = 0$.

| Review and complete | You try |
|---|---|
| 16. Given the function defined by $f(x) = -4x + 1$, | 17. Given the function defined by $f(x) = -2x - 1$, |
|   a. Determine the *y*-intercept. |   a. Determine the *y*-intercept. |
|     To find the *y*-intercept, substitute 0 for *x*. | |
|     $f(0) = -4(0) + 1$ | |
|     $f(0) = 1 \implies (0,1)$ | |
|   b. Determine the *x*-intercept. |   b. Determine the *x*-intercept. |
|     $f(x) = -4x + 1$ | |
|     $0 = -4x + 1$     Add 4*x* to both sides. | |
|     $4x = 1$         Divide by sides by 4 | |

$$\frac{4x}{4} = \frac{1}{4} \quad \Rightarrow \quad x = \frac{1}{4} \quad \Rightarrow \quad \left(\frac{1}{4}, 0\right)$$

c. Graph the function.

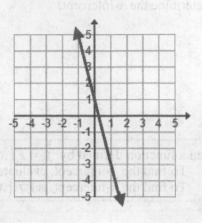

c. Graph the function.

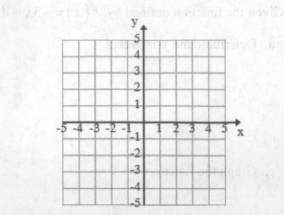

## Video 7: Identifying the *x*- and *y*-Intercepts from the Graph of a Function (0:50 min)

▶ Play the video and work along.

Estimate the *x*- and *y*-intercepts of the function.

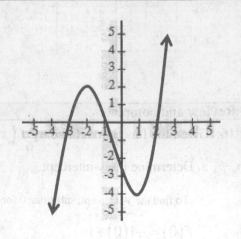

*x*-intercept(s):

*y*-intercept:

| Review and complete | You try |
|---|---|
| 18. True or false: A function can have two *y*-intercepts. | |
| 19. True or false: A function can have two *x*-intercepts. | |

20. Estimate the *x*- and *y*-intercepts of the function given by the graph.

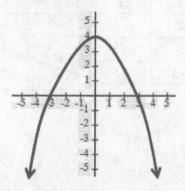

The *x*-intercept(s) is where the graph crosses the *x*-axis.

Notice there are two *x*-intercepts located at $(-3,0)$ and $(3,0)$.

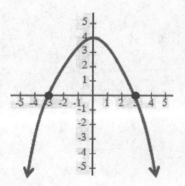

The *y*-intercept is where the graph crosses the *y*-axis.

Notice there is only one *y*-intercept located at $(0,4)$.

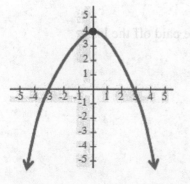

21. Estimate the *x*- and *y*-intercepts of the function given by the graph.

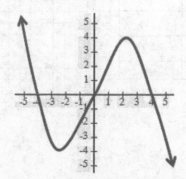

**Answers:** Section 2.1

2a. 34 in.    2b. day 19    3. $x$; $y$-axis    4. quadrants    5. III    6. negative

7. Quadrant II    8a. $y$-axis    8b. $x$-axis    10. No

12.

| x | y |
|---|---|
| 0 | −4 |
| 3 | −2 |
| −3 | −6 |

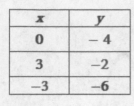

14.

| x | y |
|---|---|
| 0 | −4 |
| 3 | −3 |
| −3 | −5 |

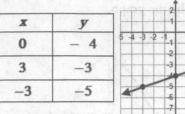

15. $(0,1)$    16. $x$-intercept: $(3,0)$; $y$-intercept: $(0,-1)$    17. $x$-intercept    18. $y$-intercept

20. $x$-intercept: $(2,0)$
    $y$-intercept: $(0,-3)$

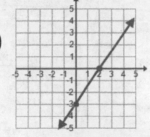

22a. $(0,250)$ Amber initially owed $250 for the dress.

22b. $(6,0)$ Six months after borrowing the money, she paid off the loan.

24. $x$-intercept: none
    $y$-intercept: $(0,1)$

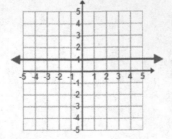

**Answers**: Section 2.2

2. Zero slope

4. $m = -\dfrac{1}{2}$

6. $m = -\dfrac{3}{4}$

8. Undefined

9. $m = 0$

10. undefined

11. $m = -\dfrac{1}{3}$

13. $m_1 = \dfrac{7}{2}; m_2 = -\dfrac{2}{7}$; perpendicular

15. The slope is approximately 200.83. This indicates that the revenue increased at a rate of $200.83 per month during this time period.

**Answers**: Section 2.3

2. $y$-intercept: $(0, -2)$, $m = -\dfrac{2}{3}$

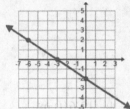

4. Parallel

5. $2x + y = 16$

6. $y = 4x - 2$; $4x - y = 2$

7. $2x + y = 16$

8. $y = 3x - 2$; $3x - y = 2$

10. $y = -\dfrac{3}{5}x - \dfrac{11}{5}$

12. $y = -\dfrac{2}{3}x + 5$

14. $y = \dfrac{3}{4}x - 5$

16. $y = -2$

**Answers**: Section 2.4

Monthly Salary Versus Sales

2a. $y = 0.04x + 1000$

2c. $(0,1000)$; Brian's base salary with $0 in sales is $1000.          2d. $2,200          4a. 12,610 mi

4b. 9,770 mi; The approximate value is very close to the actual value. It differs by only 70 mi.

4c. The slope $m=142$ means that the average number of miles driven per year increases by 142 mi each year.

4d. The y-intercept, $(0,9060)$, means that the average mileage was 9,060 mi in 1980 when $x=0$.

6a. $y=-0.25x+7$      6b. 3.25 ft

6c. $m=-0.25$; The slope means that the water level decreases at a rate of 0.25 ft/day.

**Answers**: Section 2.5

1. domain    2. range    3b. $\{3,1,5,-2\}$

4a. $\{(\text{Maine, }1820),\ (\text{Nebraska,}1823),\ (\text{Utah,}1847),\ (\text{Hawaii,}1959),\ (\text{Alaska, }1959)\}$

4b. Domain: $\{\text{Maine, Nebraska, Utah, Hawaii, Alaska}\}$; Range: $\{1820,1823,1847,1959\}$

6. Domain: $(-\infty,\infty)$; Range: $(-\infty,3]$

**Answers**: Section 2.6

2a. not a function          2b. not a function                4. function          6a. $f(-5)=-9$

6b. $f(w)=-w^2-3w+1$   8a. 1   8b. $x=-3$ and $x=5$   8c. $(-\infty,3]$       9a. zero

9b. negative                10. $(-\infty,4)\cup(4,\infty)$   11. $(-\infty,2]$

13. $P(10)=66.7\%$; $P(20)=88.9\%$; $P(20)$ means that if Brian studies for 20 hr, he will get a score of
    88.9%.

**Answers**: Section 2.7

1. is    2. is not    4. 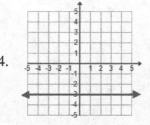    6.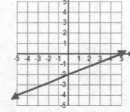

Video and Study Guide to accompany Intermediate Algebra, 5th ed., Miller/O'Neill/Hyde

8.

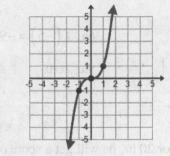

| x | $f(x) = x^3$ |
|----|------|
| -3 | -27 |
| -2 | -8 |
| -1 | -1 |
| 0 | 0 |
| 1 | 1 |
| 2 | 8 |
| 3 | 27 |

9.  downward    10.  upward      11.  linear      12.  constant

13. quadratic      15.  quadratic      17a. $(0,-1)$   17b. $\left(-\dfrac{1}{2},0\right)$      17c.

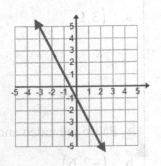

18. false      19.  true      21. $x$-intercepts: $(-4,0)$, $(0,0)$, and $(4,0)$; $y$-intercept: $(0,0)$

**Concept 1**: Solutions to System of Linear Equations

| Video 1: Determining if an Ordered Pair is a Solution to a System of Linear Equations (2:35 min) |
|---|

▶ Play the video and work along.

Determine if the given ordered pair is a solution to the system.

$$x + y = 4$$
$$2x - y = 5$$

a. $(3, 1)$

⏸ Pause the video and try this yourself.

b. $(-2, 6)$

▶ Play the video and check your answer.

| Review and complete | You try |
|---|---|
| Two or more linear equations form a **system of linear equations**. | |
| 1. An ordered pair is a **solution** to a system of equations if it satisfies _____ equations in the system. | |
| 2. Determine if $(3, 7)$ is a solution to the system | 3. Determine if $(2, 4)$ is a solution to the system |
| $$3x - y = 2$$ $$y = x + 2$$ | $$3x - y = 2$$ $$y = x + 2$$ |
| Equation 1:       Equation 2: $$3(3) - (7) \overset{?}{=} 2 \qquad (7) \overset{?}{=} (3) + 2$$ $$9 - 7 = 2 \checkmark \qquad\quad 7 \neq 5 \text{ False}$$ Because the ordered pair $(3, 7)$ does not satisfy *both* equations, the ordered pair _____ (**choose one**: is or is not) a solution to the system. | |

**Concept 2:** Solving Systems of Linear Equations by Graphing

| Video 2: Solving a System of Linear Equations by Using the Graphing Method (2:50 min) |
| --- |

▶ Play the video and work along.

Graph the system of equations and determine the solution set.

$$x + y = 4$$
$$2x - y = 5$$

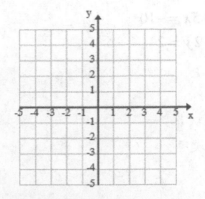

| Review and complete | You try |
| --- | --- |

To solve a system of linear equations by graphing, graph the individual equations in the system. The solution set to the system consists of all points of intersection of the graphs.

| | |
| --- | --- |
| 4. Graph the system of equations and determine the solution set.<br><br>$$2x + y = 1$$<br>$$y = x - 2$$<br><br>Graph each equation and determine the point(s) of intersection.<br><br>Equation 1: Slope-intercept form: $y = -2x + 1$<br>Equation 2: Slope-intercept form: $y = x - 2$<br><br>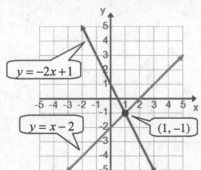<br><br>The solution set is _____. | 5. Graph the system of equations and determine the solution set.<br><br>$$3x - y = -2$$<br>$$x + 2y = -10$$<br><br><br><br><br><br><br>The solution set is _____. |

Video and Study Guide to accompany Intermediate Algebra, 5<sup>th</sup> ed., Miller/O'Neill/Hyde
Copyright © 2018 McGraw-Hill Education

## Video 3: Solving a System of Linear Equations by Using the Graphing Method (2:31 min)

■ Before starting the video, try this yourself.

Graph the system of equations and determine the solution set.

$$5x = -10$$
$$-x + 2y = 2$$

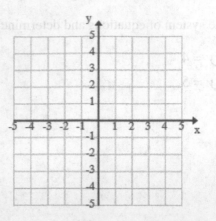

▶ Play the video and check your answer.

| Review and complete | You try |
|---|---|
| 6. Graph the system of equations and determine the solution set. | 7. Graph the system of equations and determine the solution set. |

6. Graph the system of equations and determine the solution set.

$$4y - 7 = 1$$
$$x - y = 1$$

Graph each equation and determine the point(s) of intersection.

Equation 1: Horizontal line: $y = 2$
Equation 2: Slope-intercept form: $y = x - 1$

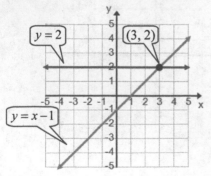

The solution set is _____.

7. Graph the system of equations and determine the solution set.

$$y = -2x - 3$$
$$-3y + 1 = -2$$

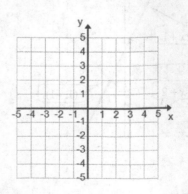

The solution set is _____.

## Video 4: Solutions to a System of Linear Equations in Two Variables: A Summary (3:32 min)

▶ Play the video and work along.

### Example 1

$$2x + y = -2$$
$$-x + 3y = 15$$ → Slope-intercept → $$y = -2x - 2$$
$$y = \frac{1}{3}x + 5$$

One unique solution

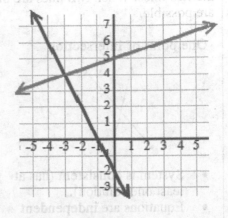

### Example 2

$$x + y = 3$$
$$2x + 2y = -4$$ → Slope-intercept → $$y = -x + 3$$
$$y = -x + 2$$

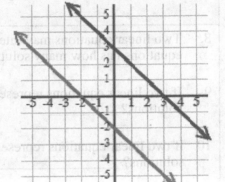

- A system is **inconsistent** if it has no solution.

- A system is **consistent** if it has one or more solutions.

### Example 3

$$x + y = 2$$
$$3x + 3y = 6$$ → Slope-intercept → $$y = -x + 2$$
$$y = -x + 2$$

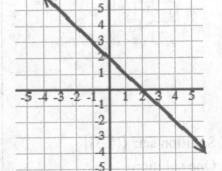

- If the equations in a system of equations represent the same line, then the equations are **dependent**.

- If the equations represent different lines, then the equations are **independent**.

Video and Study Guide to accompany Intermediate Algebra, 5th ed., Miller/O'Neill/Hyde

| Review and complete | You try |
|---|---|

A solution to a system of two linear equations may be interpreted graphically as the point of intersection of the two lines. When two lines are drawn in a rectangular coordinate system, three geometric relationships are possible.

**One point of intersection**

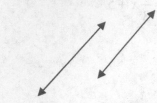

- System is **consistent** (has at least one solution).
- Equations are **independent** (equations represent two different lines).

**Parallel lines**

- System is **inconsistent** (has no solution).
- Equations are **independent** (equations represent two different lines).

**Coinciding lines (Same line)**

- System is **consistent** (has at least one solution).
- Equations are **dependent** (equations represent the *same* line).

---

8. Two linear equations may intersect in exactly one point. In such a case, the related system of equations has how many solutions? _____

9. If two linear equations represent parallel lines, then the related system of equations has how many solutions? _____

10. If two linear equations represent the same line, then the related system of equations has how many solutions? _____

---

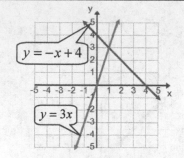

$y = -x + 4$

$y = 3x$

Solution set: $\{(1,3)\}$

One solution

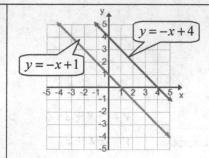

$y = -x + 4$

$y = -x + 1$

Solution set: $\{\ \}$

No solution

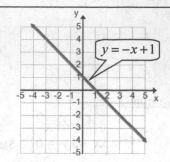

$y = -x + 1$

Solution set: (All points on the common line)

$\{(x, y) \mid y = -x + 1\}$

Video and Study Guide to accompany Intermediate Algebra, 5ᵗʰ ed., Miller/O'Neill/Hyde
Copyright © 2018 McGraw-Hill Education

11. Graph the system of equations and determine the solution set.

$$x + 3y = 3$$

$$\frac{1}{3}x + y = -2$$

Equation 1: Slope-intercept form: $y = -\frac{1}{3}x + 1$

Equation 2: Slope-intercept form: $y = -\frac{1}{3}x - 2$

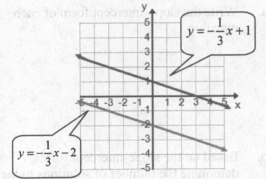

a. The lines each have a slope of _____, but different $y$-intercepts.

b. The lines are parallel and have no point of intersection. Thus, the number of solutions is _____.

c. The solution set is _____.

12. Graph the system of equations and determine the solution set.

$$\frac{1}{2}x + \frac{1}{4}y = \frac{3}{4}$$

$$y - 3 = -2x$$

a. The slope-intercept form of Equation 1 is: _____.

b. The slope-intercept form of Equation 2 is: _____.

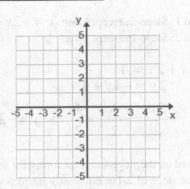

c. The lines are (**choose one**: parallel or coinciding).

d. How many solutions does the system have? _____

e. The solution set is _____.

## Video 5: Solving a System of Linear Equations Where the System is Inconsistent (2:20 min)

▶ Play the video and work along.

Graph the system of equations and determine the solution set. Also indicate if the system is inconsistent or if the equations are dependent.

$$y = -3x - 2$$

$$x + \frac{1}{3}y = 0$$

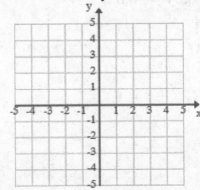

| Review and complete | You try |
| --- | --- |

**13.** If the equations in a system of linear equations represent parallel lines, then the lines never intersect. In such a case,

    a.   The system has how many solutions? _____

    b.   The solution set is _____, and the system is said to be _____.

---

**14.** Determine the solution set.

$$y = -2x + 4$$
$$4x + 2y = 4$$

Equation 1: Slope-intercept form: $y = -2x + 4$
Equation 2: Slope-intercept form:

$$4x + 2y = 4$$
$$2y = -4x + 4$$
$$\frac{2y}{2} = \frac{-4x}{2} + \frac{4}{2} \implies y = -2x + 2$$

- Based on the slope-intercept forms, the lines have the same slope of 2, but different $y$-intercepts.

- The lines are parallel and do not intersect. Thus, there is no solution.

The solution set is $\{\ \}$.

**15.** Determine the solution set.

$$2x + 3y = 6$$
$$6y = -4x - 6$$

    a.   Write the slope-intercept form of each equation.

    b.   Based on the slope-intercept forms, determine the number of solutions to the system.

    c.   Write the solution set to the system.

---

## Video 6: Solving a System of Dependent Linear Equations (2:37 min)

▶ Play the video and work along.

Graph the system of equations and determine the solution set.

$$x + y = -3$$
$$0.2x + 0.2y = -0.6$$

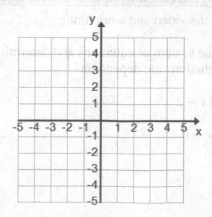

| Review and complete | You try |
|---|---|
| 16. If the equations in a system of linear equations represent the *same* line, then the lines intersect at all points on the common line. In such a case, | |

16. If the equations in a system of linear equations represent the *same* line, then the lines intersect at all points on the common line. In such a case,

    a.   The system has how many solutions? _____

    b.   The equations are said to be _____.

---

**17.** Determine the solution set.

$$15x - 15y = 20$$
$$3x = y + 4$$

Equation 1: Slope-intercept form:

$$15x - 5y = 20$$
$$-5y = -15x + 20$$
$$\frac{-5y}{-5} = \frac{-15x}{-5} + \frac{20}{-5} \Rightarrow y = 3x - 4$$

Equation 2: Slope-intercept form:

$$3x = y + 4 \Rightarrow y = -3x - 4$$

- The lines have the same slope of –3, and the same *y*-intercept $(0, -4)$.

- The equations represent the same line. All points on the common line are solutions (infinitely many solutions).

The solution set is $\{(x, y) \mid y = -3x + 4\}$.

---

**18.** Determine the solution set.

$$5x + 3y = 9$$
$$\frac{5}{3}x + y = 3$$

    a.   Write the slope-intercept form of each equation.

    b.   Based on the slope-intercept forms, determine the number of solutions to the system.

    c.   Write the solution set to the system.

**Concept 1**: The Substitution Method

| Video 1: Solving a System of Equations by Using the Substitution Method (4:05 min) |
|---|

▶ Play the video and work along.

Solve the system by using the substitution method.

$$2x + y = 1$$
$$4x + 3y = 1$$

**Step 1:** Isolate one of the variables from one of the equations.

**Step 2:** Substitute the quantity found in step 1 into the other equation.

**Step 3:** Solve the resulting equation.

**Step 4:** Substitute the value of the variable from step 3 into one of the original equations. Then solve for the remaining variable.

**Step 5:** Check the solution in both original equations.

| Review and complete | You try |
|---|---|

1. Solve the system by using the substitution method.

$$12x - 2y = 0$$
$$-7x + y = -1$$

**Step 1:** The $y$-variable in the second equation is the easiest variable to isolate because its coefficient is 1.

$$12x - 2y = 0$$
$$-7x + y = -1 \implies y = 7x - 1$$

**Step 2:** Substitute the quantity $7x - 1$ for $y$ in the other equation.

$$12x - 2(7x - 1) = 0$$

**Step 3:** Solve for $x$.

$$12x - 14x + 2 = 0 \quad \text{Simplify.}$$
$$-2x = -2 \quad \text{Divide by -2.}$$
$$x = 1$$

2. Solve the system by using the substitution method.

$$6x + 3y = 9$$
$$2x - y = -3$$

**Step 4:** Substitute the known value for *x* into one of the
original equations to solve for *y*.

$$y = 7x - 1$$
$$y = 7(1) - 1$$
$$y = 7 - 1$$
$$y = 6$$

**Step 5:** Check the ordered pair, $(1,6)$ in both original
equations.

$$12(1) - 2(6) = 0 \qquad -7(1) + (6) = -1$$

$$12 - 12 = 0$$

$$\qquad\qquad\qquad -7 + 6 = -1$$

$$0 = 0$$

$$\qquad\qquad\qquad -1 = -1$$

The ordered pair, checks in both equations.

Therefore, the solution to the system is $\{(1,6)\}$.

## Video 2: Solving a System of Equations by Using the Substitution Method (2:58 min)

■ Before starting the video, try this yourself.

Solve the system by using the substitution method.

$$2x - 3y = 4$$
$$4x + 5y = 8$$

Play the video and check your answer.

| Review and complete | You try |
|---|---|
| 3. Solve the system by using the substitution method.<br><br>$2x - 3y = 4$<br><br>$4x - 5y = 6$ | 4. Solve the system by using the substitution method.<br><br>$2x + 3y = 7$<br><br>$5x + 2y = 12$ |

Since we do not have a coefficient of 1 or -1 on any of the variables, we choose to solve the first equation for $x$.

$$2x - 3y = 4 \quad \rightarrow \quad 2x = 3y + 4$$

$$4x - 5y = 6 \qquad\qquad x = \frac{3}{2}y + 2$$

Substitute $\frac{3}{2}y + 2$ for $x$ in the other equation.

$$4x - 5y = 8$$

$$4\left(\frac{3}{2}y + 2\right) - 5y = 6$$

$$6y + 8 - 5y = 6$$

$$y + 8 = 6$$

$$y = -2$$

Substitute $y = -2$ into any of the original equations to find $x$.

$$x = \frac{3}{2}y + 2 \;\Rightarrow\; x = \frac{3}{2}(-2) + 2$$

$$\Rightarrow x = -3 + 2$$

$$\Rightarrow x = -1$$

The solution to the system is $\{(-1, -2)\}$.

## Video 3: Solving a System of Equations by Using the Substitution Method (1:40)

■ Before starting the video, try this yourself.

Solve the system by using the substitution method.

$$-4x - 5y = 2$$

$$3x = -9$$

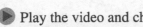 Play the video and check your answer.

| Review and complete | You try |
|---|---|
| 5.  Solve the system by using the substitution method. | 6.  Solve the system by using the substitution method. |

5.  $5x+2y=6$

$\quad\ \ 3y=9$

Since the second equation has only one variable, it is easy to isolate $y$ by dividing both sides by 3.

$5x+2y=6$

$3y=9 \quad \rightarrow \quad \dfrac{3y}{3}=\dfrac{9}{3} \quad \rightarrow \quad y=3$

6.  $3x-2y=6$

$\quad\ \ 4x=8$

Substitute 3 for $y$ in the first equation.

$5x+2(3)=6$

$5x+6=6 \qquad$ Subtract 6 from both sides.

$5x=0 \qquad\ $ Divide both sides by 5.

$x=0$

Therefore, the solution to the system is $\{(0,3)\}$.

**Concept 2:**  Solving Inconsistent Systems and Systems of Dependent Equations

| Video 4: Solving a System of Equations in which the System is Inconsistent (2:11 min) |
|---|

 Play the video and work along.

Solve the system by using the substitution method.

$x-3y=2$

$\quad 2x=6y+8$

$$x - 3y = 2 \longrightarrow y = \frac{1}{3}x - \frac{2}{3}$$

**slope-intercept form**

$$2x = 6y + 8 \longrightarrow y = \frac{1}{3}x - \frac{4}{3}$$

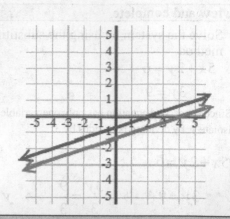

| Review and complete | You try |
|---|---|
| 7. Solve the system by using the substitution method.<br>$$-5x + y = -10$$<br>$$2y = 10x - 5$$ | 8. Solve the system by using the substitution method.<br>$$x + 4y = 8$$<br>$$3x = 3 - 12y$$ |

Solve the first equation for $y$ since its coefficient is 1.

$$-5x + y = -10 \quad \rightarrow \quad y = 5x - 10$$
$$2y = 10x - 5$$

Substitute $5x - 10$ for $y$ in the second equation.

$$2(5x - 10) = 10x - 5 \quad \text{Simplify.}$$
$$10x - 20 = 10x - 5 \quad \text{Subtract } 10x \text{ from both sides.}$$
$$-20 = -5$$

This is a contradiction which means there is no solution for $x$ and consequently no ordered pair that will be a solution to both equations. Therefore, the lines are parallel.

No solution; $\{ \ \}$

### Video 5: Solving a System of Dependent Equations (2:59 min)

▶ Play the video and work along.

Solve the system by using the substitution method.
$$\frac{1}{2}x + \frac{1}{4}y = 1$$
$$2(x + y) = 4 + y$$

$$\frac{1}{2}x+\frac{1}{4}y=1 \longrightarrow y=-2x+4$$

**slope-intercept form**

$$2(x+y)=4+y$$

$$\longrightarrow y=-2x+4$$

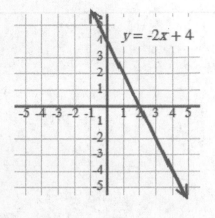

| Review and complete | You try |
|---|---|
| 9. Solve the system by using the substitution method.<br><br>$$x-4y=8$$<br>$$\frac{1}{16}x-\frac{1}{4}y=\frac{1}{2}$$<br><br>Clear fractions from the second equation by multiplying each side by the least common denominator, 16.<br><br>$$16\cdot\left(\frac{1}{16}x-\frac{1}{4}y\right)=16\cdot\left(\frac{1}{2}\right)$$<br>$$x-4y=8$$<br><br>Now solve this equation for $x$.<br><br>$$x=4y+8$$<br><br>Substitute $4y+8$ for $x$ in the first equation.<br><br>$(4y+8)-4y=8$  Simplify and solve for $y$.<br><br>$8=8$  This is called an identity.<br><br>When the variables are eliminated from the equation and the resulting equation is an identity, this means that the two equations represent the same line. | 10. Solve the system by using the substitution method.<br><br>$$2x-y=6$$<br>$$\frac{1}{6}x-\frac{1}{12}y=\frac{1}{2}$$ |

Infinitely many solutions; $\{(x,y)|x-4y=8\}$

## Concept 1: The Addition Method

| Video 1: Solving a System of Linear Equations by Using the Addition Method (3:35 min) |
| --- |

▶ Play the video and work along.

Solve the system by using the addition method.

$$3x = 2y - 5$$
$$2x + y = 6$$

**Step 1:** Write both equations in standard form, $Ax + By = C$.

**Step 2:** Clear fractions or decimals (optional).

**Step 3:** Multiply one or both equations by nonzero constants to create opposite coefficients for one of the variables.

**Step 4:** Add the equations from step 3.

**Step 5:** Solve for the remaining variable.

**Step 6:** Substitute the known value into one of the original equations.

**Step 7:** Check the ordered pair in both original equations.

| Video 2: Solving a System of Equations by Using the Addition Method (3:14 min) |
| --- |

■ Before starting the video, try this yourself.

Solve the system by using the addition method.

$$5x - 4y = 2$$
$$-3x + 7y = -15$$

▶ Play the video and check your answer.

| Review and complete | You try |
|---|---|
| 1. Solve the system by using the addition method. | 2. Solve the system by using the addition method. |

**Review and complete**

1. Solve the system by using the addition method.

$$-5x = 6y - 4$$
$$5y = 1 - 3x$$

**Step 1:** Write both equations in standard form.

$$-5x = 6y - 4 \implies -5x - 6y = -4$$
$$5y = 1 - 3x \implies 3x + 5y = 1$$

**Step 3:** Multiply one or both equations by appropriate constants to obtain opposite coefficients for either the $x$ or $y$ variable.

Multiply the first equation by 3 and the second equation by 5.

$$-5x - 6y = -4 \quad \rightarrow \quad -15x - 18y = -12$$
$$3x + 5y = 1 \quad \rightarrow \quad 15x + 25y = 5$$

**Step 4:** Add the resulting equations from step 3.

$$-15x - 18y = -12$$
$$\underline{15x + 25y = 5}$$
$$7y = -7$$

**Step 5:** Solve for the remaining variable.

$$\frac{7y}{7} = \frac{-7}{7} \quad \rightarrow \quad y = -1$$

**Step 6:** Substitute the known value into one of the original equations.

$$-5x = 6(-1) - 4$$
$$-5x = -6 - 4$$
$$-5x = -10$$
$$x = 2$$

The ordered pair $(2, -1)$ satisfies both equations.

So, the solution to the system is $\{(2, -1)\}$.

**You try**

2. Solve the system by using the addition method.

$$2x = 3y - 5$$
$$5x - 2y = -40$$

## Video 3: Solving a System of Equations by Using the Addition Method (4:01 min)

■ Before starting the video, try this yourself.

Solve the system by using the addition method.

$$3(x+2y) = y - 1200$$
$$-\frac{1}{10}x = 40 - \frac{1}{20}y$$

▶ Play the video and check your answer.

| Review and complete | You try |
|---|---|
| 3.  Solve the system by using the addition method.<br><br>$$3(x+y) = y + 2x + 2$$<br>$$x = -\frac{1}{3}y + \frac{16}{3}$$<br><br>Let's clean up the equations.<br><br>Clear the parenthesis and fractions, and then write both equations in standard form.<br><br>$$3(x+y) = y + 2x + 2 \rightarrow 3x + 3y = y + 2x + 2$$<br>$$\rightarrow x + 2y = 2$$<br><br>$$x = -\frac{1}{3}y + \frac{16}{3} \rightarrow 3 \cdot (x) = 3 \cdot \left(-\frac{1}{3}y + \frac{16}{3}\right)$$<br><br>$$\rightarrow 3x = -y + 16 \rightarrow 3x + y = 16$$<br><br>$$x + 2y = 2$$<br>$$3x + y = 16$$<br><br>Obtaining opposite coefficients for either the $x$ or $y$ variable. | 4.  Solve the system by using the addition method.<br><br>$$-7(x-y) = 16 + 3y$$<br>$$\frac{1}{2}x + \frac{5}{2}y = 10$$ |

Video and Study Guide to accompany Intermediate Algebra, 5<sup>th</sup> ed., Miller/O'Neill/Hyde

Let's multiply the first equation by -3.

$$x + 2y = 2 \quad \rightarrow \quad -3x - 6y = -6$$
$$3x + y = 16$$

Add the equations.

$$-3x - 6y = -6$$
$$\underline{3x + y = 16}$$
$$-5y = 10$$
$$y = -2$$

Substitute the known value into one of the original equations as given or rewritten in standard form.

$$x + 2y = 2 \rightarrow x + 2(-2) = 2 \rightarrow x - 4 = 2$$
$$\rightarrow x = 6$$

Check the ordered pair $(6, -2)$ in both original equations.

| $(6) + 2(-2) = 2$ | $3(6) + (-2) = 16$ |
|---|---|
| $6 - 4 = 2$ | $18 - 2 = 16$ |
| $2 = 2$ | $16 = 16$ |

Because the ordered pair satisfies both equations, the solution set to the system is

$$\{(6, -2)\}.$$

## Video 4: Solving a System of Equations in which the Solutions Are Fractions (2:43 min)

⏹ Before starting the video, try this yourself.

Solve the system by using the addition method.

$$3c - 14d = 7$$
$$9c + 2d = 11$$

⏸ Play the video and check your answer.

| Review and complete | You try |
|---|---|
| 5. Solve the system by using the addition method. | 6. Solve the system by using the addition method. |

**Review and complete**

5. Solve the system by using the addition method.

$$6x + 8y = 5$$
$$3x - 2y = 1$$

To get opposite coefficients on the $y$-variable, multiply the second equation by 4.

$$6x + 8y = 5$$
$$3x - 2y = 1 \quad \rightarrow \quad 12x - 8y = 4$$

Add the equations and solve for $x$.

$$6x + 8y = 5$$
$$\underline{12x - 8y = 4}$$
$$18x \qquad = 9$$
$$x \qquad = \frac{1}{2}$$

Substitute the known value into one of the original equations to find $y$.

$$6\left(\frac{1}{2}\right) + 8y = 5 \quad \rightarrow \quad 3 + 8y = 5 \quad \rightarrow \quad 8y = 2$$

$$\rightarrow \quad y = \frac{2}{8} = \frac{1}{4}$$

Check the ordered pair $\left(\frac{1}{2}, \frac{1}{4}\right)$ in both original equations.

$$6\left(\frac{1}{2}\right) + 8\left(\frac{1}{4}\right) = 5 \quad \bigg| \quad 3\left(\frac{1}{2}\right) - 2\left(\frac{1}{4}\right) = 1$$

$$\qquad\qquad\qquad\qquad \frac{3}{2} - \frac{1}{2} = 1$$

$$3 + 2 = 5$$

$$\qquad\qquad\qquad\qquad \frac{1}{2} = \frac{1}{2}$$

$$5 = 5$$

The solution to the system is $\left\{\left(\frac{1}{2}, \frac{1}{4}\right)\right\}$.

**You try**

6. Solve the system by using the addition method.

$$2c + 4d = 1$$
$$3c + 5d = 3$$

**Concept 2:** Solving Inconsistent Systems and Systems of Dependent Equations

## Video 5: Solving a System of Dependent Equations (1:37 min)

▶ Play the video and work along.

Solve the system by using the addition method.

$$-y - 2x = -28$$

$$x + \frac{1}{2}y = 14$$

---

| Review and complete | You try |
|---|---|
| 7.  Solve the system by using the addition method. | 8.  Solve the system by using the addition method. |

7.  Solve the system by using the addition method.

$$10x - 15y = 5$$

$$x - \frac{3}{2}y = \frac{1}{2}$$

To clear fractions from the second equation, multiply both sides of the equation by the least common denominator, 2.

$$10x - 15y = 5$$

$$2\left(x - \frac{3}{2}y = \frac{1}{2}\right) \quad \rightarrow \quad 2x - 3y = 1$$

Now multiply the second equation by -5 to get opposite coefficients on the $y$-terms in the system.

$$10x - 15y = 5$$

$$-5\left(2x - 3y = 1\right) \quad \rightarrow \quad -10x + 15y = -5$$

Add the equations.

$$\begin{array}{r} 10x - 15y = 5 \\ \underline{-10x + 15y = -5} \\ 0 = 0 \end{array}$$

The identity $0 = 0$ indicates that the equations in the system are dependent. The solution set consists of an infinite number of ordered pairs $(x, y)$ that fall on the common line of intersection, $10x - 15y = 5$,

or equivalently, $\left\{(x, y) \mid 10x - 15y = 5\right\}$

8.  Solve the system by using the addition method.

$$3x + y = 1$$

$$-x - \frac{1}{3}y = -\frac{1}{3}$$

---

**Video 6: Solving a System of Equations by Using the Addition Method (System is Inconsistent) (1:56 min)**

▶ Play the video and work along.

Solve the system by using the addition method.

$$0.3x = 0.4y - 1$$
$$6x = 8y$$

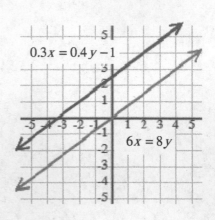

---

| **Review and complete** | **You try** |
|---|---|
| 9. Solve the system by using the addition method. | 10. Solve the system by using the addition method. |

<table>
<tr><td>

$$0.2x - 0.1y = -1.2$$
$$2x - y = 6$$

Clear the decimals from the first equation by multiplying both sides by 10.

$$0.2x - 0.1y = -1.2 \quad \rightarrow \quad 2x - y = -12$$
$$2x - y = 6$$

Multiply the resulting first equation by -1 to get opposties as coefficients for the *y*-variables.

$$-1(2x - y = -12) \quad \rightarrow \quad -2x + y = 12$$
$$2x - y = 6$$

</td><td>

$$0.4x + y = 1.2$$
$$2x + 5y = 10$$

</td></tr>
</table>

Add the equations.

$$-2x + y = 12$$
$$\underline{2x - y = 6}$$
$$0 = 18$$

This is a contradiction which means there is no solution for *x* and, consequently, no ordered pair that satisfies both equations.

Therefore, the lines in the system are parallel, and the system is inconsistent.

No solution; $\left\{ \quad \right\}$

**Concept 1**: Applications Involving Cost

| Video 1: An Application of Systems of Linear Equations (Popcorn and Drink Sales) (4:15 min) |
| --- |

▶ Play the video and work along.

At a movie theater, Maria bought 2 large popcorns and 3 drinks for $16. Annie bought 3 large popcorns and 5 drinks for $25.50. Determine the cost of one large popcorn and one drink.

**Check:**

2 popcorns + 3 drinks = $16.00                     3 popcorns + 5 drinks = $25.50

| Review and complete | You try |
| --- | --- |
| 1.  Adam and Eric bought school supplies. Adam spent $10.65 on 4 notebooks and 5 pens. Eric spent $7.50 on 3 notebooks and 3 pens. What is the cost of 1 notebook, and what is the cost of 1 pen? | 2.  Melissa has 2 cups of popcorn and 8 oz of soda for a total of 216 calories. Lenny has 1 cup of popcorn and 12 oz of soda for a total of 204 calories. Determine the number of calories per cup of popcorn and per ounce of soda. |

Let $n$ represent the cost of a notebook.
Let $p$ represent the cost of a pen.

$$4n + 5p = 10.65$$
$$3n + 3p = 7.50$$

Solve using the addition method. Multiply the first equation by -3 and the second equation by 5.

$$-3(4n + 5p = 10.65) \rightarrow -12n - 15p = -31.95$$
$$5(3n + 3p = 7.50) \rightarrow 15n + 15p = 37.5$$

$-12n-15p=-31.95$

$\underline{\phantom{-}15n+15p=37.5}$    Add the equations.

$\phantom{-}3n\phantom{+15p}=5.55$

$\phantom{-}n\phantom{+15p}=1.85$

Substitute 1.85 into one of the original equations to solve for p.

$3(1.85)+3p=7.50 \;\rightarrow\; 5.55+3p=7.50$

$\rightarrow\; 3p=1.95 \;\rightarrow\; p=.65$

The cost of 1 pen is $0.65. The cost of 1 notebook is $1.85.

## Concept 2: Applications Involving Mixtures

### Video 2: Setting up a System of Equations to Solve a Mixture Application (4:48 min)

▶ Play the video and work along.

How much 45% acid solution should be mixed with a 30% acid solution to make 450 mL of a 40% acid solution?

| 45% solution | | 30% solution | | 40% solution |
|---|---|---|---|---|
|  | + |  | = | 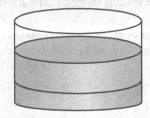 |

|  |  |  |  |
|---|---|---|---|
|  |  |  |  |
|  |  |  |  |

## Video 3: Solving a Mixture Application Using a System of Linear Equations (2:40 min)

■ Before starting the video, try this yourself.

How much 45% acid solution should be mixed with a 30% acid solution to make 450 mL of a 40% acid solution?

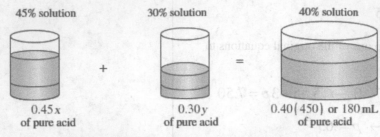

| | 45% acid | 30% acid | 40% acid |
|---|---|---|---|
| Amount of solution | $x$ | $y$ | 450 |
| Amount of pure acid | $0.45x$ | $0.30y$ | $0.40(450)$ or 180 |

▶ Play the video and check your answer.

| Review and complete | You try |
|---|---|

**3.** How many gallons each of 65% acid and 20% acid should be mixed to get 120 gallons of 50% acid?

Let $x$ = the number of gal of the 65% solution.
Let $y$ = the number of gal of the 20% solution.

| | 65% acid | 20% acid | 50% acid |
|---|---|---|---|
| Amount of solution | $x$ | $y$ | 120 |
| Amount of pure acid | $0.65x$ | $0.20y$ | $0.50(120)$ or 60 |

$x + y = 120$

$0.65x + 0.20y = 0.50(120)$

There is a total of 120 gal of acid solution.
The amount of pure acid in each solution is given by the second equation.

**4.** How many milliliters each of 30% acid solution and 10% acid should be mixed to get 100 mL of a 12% acid solution?

| | ___% acid | ___% acid | ___% acid |
|---|---|---|---|
| Amount of solution | | | |
| Amount of pure acid | | | |

First, clear the decimals from the second equation by multiplying both sides by 100.

$$x + y = 120$$

$$0.65x + 0.20y = 60 \quad \rightarrow \quad 65x + 20y = 6000$$

Now use the substitution method to solve the system. Solve the first equation for x.

$$x + y = 120 \quad \rightarrow \quad x = 120 - y$$
$$65x + 20y = 6000$$

Substitute $120 - y$ for $x$ in the other equation.

$$65x + 20y = 6000$$
$$65(120 - y) + 20y = 6000$$
$$7800 - 65y + 20y = 6000$$
$$-45y = -1800$$
$$y = 40$$

Substitute 40 in one of the original equations to find $x$.

$$x + (40) = 120 \quad \rightarrow \quad x = 80$$

Mix 80 gal of the 65% acid solution and 40 gal of the 20% acid solution.

**Concept 3:** Applications Involving Principal and Interest

---

## Video 4: Solving a Finance Application Using a System of Linear Equations (4:38 min)

▶ Play the video and work along.

Sandy invested a total of $7000. She put part of the money in an account earning 5% simple interest and the rest of the money in a mutual fund that earns 8% annual interest. If her total earnings is $485 after one year, find the amount invested in each account.

| | 5% account | 8% account | total |
|---|---|---|---|
| Principal invested | | | |
| Interest earned | | | |

| Review and complete | You try |
|---|---|

**Review and complete**

5. Jody invested $20,000. She put part of it at 3% simple interest and the rest at 4% simple interest. At the end of the first year, the total interest from both accounts was $725. Find the amount invested in each account.

Let $x$ represent the amount invested in the 3% account.
Let $y$ represent the amount invested in the 4% account.

|  | 3% account | 4% account | total |
|---|---|---|---|
| Principal invested | x | y | $20,000 |
| Interest earned | $0.03x$ | $0.04y$ | $725 |

The first row of the table gives us the equation

$$x + y = 20,000 \qquad \textbf{PRINCIPAL INVESTED}$$

The second row of the table gives us the equation

$$0.03x + 0.04y = 725 \quad \textbf{INTEREST EARNED}$$

Clear the decimals from the second equation by multiplying both sides by 100.

$$x + y = 20,000$$
$$0.03x + 0.04y = 725 \quad \rightarrow \quad 3x + 4y = 72,500$$

Solve the system by using the addition method. Multiply the first equation by -3 to eliminate x.

$$x + y = 20,000 \quad \rightarrow \quad -3x - 3y = -60,000$$
$$3x + 4y = 72,500$$

$$\begin{aligned} -3x - 3y &= -60,000 \\ 3x + 4y &= 72,500 \\ \hline y &= 12,500 \end{aligned}$$    Add the equations.

Substitute the known value into one of the original equations to find $x$.

$$x + (12,500) = 20,000 \quad \rightarrow \quad x = 7500$$

Jody invested $7500 in the 3% account and $12,500 in the 4% account.

**You try**

6. Mitch won $2000 at the racetrack. He invested part of it at 6% simple interest and the rest at 7%. His total annual interest from the two investments was $135. How much did he invest at each rate?

|  | ___% account | ___% account | total |
|---|---|---|---|
| Principal invested |  |  |  |
| Interest earned |  |  |  |

## Concept 4: Applications Involving Uniform Motion

**Video 5: Solving an Application Using Uniform Motion with a System of Linear Equations (5:22 min)**

▶ Play the video and work along.

Suppose a plane leaves Fargo, ND, headed for Orlando, FL, 2400 miles away. Going to Orlando, the plane encounters a "tailwind" and gets to Orlando in 4 hr. Returning to Fargo, the same wind is now an opposing "headwind," and the plane takes 5 hr to return. Find the speed of the airplane in still air and speed of the wind.

|  | Distance | Rate | Time |
|---|---|---|---|
| With a tailwind |  |  |  |
| Against a headwind |  |  |  |

Check:

|  | Distance | Rate | Time |
|---|---|---|---|
| With a tailwind |  |  |  |
| Against a headwind |  |  |  |

| **Review and complete** | **You try** |
|---|---|

**7.** In her kayak, Bonnie can travel 31.5 mi in 7 hr going downstream with the current. The return trip against the current takes 9 hr. Find the speed of the kayak in still water and the speed of the current.

**8.** Kim rides a total of 48 km in the bicycle portion of a triathlon. The course is an "out and back" route. It takes her 3 hr on the way out against the wind. The ride back takes her 2 hr with the wind. Find the speed of the wind and Kim's speed riding her bike in still air.

Let $k$ represent the speed of the kayak in still water.
Let $c$ represent the speed of the current.

|  | Distance | Rate | Time |
|---|---|---|---|
| With the current | 31.5 | $k + c$ | 7 |
| Against the current | 31.5 | $k - c$ | 9 |

|  |  |  |  |
|---|---|---|---|
|  |  |  |  |
|  |  |  |  |

We use the fact that distance is rate times time to get two equations for the system.

$$31.5 = 7(k + c)$$
$$31.5 = 9(k - c)$$

Simplify each equation in the system.

Divide by 7.

$$31.5 = 7(k + c) \quad \rightarrow \quad 4.5 = k + c$$
$$31.5 = 9(k - c) \quad \rightarrow \quad 3.5 = k - c$$

Divide by 9.

Now apply the addition method to solve the system.

$$4.5 = k + c$$
$$\underline{3.5 = k - c}$$
$$8 = 2k$$
$$4 = k$$

Substitute the known value into one of the original equations to solve for $c$.

$$4.5 = (4) + c \quad \rightarrow \quad .5 = c$$

Bonnie's speed in still water is 4 mph and the speed of the current is 0.5 mph.

**Concept 5:** Applications Involving Geometry

## Video 6: Using a System of Equations for an Application in Geometry (2:35 min)

▶ Play the video and work along.

Two angles are complementary. The measure of one angle is 5° more than four times the other. Find the measure of each angle.

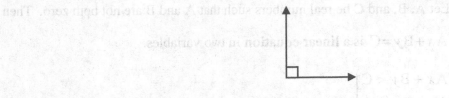

| Review and complete | You try |
|---|---|
| 9. The measure of one angle is 5 times the measure of another. If the two angles are supplementary, find the measures of the angles. | 10. Two angles are complementary. One angle measures 15° more than 2 times the measure of the other. What are the measures of the two angles? |

Let $x$ represent smaller angle.
Let $y$ represent the larger angle.

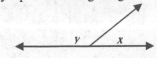

$$y = 5x$$
$$x + y = 180$$

The larger angle is 5 times the measure of the smaller angle.
The two angles are supplementary.

We solve the system using the substitution method.
Substitute $5x$ for $y$ in the second equation.

$$x + (5x) = 180$$
$$6x = 180$$
$$x = 30$$

Substitute the known value into one of the original equations.

$$y = 5(30) \rightarrow y = 150$$

The angles are 30° and 150°.

Video and Study Guide to accompany Intermediate Algebra, 5th ed., Miller/O'Neill/Hyde

**Concept 1**: Graphing Linear Inequalities in Two Variables

**Video 1: Introduction to Linear Inequalities in Two Variables (2:30 min)**

▶ Play the video and work along.

---

**Definition of Linear Equations and Inequalities in Two Variables**

Let A, B, and C be real numbers such that A and B are not both zero. Then

$Ax + By = C$ is a **linear equation** in two variables.

$$\left.\begin{array}{l} Ax + By < C \\ Ax + By \leq C \\ Ax + By > C \\ Ax + By \geq C \end{array}\right\}$$ are **linear *inequalities*** in two variables.

---

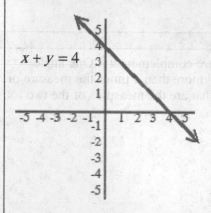

$x + y = 4$

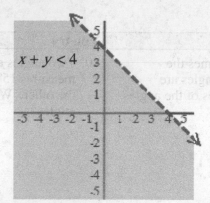

$x + y < 4$

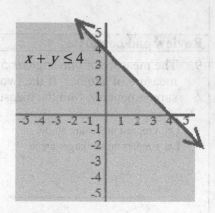

$x + y \leq 4$

**Video 2: Solving a Linear Inequality in Two Variables (2:29 min)**

▶ Play the video and work along.

Graph the solution set.

$2x + y \geq 3$

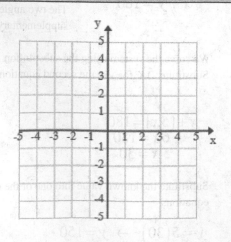

## Video 3: Solving a Linear Inequality in Two Variables (2:47 min)

■ Before starting the video, try this yourself.

Graph the solution set.

$-3y > 5x$

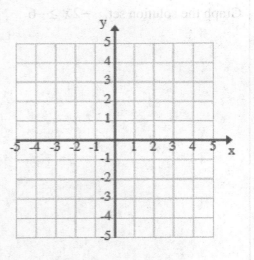

▶ Play the video and check your answer.

| Review and complete | You try |
|---|---|

1. Graph the solution set. $-2y < 6x$

Solve for $y$ by diving both sides by –2. Remember to reverse the inequality sign.

$$\frac{-2y}{-2} > \frac{6x}{-2} \quad \rightarrow \quad y > -3x$$

Graph the line defined by the related equation, $y = -3x$. The line has a $y$-intercept of (0, 0) and slope -3. Draw a dashed line through the points because the inequality is strict.

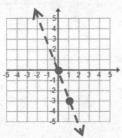

**Shading:** The points in the solution set are the points where $y > -3x$. These are points in the plane where the $y$-values are greater than those on the boundary line.

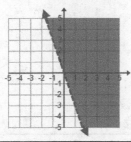

2. Graph the solution set. $x - 3y < 8$

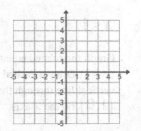

Video and Study Guide to accompany Intermediate Algebra, 5th ed., Miller/O'Neill/Hyde

## Video 4: Solving a Linear Inequality in Two Variables (1:37 min)

■ Before starting the video, try this yourself.

Graph the solution set.   $-2x \geq -6$

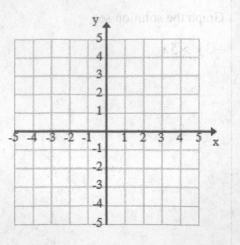

▶ Play the video and check your answer.

| Review and complete | You try |
|---|---|

3.   Graph the solution set.

$$x+6<7$$

Although we can't solve for $y$, we can simplify the inequality by subtracting 6 from both sides.

$$x<1$$

Graph the related equation $x=1$. Draw a dashed vertical line through the point $(1,0)$ on the $x$-axis.

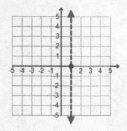

The solution is all ordered pairs whose $x$-coordinates are 1 or less, which are the points to the left of the line.

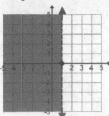

You can verify this graph by using test points.

4.   Graph the solution set.

$$y+1>-1$$

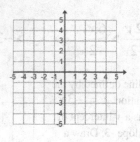

Video and Study Guide to accompany Intermediate Algebra, 5ᵗʰ ed., Miller/O'Neill/Hyde

## Concept 2: Compound Linear Inequalities in Two Variables

### Video 5: Graphing the Intersection of Two Linear Inequalities (2:46 min)

▶ Play the video and work along.

Graph the solution set to the compound inequality.

$$y \geq \frac{1}{3}x - 1 \quad \text{and} \quad -2x + y < 2$$

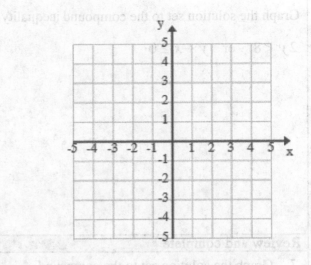

---

### Review and complete

5.  Graph the solution set to the compound inequality.

$$y < 3 \quad \text{and} \quad x + 2y < 6$$

$y < 3$: Draw a dashed horizontal line through the point $(0,3)$ on the $y$-axis. The solutions to this inequality are points below the horizontal line.

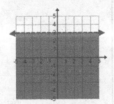

$x + 2y < 6$: $y < -\frac{1}{2}x + 3$

Draw a dashed line through the points (0, 3) and (2, 2) (down 1, right 2 from the y-intercept). The solutions to this inequality are points below the line.

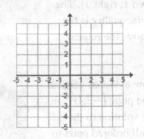

The solution to the system of inequalities is the intersection of these sets as shown.

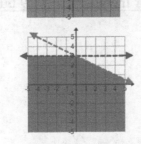

### You try

6.  Graph the solution set to the compound inequality.

$$x + y < 4 \quad \text{and} \quad 3x + y < 9$$

---

## Video 6: Graphing the Union of Two Linear Inequalities (2:31 min)

▶ Play the video and work along.

Graph the solution set to the compound inequality.

$2y < 8$ or $y + x \geq 0$

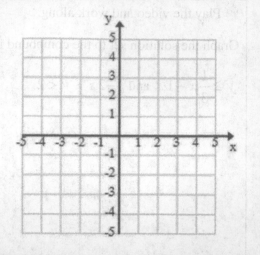

| Review and complete | You try |
|---|---|
| 7. Graph the solution set to the compound inequality. | 8. Graph the solution set to the compound inequality. |

$$x + 3y \geq 3 \quad \text{or} \quad x < -2$$

$x + 3y \geq 3$ is $y \geq -\dfrac{1}{3}x + 1$.

Draw a solid line through $(0, 1)$ and $(3, 0)$ (down 1, right 3). The solution to the inequality is the set of points above the related line.

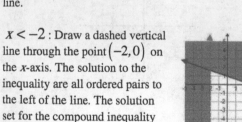

$x < -2$ : Draw a dashed vertical line through the point $(-2, 0)$ on the $x$-axis. The solution to the inequality are all ordered pairs to the left of the line. The solution set for the compound inequality is the union of the two graphs.

$$3x + 2y \geq 4 \quad \text{or} \quad x - y < 3$$

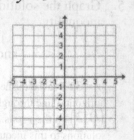

## Video 7: Graphing the Intersection of Two Linear Inequalities (1:29 min)

▶ Play the video and work along.

Graph the solution set to the compound inequality.

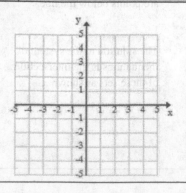

$x \geq 0$    and    $y > 0$

| Review and complete | You try |
|---|---|
| 9.  Graph the solution set to the compound inequality. | 10.  Graph the solution set to the compound inequality. |

9.  Graph the solution set to the compound inequality.

$$x < 3 \quad \text{and} \quad y > -1$$

$x < 3$: Draw a dashed vertical line through the point $(3, 0)$ on the $x$-axis. The solutions are all ordered pairs to the left of the vertical line.

$y > -1$: Draw a dashed horizontal line through the point $(0, -1)$ on the $y$-axis. The solutions are all ordered pairs above the horizontal line.

The solution to the system of inequalities is the intersection (overlap) of these sets as shown.

10.  Graph the solution set to the compound inequality.

$$y < 1 \quad \text{and} \quad x > -3$$

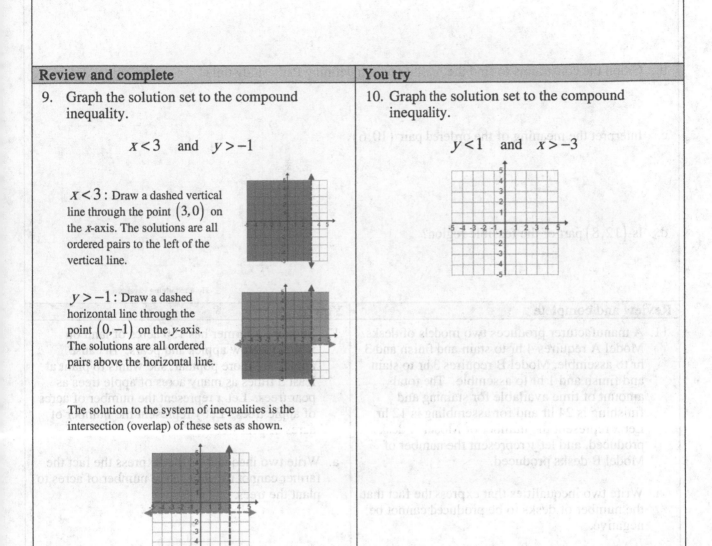

**Concept 3:** Graphing a Feasible Region

| Video 8: Writing Constraints and Graphing a Feasible Region (4:00 min) |
|---|

▶ Play the video and work along.

Pat has two tests: one for psychology and one for history. The tests are on the same day. The classes meet in consecutive hours, so she has no time to study between classes. Pat estimates that she has a maximum of 18 h of time to study. She needs to divide her time to study for the two tests.

Let $x$ represent the number of hours Pat studies psychology.
Let $y$ represent the number of hours Pat studies history.

a. Write a set of inequalities to describe the constraints on Pat's time.

b. Graph the constraints to find the feasible region defining Pat's study time.

c. Interpret the meaning of the ordered pair $(10,6)$.

d. Is $(12,8)$ part of the feasible region?

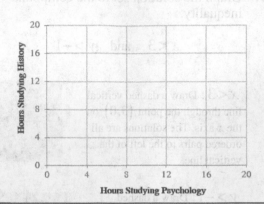

| Review and complete | You try |
|---|---|
| 11. A manufacturer produces two models of desks. Model A requires 4 hr to stain and finish and 3 hr to assemble. Model B requires 3 hr to stain and finish and 1 hr to assemble. The total amount of time available for staining and finishing is 24 hr and for assembling is 12 hr. Let $x$ represent the number of Model A desks produced, and let $y$ represent the number of Model B desks produced. | 12. Suppose a farmer has 150 acres of land on which to grow apples and pears. Because apples are more popular, she wants to plant at least 5 times as many acres of apple trees as pear trees. Let $x$ represent the number of acres of apple trees. Let $y$ represent the number of acres of pear trees. |
| a. Write two inequalities that express the fact that the number of desks to be produced cannot be negative.<br><br>$x \geq 0,\ y \geq 0$ | a. Write two inequalities that express the fact the farmer cannot use a negative number of acres to plant the trees. |
| b. Write an inequality in terms of the number of desks that can be produced if the total time for staining and finishing is at most 24 hr.<br><br>"At most" implies that the total time is less than or equal to 24 hr. It takes 4 hr to stain and finish Model A desks, and 3 hr to stain and finish Model B desks.<br><br>$4x + 3y \leq 24$ | b. Write an inequality that expresses that the total number of acres use for growing the trees is at most 150. |
| c. Write an inequality in terms of the number of desks that can be produced if the total time for assembly is no more than 12 hr. | c. Write an inequality that expresses the fact that the farmer wants to plant at least 5 times as many acres of apple trees as pear trees. |

"No more than" implies that the total time is less than or equal to 12 hr. It takes 3 hr to assemble Model A desks, and 1 hr to assemble Model B desks.

$$3x + y \leq 12$$

d. Graph the constraints to find a feasible region.

The first two conditions, $x \geq 0$ and $y \geq 0$, represent the set of points in the first quadrant.

Solve the third and fourth inequalities for $y$.

$$4x + 3y \leq 24 \qquad\qquad 3x + y \leq 12$$
$$3y \leq -4x + 24 \qquad\qquad y \leq -3x + 12$$
$$y \leq -\frac{4}{3}x + 8$$

Draw a solid line through these related equations and shade below each line.

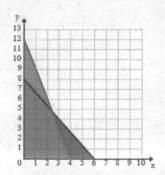

The feasible region is the intersection of the inequalities.

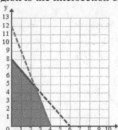

e. Is the point $(3,1)$ in the feasible region? What does the point $(3,1)$ represent in the context of this problem?

Yes, the point $(3,1)$ is in the feasible region.

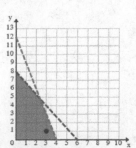

The $x$-coordinate represents the number of Model A desks produced and $y$-coordinate represents the number of

d. Graph the constraints to find a feasible region.

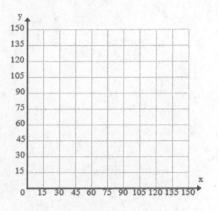

e. Is the point $(105,30)$ in the feasible region? What does the point $(105,30)$ represent in the context of this problem?

Model B desks produced.

Therefore, the point $(3,1)$ means that 3 Model A desks and 1 Model B desk are produced.

**Concept 1**: Solutions to Systems of Linear Equations in Three Variables

| Video 1: Introduction to Linear Equations in Three Variables (2:30 min) |
| --- |

▶ Play the video and work along.

**Definition of Linear Equations in Two Variables**

Let A, B, and C be real numbers. Then $Ax + By = C$ is a linear equation in *two* variables. (A, B are not both 0)

$$x + y = 4$$

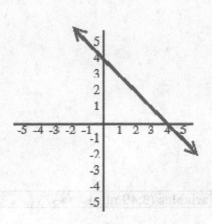

**Definition of Linear Equations in Three Variables**

Let A, B, C and D be real numbers. Then $Ax + By + Cz = D$ is a linear equation in *two* variables. (A, B, and C are not all 0)

$$4x + 2y + 3z = 12$$

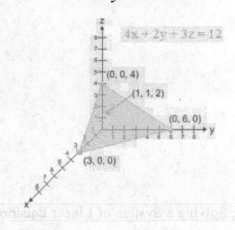

More solutions

$(3, 0, 0)$

$(0, 6, 0)$

$(0, 0, 4)$

| Review and complete | You try |
| --- | --- |
| 1. Is $(1,1,3)$ a solution of the system<br><br>$-3x - 3y - 6z = -24$<br><br>$-9x - 6y + 3z = -45$ ?<br><br>$9x + 3y - 9z = 33$<br><br>Check $(1,1,3)$<br><br>$-3(1) - 3(1) - 6(3) = -24$<br><br>$-9(1) - 6(1) + 3(3) \neq -45$<br><br>$9(1) + 3(1) - 9(3) \neq 33$<br><br>$(1,1,3)$ is not a solution. | 2. Is $(6,1,-1)$ a solution of the system<br><br>$x + y - z = 8$<br><br>$x - 2y + z = 3$ ?<br><br>$x + 3y + 2z = 7$ |

**Concept 2:** Solving Systems of Linear Equations in Three Variables

| Video 2: Solving a System of Linear Equations in Three Variables (5:16 min) |
| --- |

▶ Play the video and work along.

Solve the system.
$$3x + 2y + z = 10$$
$$4x - 3y + 2z = 4$$
$$2x + y - 4z = -8$$

| Video 3: Solving a System of Linear Equations in Three Variables (3:49 min) |
| --- |

■ Before starting the video, try this yourself.

Solve the system.
$$2x \qquad + z = 7$$
$$5y - 3z = -7$$
$$-3x - 4y \qquad = -4$$

▶ Play the video and check your answer.

| Review and complete | You try |
|---|---|
| 3.  Solve the system.<br><br>A $2x + \quad 2z = 18$<br><br>B $\quad y - 7z = -37$<br><br>C $x - 6y \quad = 16$ | 4.  Solve the system.<br><br>$x + y \quad = 3$<br><br>$2x \quad - z = 5$<br><br>$3y + z = 2$ |

Eliminate $x$ from equations A and C.

$(-2C) \quad -2x + 12y = -32$

$\underline{(A) \qquad 2x \quad + 2z = 18}$

$(D) \qquad 12y + 2z = -14$

Use equations D and B to solve for $z$.

$(D) \qquad 12y + 2z = -14$

$\underline{(-12B) \ -12y + 84z = 444}$

$\qquad\qquad\qquad 86z = 430$

$\qquad\qquad\qquad\quad z = 5$

Use equation B to solve for $y$.

$(B) \quad y - 7(5) = -37$

$\qquad\quad y - 35 = -37$

$\qquad\qquad\quad y = -2$

Now use equation C to solve for $x$.

$(C) \qquad x - 6y = 16$

$\qquad\quad x - 6(-2) = 16$

$\qquad\qquad x + 12 = 16$

$\qquad\qquad\qquad x = 4$

The solution set is $\{(4, -2, 5)\}$.

|  |  |
|---|---|
|  |  |

**Concept 3:** Applications of Linear Equations in Three Variables

---

**Video 4: Using a System of Three Equations to Solve an Application in Geometry (4:56 min)**

▶ Play the video and work along.

The measure of the largest angle in a triangle is 3° less than 4 times the measure of the smallest angle. The middle angle is 12° less than the measure of the largest angle. Find the measure of each angle.

Let $x$ represent the measure of the smallest angle.
Let $y$ represent the measure of the middle angle.
Let $z$ represent the measure of the largest angle.

---

| **Review and complete** | **You try** |
|---|---|
| 5. The largest angle of a triangle measures 4° less than 5 times the measure of the smallest angle. The middle angle measures twice that of the smallest angle. Construct the system that represents this situation. | 6. Solve the system you constructed in the previous problem, #5. |
| Let $x$ represent the measure of the smallest angle. | $5x \quad\quad - z = \quad 4$ |
| Let $y$ represent the measure of the middle angle. | $2x - y \quad\ = \quad 0$ |
| Let $z$ represent the measure of the largest angle. | $x + y + z = 180$ |
| $z = 5x - 4$ | |
| $y = 2x$ | |

Now use the fact that the sum of the angles of a triangle is 180° to write the third equation in the system.

$$z = 5x - 4$$
$$y = 2x$$
$$x + y + z = 180$$

## Video 5: Using a System of Three Equations to Solve a Consumer Application (4:46 min)

▶ Play the video and work along.

At a county fair, a family of four bought 1 hot dog, 3 hamburgers, and 3 sodas for $19. A couple bought two hamburgers and two sodas for $11, and a group of teenagers bought 6 hot dogs, 8 hamburgers, and 14 sodas for $68. Find the unit cost for a hot dog, a hamburger, and a soda.

Let $x$ represent the cost of one hot dog.
Let $y$ represent the cost of one hamburger.
Let $z$ represent the cost of one soda.

| Review and complete | You try |
|---|---|
| 7. At a rock concert, hats, long-sleeved and short-sleeved t-shirts are available for sale. Two hats, three short-sleeved shirts, and one long-sleeved shirt cost $144. Two short-sleeved shirts and two long-sleeved shirts cost $76. Five hats and one short-sleeved shirt cost $125. Construct a system of equations to find the cost of each item. | 8. Solve the system constructed in problem 7 to determine that cost of each item. $$2x + 3y + z = 144$$ $$2y + 2z = 76$$ $$5x + y = 125$$ |

Let $x$ represent the cost of one hat.
Let $y$ represent the cost of one short-sleeved shirt.
Let $z$ represent the cost of one long-sleeved shirt.

$$2x + 3y + z = 144$$
$$2y + 2z = 76$$
$$5x + y = 125$$

**Concept 4:** Solving Inconsistent Systems and Systems of Dependent Equations

**Video 6: Solutions to a System of Linear Equations in Three Variables (1:02 min)**

▶ Play the video and work along.

One unique solution (planes intersect at one point)

- The system is consistent.
- The equations are independent.

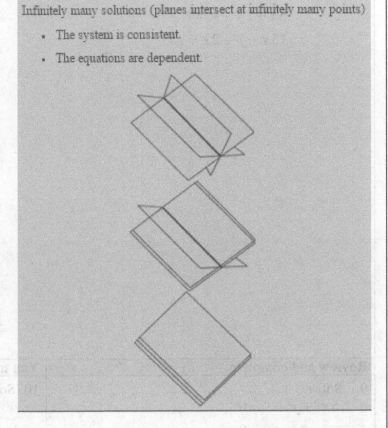

Infinitely many solutions (planes intersect at infinitely many points)

- The system is consistent.
- The equations are dependent.

No solution (the three planes do not all intersect)

- The system is inconsistent.
- The equations are independent.

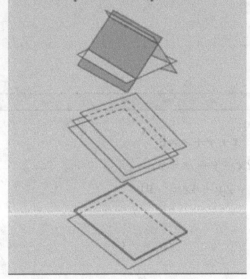

## Video 7: Solving a System of Three Linear Equations: Dependent Equations (3:12 min)

▶ Play the video and work along.

Solve.
$$4x - y + z = 8$$
$$3x + 2y - z = 1$$
$$15x - y + 2z = 25$$

| Review and complete | You try |
| --- | --- |
| 9. Solve. | 10. Solve. |

9. Solve.
$$5x + y = 0$$
$$4y - z = 0$$
$$5x + 5y - z = 0$$

Use equations A and C to eliminate $x$.

(A)      $5x + y = 0$
(-1C)  $\underline{-5x - 5y + z = 0}$
(D)        $-4y + z = 0$

Use equations B and D to eliminate z.
(B)    $4y - z = 0$
(D)  $\underline{-4y + z = 0}$
          $0 = 0$

An identity results which means that there are infinitely many solutions to the system. The equations are dependent.

10. Solve.
$$x + y + z = 8$$
$$2x - y + z = 6$$
$$-5x - 2y - 4z = -30$$

**Video 8: Solving a System of Three Linear Equations: Inconsistent System (2:34 min)**

▶ Play the video and work along.

Solve.

$$5x = 3y - 4z + 1$$

$$3(x - 2y) = -8z - 7x + 4$$

$$3x + 2y + 7z = 1$$

| Review and complete | You try |
|---|---|
| 11. Solve.<br><br>$6x - 2y + 2z = 2$<br><br>$4x + 8y - 2z = 5$<br><br>$2x + 4y - z = 2$<br><br><br>Use equations B and C to eliminate $x$.<br>(B)   $4x + 8y - 2z = 5$<br>(-2C)   $\underline{-4x + 8y + 2z = -4}$<br>$\qquad\qquad 0 = 1$<br><br><br>Because a contradiction results, the system has no solution. | 12. Solve.<br><br>$-y + 2z = 2$<br><br>$x + y + z = 2$<br><br>$-x \qquad - 3z = 2$ |

**Concept 1**: Introduction to Matrices

| Video 1: Introduction to Matrices (1:35 min) |
| --- |

▶ Play the video and work along.

> ### Definition of a Matrix and Order of a Matrix
>
> A **matrix** is a rectangular array of numbers (the plural of matrix is "matrices.") The **order of a matrix** is determined by the number of rows and columns. A matrix with $m$ rows and $n$ columns is a $m \times n$ matrix.

Determine the order of each matrix.

$$A = \begin{bmatrix} 3 & 2 & 5 & 1 \\ 6 & \sqrt{7} & -3 & \pi \end{bmatrix}$$

$$B = \begin{bmatrix} 5 & 2 & 1 \\ 6 & 3 & -9 \\ 2.2 & 4 & 0 \end{bmatrix}$$

⏸ Pause the video and try these yourself.

$$C = \begin{bmatrix} 2 & -8 & \dfrac{1}{2} \end{bmatrix}$$

$$D = \begin{bmatrix} 2 \\ -5 \end{bmatrix}$$

▶ Play the video and check your answers.

| Review and complete | You try |
| --- | --- |
| 1. Determine the order of the matrix. $$M = \begin{bmatrix} 3 & 8 & 9 & 0 & 1 \\ 1 & 2 & 5 & 6 & 7 \\ 2 & 2 & 4 & -9 & 1 \end{bmatrix}$$ There are three rows and five columns, so this matrix is a $3 \times 5$ matrix. | 2. Determine the order of the matrix. $$N = \begin{bmatrix} 5 \\ 9 \\ 1 \\ -2 \end{bmatrix}$$ |

| Video 2: Introduction to Coefficient Matrices and Augmented Matrices (2:30 min) |
| --- |

▶ Play the video and work along.

For each system of linear equations, identify the coefficient matrix and the augmented matrix.

$$3x = 2y - 2$$
$$4x + y = 12$$

coefficient matrix:

augmented matrix:

**⏸ Pause the video and try this yourself.**

$$3x \quad + \quad z = 9$$
$$x + 6y - 5z = 13$$
$$\qquad 5y - 2z = 1$$

coefficient matrix

augmented matrix:

**▶ Play the video and check your answer.**

| Review and complete | You try |
|---|---|
| 3.  Write the augmented matrix for the linear system.<br><br>$\qquad x + \phantom{6}y - z = 12$<br>$2x + 6y \qquad = \phantom{-}1$<br>$\qquad\qquad y - z = -5$<br><br>$\begin{bmatrix} 1 & 1 & -1 & 12 \\ 2 & 6 & 0 & 1 \\ 0 & 1 & -1 & -5 \end{bmatrix}$ | 4.  Write the augmented matrix for the linear system.<br><br>$\qquad x + \phantom{2}y - \phantom{2}z = 8$<br>$\qquad x - 2y + \phantom{2}z = 3$<br>$\qquad x + 3y + 2z = 7$ |

**Video 3: Writing a System of Equations from an Augmented Matrix (2:45 min)**

**▶ Play the video and work along.**

Write a system of linear equations represented by each augmented matrix.

$$\begin{bmatrix} 3 & -7 & -4 \\ 4 & 2 & 6 \end{bmatrix}$$

**⏸ Pause the video and try these yourself.**

$$\begin{bmatrix} -3.1 & 2.4 & 9.1 & 1.7 \\ 0 & 1.3 & 8.7 & 2.6 \\ 6.1 & 2.2 & -6.6 & 10.5 \end{bmatrix}$$

$$\begin{bmatrix} 1 & 0 & 0 & \bigm| & \dfrac{1}{2} \\ 0 & 1 & 0 & \bigm| & \dfrac{1}{3} \\ 0 & 0 & 1 & \bigm| & 4 \end{bmatrix}$$

▶ Play the video and check your answers.

| Review and complete | You try |
|---|---|
| 5.  Write a system of linear equations represented by the augmented matrix.<br><br>$$\begin{bmatrix} 0 & 3 & \bigm| & 30 \\ -2 & 5 & \bigm| & 4 \end{bmatrix}$$<br><br>$$3y = 30$$<br>$$-2x + 5y = 4$$ | 6.  Write a system of linear equations represented by the augmented matrix.<br><br>$$\begin{bmatrix} 2 & 0 & \bigm| & 12 \\ 1 & -6 & \bigm| & 0 \end{bmatrix}$$ |

**Concept 2:** Solving Systems of Linear Equations by Using the Gauss-Jordan Method

**Video 4: Elementary Row Operations (4:32 min)**

▶ Play the video and work along.

$$4x + 2y = 26$$

$$-\dfrac{1}{2}x + y = 3$$

**Summary of Elementary Row Operations**

The following operations on an augmented matrix result in an equivalent augmented matrix.
  *  Interchange two rows.
  *  Multiply every element in a row by a nonzero real number.
  *  Add a multiple of one row to another row.

| Review and complete | You try |
|---|---|

7. Consider the matrix given. Do the operations result in an equivalent augmented matrix?

$$\begin{bmatrix} 3 & 4 & | & -7 \\ 2 & -1 & | & 12 \end{bmatrix}$$

   a. Add rows 1 and 2.
   b. Add 3 to every element of row 2.
   c. Multiplying every element of row 2 by 3.
   d. Interchanging rows 1 and 2.

## Video 5: Introduction to the Gauss-Jordan Method to Solve Systems of Linear Equations (1:59 min)

▶ Play the video and work along.

Use the Gauss-Jordan method to solve the system of equations.

$2x + 4y = -2$
$5x + 9y = -6$

Write augmented matrix.

$$\begin{bmatrix} 2 & 4 & | & -2 \\ 5 & 9 & | & -6 \end{bmatrix}$$

Get reduced row-echelon form.

$$\begin{bmatrix} 1 & 0 & | & -3 \\ 0 & 1 & | & 1 \end{bmatrix}$$

Write the equivalent system of equations.

Simplify.

$-3x \quad\quad + z = -1$
$x + 2y + 3z = 5$
$2x - 5y + z = 9$

Write augmented matrix.

$$\begin{bmatrix} 3 & 0 & 2 & | & -1 \\ 1 & 2 & 3 & | & 5 \\ 2 & -5 & 1 & | & 9 \end{bmatrix}$$

Get reduced row-echelon form.

$$\begin{bmatrix} 1 & 0 & 0 & | & 1 \\ 0 & 1 & 0 & | & -1 \\ 0 & 0 & 1 & | & 2 \end{bmatrix}$$

**Video 6: Solving a System of Linear Equations by Using the Gauss-Jordan Method (4:24 min)**

● Before starting the video, try this yourself.

Solve the system by using the Gauss-Jordan method.

$$4x + 12y = -4$$
$$2x + y = -7$$

▶ Play the video and check your answer.

| Review and complete | You try |
|---|---|
| 8. Solve the system by using the Gauss-Jordan method. | 9. Solve the system by using the Gauss-Jordan method. |

8.

$$2x - y = 0$$
$$x + y = 3$$

The augmented matrix is $\begin{bmatrix} 2 & -1 & \vert & 0 \\ 1 & 1 & \vert & 3 \end{bmatrix}$.

$R1 + R2 \Rightarrow R1 \quad \begin{bmatrix} 3 & 0 & \vert & 3 \\ 1 & 1 & \vert & 3 \end{bmatrix}$

$\dfrac{1}{3} \cdot R1 \Rightarrow R1 \quad \begin{bmatrix} 1 & 0 & \vert & 1 \\ 1 & 1 & \vert & 3 \end{bmatrix}$

$-1 \cdot R1 + R2 \Rightarrow R2 \quad \begin{bmatrix} 1 & 0 & \vert & 1 \\ 0 & 1 & \vert & 2 \end{bmatrix}$

The solution of the system is $(1, 2)$.

9.

$$x - y = 4$$
$$2x + y = 5$$

| | |
|---|---|

**Video 7: Solving a System of Three Linear Equations by Using the Gauss-Jordan Method (6:07 min)**

▶ Play the video and work along.

Solve the system by using the Gauss-Jordan method.

$$2x \quad + z = y + 5$$
$$x + y - z = -2$$
$$-x + 2y + 2z = 1$$

| **Review and complete** | **You try** |
|---|---|
| 10. Solve the system by using the Gauss-Jordan method. $$x + y + z = 2$$ $$x - y + z = 4$$ $$x + 4y + 2z = 1$$ The augmented matrix is $\begin{bmatrix} 1 & 1 & 1 & 2 \\ 1 & -1 & 1 & 4 \\ 1 & 4 & 2 & 1 \end{bmatrix}$. | 11. Solve the system by using the Gauss-Jordan method. $$6x - 2y + 2z = 10$$ $$2x + 3y + 2z = 7$$ $$x + 2y - z = 4$$ |

$$
\begin{array}{l} -1R1+R2 \Rightarrow R2 \\ -1R1+R3 \Rightarrow R3 \end{array}
\left[\begin{array}{ccc|c} 1 & 1 & 1 & 2 \\ 0 & -2 & 0 & 2 \\ 0 & 3 & 1 & -1 \end{array}\right]
$$

$$
-\frac{1}{2}R2 \Rightarrow R2
\left[\begin{array}{ccc|c} 1 & 1 & 1 & 2 \\ 0 & 1 & 0 & -1 \\ 0 & 3 & 1 & -1 \end{array}\right]
$$

$$
\begin{array}{l} -1R2+R1 \Rightarrow R1 \\ -3R2+R3 \Rightarrow R3 \end{array}
\left[\begin{array}{ccc|c} 1 & 0 & 1 & 3 \\ 0 & 1 & 0 & -1 \\ 0 & 0 & 1 & 2 \end{array}\right]
$$

$$
-1R3+R1 \Rightarrow R1
\left[\begin{array}{ccc|c} 1 & 0 & 0 & 1 \\ 0 & 1 & 0 & -1 \\ 0 & 0 & 1 & 2 \end{array}\right]
$$

The solution set is $\{(1,-1,2)\}$.

## Video 8: Solving a System of Linear Equations Using the Gauss-Jordan Method Dependent Equations (2:22 min)

▶ Play the video and work along.

Solve the system by using the Gauss-Jordan method.

$$
x + 4y = -8
$$
$$
-\frac{1}{8}x - \frac{1}{2}y = 1
$$

| Review and complete | You try |

12. Solve the system by using the Gauss-Jordan method.

$$x + 3y = -2$$
$$2x + 6y = -4$$

$$\begin{bmatrix} 1 & 3 & | & -2 \\ 2 & 6 & | & -4 \end{bmatrix} \quad -2R1+R2 \Rightarrow R2 \quad \begin{bmatrix} 1 & 3 & | & -2 \\ 0 & 0 & | & 0 \end{bmatrix}$$

The equations are dependent since the last row corresponds to an identity, $0 = 0$. There are infinitely many solutions to the system.

13. Solve the system by using the Gauss-Jordan method.

$$-x + y = -2$$
$$3x - 3y = 6$$

---

**Video 9: Solving an Inconsistent System by Using the Gauss-Jordan Method (2:24 min)**

▶ Play the video and work along.

Solve the system by using the Gauss-Jordan method.

$$5x + 10y = 5$$
$$-2x - 4y = 0$$

---

**Review and complete** | **You try**

14. Solve the system by using the Gauss-Jordan method.

$$x + y = 5$$
$$-2x - 2y = 2$$

$$\begin{bmatrix} 1 & 1 & | & 5 \\ -2 & -2 & | & 2 \end{bmatrix}$$

$2R1 + R2 \Rightarrow R2 \begin{bmatrix} 1 & 1 & | & 5 \\ 0 & 0 & | & 12 \end{bmatrix}$

The system is inconsistent since the last row corresponds to a contradiction, $0 = 12$. There are no solutions to the system.

15. Solve the system by using the Gauss-Jordan method.

$$-10x + 5y = 5$$
$$x - \frac{1}{2}y = 4$$

**Answers:** Section 3.1

1. both (or all)    2. is not    3. Yes    4. $\{(1,-1)\}$    5. $\{(-2,-4)\}$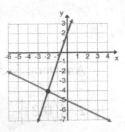

6. $\{(3,2)\}$    7. $\{(-2,1)\}$    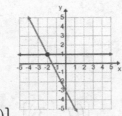    8. One    9. Zero

10. Infinitely many    11a. $-\dfrac{1}{3}$    11b. zero    11c. $\{\ \}$    12a. $y=-2x+3$

12b. $y=-2x+3$    12c. coinciding    12d. Infinitely many    12e. $\{(x,y)\,|\,y=-2x+3\}$

13a. Zero    13b. $\{\ \}$; inconsistent    15a. $y=-\dfrac{2}{3}x+2$; $y=-\dfrac{2}{3}x-1$

15b. Zero    15c. $\{\ \}$    16a. Infinitely many    16b. dependent

18a. $y=-\dfrac{5}{3}x+3$; $y=-\dfrac{5}{3}x+3$    18b. Infinitely many    18c. $\left\{(x,y)\,\middle|\,y=-\dfrac{5}{3}x+3\right\}$

**Answers:** Section 3.2

2. $\{(0,3)\}$     4. $\{(2,1)\}$     6. $\{(2,0)\}$     8. No solution; $\{\ \ \}$

10. Infinitely many solutions; $\{(x,y)\,|\,2x-y=6\}$

**Answers:** Section 3.3

2. $\{(-10,-5)\}$     4. $\{(0,4)\}$     6. $\left\{\left(\dfrac{7}{2},-\dfrac{3}{2}\right)\right\}$

8. Infinitely many solutions; $\{(x,y)\,|\,3x+y=1\}$     10. No solution; $\{\ \ \}$

**Answers:** Section 3.4

2. One cup of popcorn has 60 Cal, and 1 oz of soda has 12 Cal.

4. Mix 10 mL of the 30% acid solution and 90 mL of the 10% acid solution
.
6. He invested $500 in the 6% account and $1500 in the 7% account.

8. Kim rides 10 km/hr in still air. The wind speed is 2 km/hr.

10. The angles are 65° and 25°.

**Answers:** Section 3.5

2. 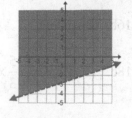    4. 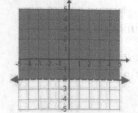    6. 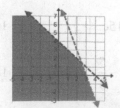    8.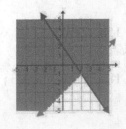

10.

12a. $x \geq 0, y \geq 0$    12b. $x + y \leq 150$    12c. $x \geq 5y$

12d.

12e. No, the point $(105, 30)$ is not in the feasible region. The point $(105, 30)$

represents 105 acres of apple trees and 30 acres of pear trees.

**Answers**: Section 3.6

2. Yes                          4.  $(2, 1, -1)$                          6.  $111°, 46°, 23°$

8.  Hat $18, short-sleeved shirt $35, long-sleeved shirt $3

10. Dependent; infinitely many solutions

12. Inconsistent; no solution

**Answers**: Section 3.7

2. $4 \times 1$

4. $\begin{bmatrix} 1 & 1 & -1 & 8 \\ 1 & -2 & 1 & 3 \\ 1 & 3 & 2 & 7 \end{bmatrix}$

6. $\begin{aligned} 2x &= 12 \\ x - 6y &= 0 \end{aligned}$

7a. Yes

7b. No              7c. Yes                          7d. Yes                          9.  $(3, -1)$

11.  $(2, 1, 0)$              13. Dependent; $\{(x, y) \mid 3x - 3y = 6\}$    15. Inconsistent; no solution

## Concept 1: Simplifying Expressions with Exponents

### Video 1: Simplifying Numerical Expressions with Exponents  (2:13 min)

▶ Play the video and work along.

Simplify.

$(-6)^2$                                                    $-6^2$

⏸ Pause the video and try this yourself.

$(-0.1)^3$                                                  $-0.1^3$

▶ Play the video and check your answer.

| Review and complete | You try |
|---|---|
| 1. To simplify an expression of the form $-a^n$, apply the exponent and then take the _____ of the result. | |
| 2. Simplify.<br>  a. $-9^2 = -(9^2) = -81$<br>  b. $(-9)^2 = (-9)(-9) = 81$<br>  c. $-2^3 = -(2^3) = -8$ | 3. Simplify.<br>  a. $(-4)^2$<br>  b. $-4^2$<br>  c. $-4^3$ |

### Video 2: Summary of Properties of Exponents  (2:59 min)

▶ Play the video and work along.

| Property/Definition | Example |
|---|---|
| $b^m b^n = b^{m+n}$ | |
| $\dfrac{b^m}{b^n} = b^{m-n}$ for $b \neq 0$ | |
| $(b^m)^n = b^{mn}$ | |
| $(ab)^m = a^m b^m$ | |
| $\left(\dfrac{a}{b}\right)^m = \dfrac{a^m}{b^m}$ for $b \neq 0$ | |
| Definition: $b^0 = 1$ for $b \neq 0$ | |
| Definition: $b^{-n} = \left(\dfrac{1}{b}\right)^n = \dfrac{1}{b^n}$ for $b \neq 0$ | |

Video and Study Guide to accompany Intermediate Algebra, 5th ed., Miller/O'Neill/Hyde

## Video 3: Definition of a b to the Zero Power (2:07 min)

▶ Play the video and work along.

Definition of $b^0$ :

Let $b$ be a nonzero real number. Then $b^0 = 1$.

Simplify.

$5^0$                           $(-5)^0$                         $-5^0$

$-6x^0$                    $(-6x)^0$

| Review and complete | You try |
|---|---|
| 4. Any nonzero quantity raised to the 0 power is _____. | |
| 5. Simplify. <br><br> a. $-9^0 = -(1) = -1$ <br><br><br> b. $7x^0 = 7(1) = 7$ | 6. Simplify. <br><br> a. $(-9)^0$ <br><br><br> b. $4 + x^0$ |

## Video 4: Definition of b to a Negative Exponent (2:06)

▶ Play the video and work along.

**Definition of $b^{-n}$**

Let $b$ be a nonzero real number and $n$ be an integer. Then

$$b^{-n} = \left(\frac{1}{b}\right)^n \text{ or } \frac{1}{b^n}$$

Simplify. Write the answers with positive exponents.

$x^{-5}$                       $4^{-1}$                       $(-2)^{-4}$

| Review and complete | You try |
|---|---|
| 7. Simplify. | 8. Simplify. |

7. Simplify.

a. $2^{-4} = \left(\dfrac{1}{2}\right)^4 = \dfrac{1}{16}$

b. $y^{-3} = \left(\dfrac{1}{y}\right)^3 = \dfrac{1}{y^3}$

8. Simplify.

a. $k^{-4}$

b. $3^{-3}$

## Video 5: Practice with the Properties of Exponents (1:36 min)

▶ Play the video and work along.

Simplify. $\dfrac{\left(z^4\right)^3 \left(z^5\right)}{\left(z^7\right)^2}$

| Review and complete | You try |
|---|---|
| 9. Simplify. $\dfrac{\left(x^3\right)^2 \left(x^9\right)^2}{\left(x^{12}\right)^2}$ | 10. Simplify. $\dfrac{\left(m^{10}\right)^2 (m)}{\left(m^8\right)^2}$ |

9. Simplify. $\dfrac{\left(x^3\right)^2 \left(x^9\right)^2}{\left(x^{12}\right)^2}$

$\dfrac{\left(x^3\right)^2 \left(x^9\right)^2}{\left(x^{12}\right)^2} = \dfrac{x^{3\cdot2} x^{9\cdot2}}{x^{12\cdot2}}$

$= \dfrac{x^6 x^{18}}{x^{24}}$

$= \dfrac{x^{24}}{x^{24}}$

$= x^{24-24}$

$= x^0$

$= 1$

## Video 6: Simplifying Expressions with Negative and Zero Exponents (2:07 min)

▶ Play the video and work along.

Simplify.

$2^{-4} + 4^{-2} + 6^0$

$\left(-\dfrac{1}{2}\right)^{-3} + \left(\dfrac{15}{11}\right)^0 + \left(\dfrac{1}{3}\right)^{-4}$

| Review and complete | You try |
|---|---|
| 11. Simplify. | 12. Simplify. |

**Review and complete**

11. Simplify.

$$4^{-2} + \left(\dfrac{1}{2}\right)^3 - \left(\dfrac{2}{13}\right)^0$$

$$= \left(\dfrac{1}{4}\right)^2 + \dfrac{1}{8} - 1$$

$$= \dfrac{1}{16} + \dfrac{2}{16} - \dfrac{16}{16}$$

$$= -\dfrac{13}{16}$$

**You try**

12. Simplify.

$$\left(\dfrac{3}{4}\right)^{-1} - 3^{-2} + \dfrac{1}{6^0}$$

## Video 7: Simplifying an Exponential Expression (2:53 min)

▶ Play the video and work along.

Simplify. Write the answer with positive exponents.

$$\dfrac{\left(3x^2 y^{-5}\right)^4}{\left(2x^{-6} y^3\right)^2}$$

## Video 8: Simplifying an Exponential Expression (2:53 min)

▶ Play the video and work along.

Simplify. Write the answer with positive exponents.

$$5c^2d^{-4}\left(\frac{10c^3}{6cd^4}\right)^{-2}$$

| Review and complete | You try |
|---|---|
| 13. Simplify. Write the answer with positive exponents. | 14. Simplify. Write the answer with positive exponents. $$7x^{-3}y^{-4}\left(\frac{3x^{-1}y^5}{9x^3y^{-2}}\right)^{-3}$$ |

$$2m^2 n^{-1}\left(\frac{4m^3}{5mn^4}\right)^{-2} = \frac{2m^2}{n}\cdot\left(\frac{4m^3}{5mn^4}\right)^{-2}$$

$$= \frac{2m^2}{n}\cdot\frac{4^{-2}m^{-6}}{5^{-2}m^{-2}n^{-8}}$$

$$= \frac{2m^2}{n}\cdot\frac{25m^{-6-(-2)}}{16\,n^{-8}}$$

$$= \frac{25m^2 m^{-4}}{8n\cdot n^{-8}}$$

$$= \frac{25m^{-2}}{8n^{-7}}$$

$$= \frac{25n^7}{8m^2}$$

## Concept 2: Scientific Notation

| Video 9: Introduction to Scientific Notation (2:50 min) |
|---|

▶ Play the video and work along.

### Definition of Scientific Notation

A positive number expressed in the form $a\times10^n$ where $1\le a<10$ and $n$ is an integer is said to be written in **scientific notation**.

Write the numbers in scientific notation.

7000                                                   0.03

| Video 10: Writing Numbers in Scientific Notation (1:33 min) |
|---|

▶ Play the video and work along.

Write the numbers in scientific notation.

895,000,000                    0.0000456                    8.5

| Review and complete | You try |
|---|---|
| 15. Write the number in scientific notation.<br><br>0.000698<br><br>$=6.98\times10^{-4}$<br><br>It is necessary to move the decimal point four places right. | 16. Write the numbers in scientific notation.<br><br>8 million |

Video and Study Guide to accompany Intermediate Algebra, 5th ed., Miller/O'Neill/Hyde
Copyright © 2018 McGraw-Hill Education

---

### Video 11: Writing Numbers in Scientific Notation (2:17 min)

▶ Play the video and work along.

Write the numbers in scientific notation.

- Pluto is approximately 5,900,000,000,000 meters from Earth.

- The sun produces 3,900,000,000,000,000,000,000,000,000,000,000,000 ergs of radiant energy per second.

- 1 electron volt (1 eV) is approximately equal to 0.000 000 000 000 000 000 16 Joules

- A dose of a certain medicine is 0.000 004 grams.

| Review and complete | You try |
|---|---|
| 17. Write the number in scientific notation.<br><br>The population of Andorra in 2016 was approximately 79,000.<br><br>$$79,000 = 7.9 \times 10^4$$<br>It is necessary to move the decimal point four places left. | 18. Write the number in scientific notation.<br><br>The population of the United States in 2014 was approximately 318,900,000. |

### Video 12: Writing a Number in Standard Form (1:51 min)

▶ Play the video and work along.

Write each number in standard form.

$1.384 \times 10^{-4}$          $3.69 \times 10^7$

⏸ Pause the video and try this yourself.

$5.892 \times 10^0$

▶ Play the video and check your answer.

| Review and complete | You try |
|---|---|
| 19. To convert a number that is in scientiic notation to standard form, move the decimal point _____ if the exponent on 10 is negative and to the _____ if the exponent is positive. | |

| 20. Write the number in standard form. | 21. Write the number in standard form. |
|---|---|
| $4.44 \times 10^5$ <br><br> $= 444,000$ <br><br> It is necessary to move the decimal point five places right. | $1.2 \times 10^{-9}$ |

### Video 13: Multiplying and Dividing Numbers in Scientific Notation (3:12 min)

▶ Play the video and work along.

Multiply or divide as indicated. Write the answer in scientific notation.

$(8.6 \times 10^3)(4.0 \times 10^{-9})$           $\dfrac{2.25 \times 10^8}{5.0 \times 10^2}$

| Review and complete | You try |
|---|---|
| 22. Multiply. <br> $\quad (7.7 \times 10^4)(1.4 \times 10^{-12})$ <br> $\quad = (7.7)(1.4) \times 10^{4+(-12)}$ <br> $\quad = 10.78 \times 10^{-8}$ <br> $\quad = (1.078 \times 10^1) \times 10^{-8}$ <br> $\quad = 1.078 \times 10^{-7}$ | 23. Multiply. <br> $\quad (3.6 \times 10^{-3})(5.2 \times 10^{12})$ |
| 24. Divide. <br> $\dfrac{9.09 \times 10^4}{1.1 \times 10^{18}} = \dfrac{9.09}{1.1} \times \dfrac{10^4}{10^{18}} \approx 8.264 \times 10^{-14}$ | 25. Divide. <br> $\dfrac{6 \times 10^{-10}}{1.9 \times 10^{-2}}$ |

### Video 14: Multiplying and Dividing Numbers in Scientific Notation on a Calculator (2:23 min)

▶ Play the video and work along.

$(8.6 \times 10^3)(4.0 \times 10^{-9})$           $\dfrac{2.25 \times 10^8}{5.0 \times 10^2}$

```
(8.6E3)*(4.0E-9)
            3.44E-5
(2.25E8)/(5.0E2)
            450000
■
```

| Review and complete | You try |
|---|---|
| 26. On a calculator display, Farook sees $4.506E-9$. What number does this represent?<br><br>The number is $4.506\times10^{-9}$ which is equivalent to<br><br>0.000000004506. | 27. On a calculator display, Evin sees $1.2E7$. What number does this represent? |
| 28. Multiply using your calculator.<br><br>$7800\times(9.56\times10^{-22})$<br><br>**7800*(9.56E-22)**<br><br>$\phantom{xxxxxxxx}$**7.4568E-18** | 29. Divide using your calculator.<br><br>$$\dfrac{4.5\times10^{18}}{2.25\times10^{-6}}$$ |

## Concept 1: Polynomials: Basic Definitions

### Video 1: Introduction to Polynomials  (3:47 min)

▶ Play the video and work along.

| Term (Expressed in the form $ax^n$) | Coefficient | Degree |
|---|---|---|
| $-5x^{12}$ | | |
| $y^4z^2$ | | |
| $11t$ | | |
| $3$ | | |

| Expression | Descending Order | Leading Coefficient | Degree of Polynomial |
|---|---|---|---|
| $6x - 3x^2 + 4$ | | | |
| $7a^2 - 9a^4$ | | | |
| $7x^3$ | | | |

| Review and complete | You try |
|---|---|
| 1. A polynomial is a sum of _____, of the form $ax^n$. | |

| Review and complete | You try |
|---|---|
| 2. True or False? <br><br> a. The coefficient of the term $x^3$ is 0. <br><br> b. The degree of the term $x^3$ is 3. <br> c. The term $-8$ has no degree. <br><br> Answers: <br> a. False. The coefficient is 1. <br> b. True. <br> c. False. The degree is 0 since $-8 = -8x^0$. | 3. True or False? <br> a. The coefficient of the term $g^2h$ is 1. <br><br><br> b. The degree of the term $x$ is 0. <br><br><br> c. The term $-8x^4y$ has degree 5. |

### Video 2: Polynomials: Degree, Descending Order, Leading Term, and Coefficients (2:05 min)

▶ Play the video and work along.

Given the polynomial $9 - 3x^4 + x^5 - 4x^2$

a. List the terms of the polynomial.

b. Write the polynomial in descending order.

c. State the degree of the polynomial and the leading coefficient.

**❚❚** Pause the video and try this yourself.

Given the polynomial $7a^2b - 8ab^5 + ab$

    a.   Identify the degree of each term.

    b.   Identify the degree of the polynomial.

**▶** Play the video and check your answer.

| Review and complete | You try |
|---|---|
| 4.   Given the polynomial $6x^2y + xy + 9$ <br><br>     a.   State the leading coefficient. <br><br>         The leading coefficient is 6. <br><br>     b.   Identify the degree of the polynomial. <br><br>         The degree of the polynomial is $2 + 1 = 3$. | 5.   Given the polynomial $-10 - x^3 + 4x^2 - x$ <br><br>     a.   State the leading term. <br><br><br>     b. Identify the degree of the polynomial. |

## Concept 2: Addition of Polynomials

| Video 3: Adding Polynomials Horizontally and Vertically (1:57 min) | |
|---|---|
| **▶** Play the video and work along. | |
| Add the polynomials. <br><br> $(-7d^5 - 4d^4 - 6) + (9d^5 + 2d^4 - 2)$ | Add the polynomials using column addition. <br><br>           $-7d^5 - 4d^4 - 6$ <br> $+ \quad\; 9d^5 + 2d^4 - 2$ <br> $\overline{\phantom{+ \quad\; 9d^5 + 2d^4 - 2}}$ |

| Review and complete | You try |
|---|---|
| 6.   Add the polynomials. <br><br> $(-x^3 + 6x^2 + x - 2) + (8x^3 + 9x + 15)$ <br> $\quad = -x^3 + 8x^3 + 6x^2 + x + 9x - 2 + 15$ <br> $\quad = 7x^3 + 6x^2 + 10x + 13$ | 7.   Add the polynomials. <br><br> $(x^3 + x^2 + 6x + 10) + (6x^3 - 3x^2 - x - 12)$ |

Video and Study Guide to accompany Intermediate Algebra, 5th ed., Miller/O'Neill/Hyde

| | |
|---|---|
| | |

### Video 4: Adding Polynomials (1:33)

▶ Play the video and work along.

Add the polynomials.

$(6.1a^2 - 8.2c + 4.7a^2c) + (-7.2a^2 - 3c - 9a^2c)$

| Review and complete | You try |
|---|---|
| 8.  Add the polynomials. | 9.  Add the polynomials. |
| $(1.4m^2 + 2.6mn - 5n^2) + (-m^2 - 5mn + 4.3n^2)$ <br><br> $= 1.4m^2 + (-m^2) + 2.6mn$ <br><br> $\qquad + (-5mn) - 5n^2 + 4.3n^2$ <br><br> $= 0.4m^2 - 2.4mn - 0.7n^2$ | $(0.24x^2 - 8x + 4.5) + (5.6x^2 + 0.5x + 1.78)$ |

### Concept 3: Subtraction of Polynomials

### Video 5: Writing the Opposite of a Polynomial (1:26 min)

▶ Play the video and work along.

Find the opposite.

$\qquad 3x^2 \qquad\qquad\qquad 7x - 3.2y - 6.8z \qquad\qquad -\frac{1}{2}m + \frac{1}{3}n$

| Review and complete | You try |
|---|---|
| | |

| 10. Find the opposite. $4x - 5y^2$ | 11. Find the opposite. $-x^2 + 7xy - 2y^2$ |
|---|---|
| $-\left(4x - 5y^2\right) = -4x + 5y^2$ | |

### Video 6: Subtracting Polynomials Horizontally and Vertically (3:01 min)

 Play the video and work along.

> **Definition of Subtraction of Polynomials**
>
> If $A$ and $B$ are polynomials, then $A - B = A + (-B)$.

Subtract the polynomials. $(-5t^3 + 4t^2 - 3t) - (2t^3 - 6t^2 + 5t)$

Subtract the polynomials using column subtraction.

$$-5t^3 + 4t^2 - 3t$$
$$- \quad 2t^3 - 6t^2 + 5t$$

| Review and complete | You try |
|---|---|
| 12. Subtract the polynomials. | 13. Subtract the polynomials. |
| $(-x^3 + 10x^2 - 12) - (-2x^3 - x^2 + 4)$ <br> $= -x^3 + 10x^2 - 12 + 2x^3 + x^2 - 4$ <br> $= -x^3 + 2x^3 + 10x^2 + x^2 - 12 - 4$ <br> $= x^3 + 11x^2 - 16$ | $(8mn^2 + 5m^2n - 10mn) - (-2mn^2 - m^2n - mn)$ |

### Video 7: Subtracting Polynomials (1:51 min)

 Play the video and work along.

Subtract the polynomials.

$(0.01c^2 - 0.03cd + 0.05d^2) - (0.05c^2 - 0.07cd - 0.02d^2)$

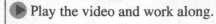

| Review and complete | You try |
|---|---|
| 14. Subtract. | 15. Subtract. |

$(0.2m^2 - 1.2m - 0.3) - (m^2 + 0.09m - 0.5)$

$= 0.2m^2 - 1.2m - 0.3 - m^2 - 0.09m + 0.5$

$= 0.2m^2 - m^2 - 1.2m - 0.09m - 0.3 + 0.5$

$= -0.8m^2 - 1.29m + 0.2$

$(8x^2 - x + 9xy) - (5x^2 - x - 7xy)$

---

### Video 8: Writing an English Phrase as an Algebraic Expression and Simplifying (1:55 min)

▶ Play the video and work along.

Subtract $\dfrac{2}{3}x^2 - \dfrac{1}{5}x$ from $x^2 + \dfrac{1}{10}x$.

Subtract $a$ from $b$.

---

| Review and complete | You try |
|---|---|

16. It is important to remember that **order matters** in subtraction. That is, subtraction is not commutative. Therefore $a - b$ _____ $b - a$ (fill in the blank with $=$ or $\neq$).

| 17. Subtract | 18. Subtract |
|---|---|

$3x^2 - 2x + \dfrac{2}{3}$ from $\dfrac{5}{2}x^2 + 3x - 1$

$\left(\dfrac{5}{2}x^2 + 3x - 1\right) - \left(3x^2 - 2x + \dfrac{2}{3}\right)$

$x^2 - \dfrac{1}{2}y^2$ from $\dfrac{4}{3}x^2 + 3y^2$

Video and Study Guide to accompany Intermediate Algebra, 5th ed., Miller/O'Neill/Hyde

$$= \frac{5}{2}x^2 + 3x - 1 - 3x^2 + 2x - \frac{2}{3}$$

$$= \frac{5}{2}x^2 - 3x^2 + 3x + 2x - 1 - \frac{2}{3}$$

$$= \frac{5}{2}x^2 - \frac{6}{2}x^2 + 5x - \frac{3}{3} - \frac{2}{3}$$

$$= -\frac{1}{2}x^2 + 5x - \frac{5}{3}$$

## Concept 4: Polynomial Functions

### Video 9: Introduction to Polynomial Functions (2:22 min)

▶ Play the video and work along.

Examples of polynomial functions.

$$f(x) = 5x + 8 \qquad g(x) = \frac{1}{2}x^3 + x^2 - 4x \qquad h(x) = 3$$

These are *not* polynomial functions.

$$k(x) = \frac{6}{x} \qquad n(x) = |x| + 3 \qquad t(x) = 4\sqrt{x} + 9x + 1$$

Given $g(x) = \frac{1}{2}x^3 + x^2 - 4x$, evaluate:

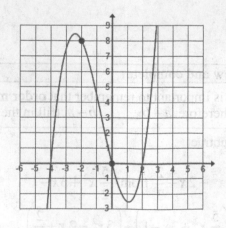

a.   $g(0)$

b.   $g(-2)$

**Review and complete**  |  **You try**

19. Circle the polynomial function(s).

$f(x) = 5x^3 + 8$        $g(x) = x^{-3} + 7$        $h(x) = 3x^{1/2}$        $m(x) = 3x$

---

20. Given $f(x) = 5x^3 + 8$, evaluate $f(-1)$.

$$f(-1) = 5(-1)^3 + 8$$
$$= 5(-1) + 8$$
$$= -5 + 8$$
$$= 3$$

21. Given $n(x) = x^3 - x - 1$, evaluate $n(0)$.

| Video 10: Writing a Polynomial Function to Express Perimeter of a Rectangle (2:06 min) |
|---|

▶ Play the video and work along.

The length of a rectangle is 7 ft more than twice the width. Let $x$ represent the width of the rectangle. Write a polynomial function $P$ that represents the perimeter of the rectangle.

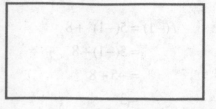

| Review and complete | You try |
|---|---|

22. A rectangle has length that is 2 units more than twice its width. If $x$ is the width, write a polynomial function $P$ that represents the perimeter of the rectangle.

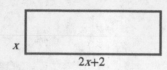

$x$

$2x+2$

$$P = 2L + 2W$$
$$P(x) = 2(x) + 2(2x + 2)$$
$$P(x) = 2x + 4x + 4$$
$$P(x) = 6x + 4$$

23. A rectangle has width that is 6 units more than half its length. If $x$ is the length, write a polynomial function $P$ that represents the perimeter of the rectangle.

| Video 11: Evaluating a Polynomial Function in an Application  (2:04 min) |
|---|

▶ Play the video and work along.

The annual expenditure for cellular service is a function of the age of the individual using the service. The annual expenditure, $E(a)$, (in dollars) can be approximated by:

$$E(a) = -0.475a^2 + 37.0a - 44.6,$$ where $a$ is an individual's age in years and $20 \le a \le 72$ .

a.  Evaluate $E(25)$ and interpret the meaning in the context of this problem.

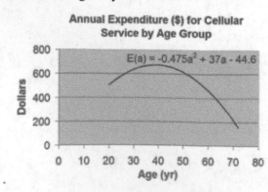

Annual Expenditure ($) for Cellular Service by Age Group

$E(a) = -0.475a^2 + 37a - 44.6$

b.  Approximate the yearly cost for cellular service for a 60 year-old.

| Review and complete | You try |
|---|---|
| 24. The percentage of American males who smoke has decreased since 1990. The function defined by $M(x) = -0.003x^2 - 0.257x + 28$ approximates the percent of males who smoke, where $x$ is the number of years since 1990. <br><br> Evaluate $M(25)$ and interpret it in this context. <br><br> $M(25) = -0.003(25)^2 - 0.257(25) + 28 \approx 20\%$ <br><br> In the year 2015, the function predicts that approximately 20% of males will be smokers. | 25. The percentage of American males who smoke has decreased since 1990. The function defined by $M(x) = -0.003x^2 - 0.257x + 28$ approximates the percent of males who smoke, where $x$ is the number of years since 1990. <br><br> Find the percent of males who smoke in the year 2020, as predicted by the function. Round to the nearest whole percent. |

**Concept 1: Multiplying Polynomials**

| Video 1: Multiplying Monomials  (2:00 min) |
| --- |

▶ Play the video and work along.

Multiply the monomials.

$$\left(-2p^3\right)\left(8p^2\right) \qquad\qquad \left(6x^2y^5\right)\left(2.1x^4yz^2\right)$$

⏸ Pause the video and try this yourself.

$$\left(\frac{1}{2}c^2d\right)\left(\frac{2}{5}d^8\right)$$

▶ Play the video and check your answer.

| Review and complete | You try |
| --- | --- |
| 1.  The _____ and _____ properties of multiplication can be used to regroup and reorder factors in a monomial multiplication. | |
| 2.  Multiply the monomials. $$\left(2mn^3\right)\left(-5m^2n\right)$$ $$= 2\cdot-5\cdot m\cdot m^2\cdot n^3\cdot n$$ $$= -10m^3n^4$$ | 3.  Multiply the monomials. $$\left(6xy^3\right)\left(\frac{1}{3}xy^4\right)$$ |

| Video 2: Multiplying a Monomial by a Polynomial (1:35 min) |
| --- |

▶ Play the video and work along.

Multiply the polynomials.

$$7y\left(3y-6\right) \qquad\qquad -6x^2y\left(2x^2-4xy+\frac{1}{2}\right)$$

| Review and complete | You try |
| --- | --- |
| 4.  Multiply the polynomials. $$5y(y^3-2xy)=5y(y^3)+5y(-2xy)$$ $$=5y^4-10xy^2$$ | 5.  Multiply the polynomials. $$3m(7mn^2-m^2n)$$ |

| | |
|---|---|
| **Video 3: Multiplying Binomials (1:10 min)** | |

▶ Play the video and work along.

Multiply the polynomials.

$(d+7)(d-4)$

| **Review and complete** | **You try** |
|---|---|
| 6.  Multiply the polynomials. | 7.  Multiply the polynomials. |
| $(x+3)(x-6) = x \cdot x + x(-6) + 3(x) + 3(-6)$ <br> $\qquad = x^2 - 6x + 3x - 18$ <br> $\qquad = x^2 - 3x - 18$ | $(x-1)(x+8)$ |

**Video 4: Multiplying Binomials (1:18 min)**

▶ Play the video and work along.

Multiply the polynomials.

$(6m+5n)(2m-3n)$

| **Review and complete** | **You try** |
|---|---|
| 8.  Multiply the polynomials. | 9.  Multiply the polynomials. |
| $(3c+d)(4c-d) = 3c(4c) + 3c(-d) + d(4c) + d(-d)$ <br> $\qquad = 12c^2 - 3cd + 4cd - d^2$ <br> $\qquad = 12c^2 + \underline{\hspace{2cm}} - d^2$ | $(2x-y)(3x-y)$ |

**Video 5: Multiplying a Binomial by a Trinomial (1:41 min)**

▶ Play the video and work along.

Multiply the polynomials.

$(x+4)(5x^2-2x-6)$

| Review and complete | You try |
|---|---|
| 10. Multiply the polynomials. | 11. Multiply the polynomials. |
| $(m+4)(2m^2-2m+1)$ <br> $\quad = m(2m^2)+m(-2m)+m(1)$ <br> $\qquad +4(2m^2)+4(-2m)+4(1)$ <br> $\quad = 2m^3-2m^2+m+8m^2-8m+4$ <br> $\quad = 2m^3+6m^2\underline{\qquad}+4$ | $(2n-1)(n^2+9n-11)$ |

## Concept 2: Special Case Products: Difference of Squares and Perfect Square Trinomials

| Video 6: Formulas for Multiplying Conjugates and Squaring Binomials (3:14 min) |
|---|
| ▶ Play the video and work along. <br><br> $(a+b)(a-b)$ $\qquad\qquad\qquad$ $(a+b)^2$ $\qquad\qquad\qquad$ $(a-b)^2$ |

| Video 7: Multiplying Conjugates  (1:48 min) |
|---|
| ▶ Play the video and work along. <br><br> Multiply. <br><br> $(z+4)(z-4)$ $\qquad\qquad$ $(z+4)(z-4)$ $\qquad\qquad$ $\left(\dfrac{1}{3}p^2-2\right)\left(\dfrac{1}{3}p^2+2\right)$ |

| Review and complete | You try |
|---|---|

12. $(a+b)(a-b)=a^2-b^2$  The product of conjugates results in a _____.

13. $(a+b)^2=a^2+2ab+b^2$  The square of a binomial results in a _____.

14. Multiply.

$$\left(\frac{1}{9}x^2-1\right)\left(\frac{1}{9}x^2+1\right)$$

$$=\frac{1}{81}x^4-1$$

15. Multiply.

$$(3t-5)(3t+5)$$

## Video 8: Squaring Binomials (2:19 min)

▶ Play the video and work along.

Multiply.

$$(2x+5)^2 \qquad\qquad \left(3y^2-7w\right)^2$$

| Review and complete | You try |

16. Squaring a _____ **cannot** be accomplished just by squaring the parts. It is important to remember that $(x+y)^2 \neq x^2+y^2$ .

17. Multiply.

$$(2x+7)^2$$
$$=(2x)^2+2(2x)(7)+(7)^2$$
$$=4x^2+28x+49$$

18. Multiply.

$$(9x-2)^2$$

## Video 9: Using Substitution and Multiplying Conjugates (1:58 min)

▶ Play the video and work along.

Multiply.    $[5+(x-y)][5-(x-y)]$

| Review and complete | You try |
|---|---|
| 19. Multiply.  $[(a-b)+3][(a-b)-3]$<br><br>Let $u=(a-b)$.<br><br>$[(a-b)+3][(a-b)-3]=(u+3)(u-3)$<br>$\qquad = u^2-9$<br>$\qquad = (a-b)^2-9$<br>$\qquad = a^2-2ab+b^2-9$ | 20. Multiply. $[a-(9+c)][a+(9+c)]$ |

## Concept 3: Applications Involving a Product of Polynomials

Video 10: Multiplying Polynomials in a Geometry Application (2:33 min)

▶ Play the video and work along.

Find a polynomial that represents the volume of the cube.

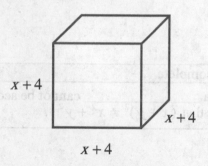

$x+4$

$x+4$

$x+4$

| Review and complete | You try |
|---|---|
| 21. A cube has side length $x-5$. Find a polynomial that represents its volume.<br><br>$V=(x-5)^3$<br>$\quad = (x-5)(x-5)(x-5)$<br>$\quad = [x^2-10x+25](x-5)$<br>$\quad = x^2(x)+x^2(-5)-10x(x)$<br>$\quad\quad -10x(-5)+25(x)+25(-5)$<br>$\quad = x^3-5x^2-10x^2+50x+25x-125$<br>$V=x^3-15x^2+75x-125$ | 22. A cube has side length $x+1$. Find a polynomial that represents its volume. |

Video 11: Translating Between English Form and Algebraic Form  (1:26 min)

▶ Play the video and work along.

Complete the table.

| English Form | Algebraic Form |
|---|---|
| The sum of the square of $a$ and the square of $b$ | ( ___ ) |
|  | $(7y^2)^2$ |
| The square of the difference of $x$ and 5 |  |

| Review and complete | You try |
|---|---|
| 23. Write a polynomial to represent the English phrase.<br><br>The square of the sum of $y$ and 7<br><br>$\left(\underline{\hspace{2cm}}\right)^2$ | 24. Write a polynomial to represent the English phrase.<br><br>The sum of the squares of $y$ and $\dfrac{1}{4}$ |

### Video 12: Writing a Polynomial Function to Represent the Area of a Triangle  (1:58 min)

▶ Play the video and work along.

The height of a triangle is 3 ft less than twice the base. Let the base of the triangle be represented by $x$.

    a.  Write a polynomial function $A$ that represents the area of the triangle as a function of $x$.

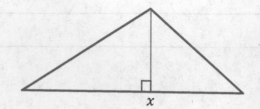

    b.  Find the area if the base is 8 in.

| Review and complete | You try |
|---|---|
| 25. The height of a triangle is 1 ft more than twice the base. Let the base of the triangle be represented by $x$.<br><br>  a.  Write a polynomial function $A$ that represents the area of the triangle as a function of $x$.<br><br>$$A = \frac{1}{2}bh$$<br>$$A(x) = \frac{1}{2}(x)(2x+1)$$<br>$$A(x) = \frac{1}{2}(2x^2 + x)$$<br>$$A(x) = x^2 + \frac{1}{2}x$$<br><br>  b.  Find the area if the base is 10 in.<br><br>$$A(10) = (10)^2 + \frac{1}{2}(10)$$<br>$$= 100 + 5 = 105 \text{ in}^2$$ | 26. The height of a triangle is 5 ft more than three times the base. Let the base of the triangle be represented by $x$.<br><br>  a.  Write a polynomial function $A$ that represents the area of the triangle as a function of $x$.<br><br><br><br>  b.  Find the area if the base is 40 in. |

## Concept 1: Division by a Monomial

### Video 1: Dividing a Polynomial by a Monomial (3:39 min)

▶ Play the video and work along.

Divide.

$$\frac{30x^3 + 15x^2 - 25}{15x^2}$$

$$\left(6a^4b^5 - 4a^3b^7 + 9a^2b^3\right) \div \left(-3a^4b^3\right)$$

| Review and complete | You try |
|---|---|
| 1. Divide. $$(12m^4n - 5m^3 + 70m^2) \div (2mn^2)$$ $$= \frac{12m^4n}{2mn^2} + \frac{-5m^3}{2mn^2} + \frac{70m^2}{2mn^2}$$ $$= \frac{6m^3}{n} - \frac{5m^2}{2n^2} + \frac{35m}{n^2}$$ | 2. Divide. $$(30x^4y^2 - 10x^2 - 2y) \div (5xy)$$ |

## Concept 2: Long Division

### Video 2: Dividing Polynomials Using Long Division (5:34 min)

▶ Play the video and work along.

Divide. $\left(4x^3 + 7x^2 - 3\right) \div \left(x - 2\right)$     Check: (divisor)(quotient) + remainder = dividend

Video and Study Guide to accompany Intermediate Algebra, 5th ed., Miller/O'Neill/Hyde
Copyright © 2018 McGraw-Hill Education

## Video 3: Long Division of Polynomials (2:33 min)

▶ Play the video and work along.

Divide. $\left(5y^2 + 2y^3 - 1\right) \div \left(2y + 1\right)$

| Review and complete | You try |
|---|---|
| 3. If necessary, terms with coefficient _____ can be inserted to make lining up terms in the division easier. | |

| 4. Divide. | 5. Divide. |
|---|---|
| $\left(x^3 + 5x - 2\right) \div \left(x - 1\right)$ | $\left(4x^3 + 7x^2 - 3\right) \div \left(x - 2\right)$ |

**4. Divide.**

$$\begin{array}{r} x^2 + x + 6 \\ x-1\overline{)x^3 + 0x^2 + 5x - 2} \\ \underline{-(x^3 - x^2)} \\ x^2 + 5x \\ \underline{-(x^2 - x)} \\ 6x - 2 \\ \underline{-(6x - 6)} \\ 4 \end{array}$$

$$x^2 + x + 6 + \frac{4}{x-1}$$

## Video 4: Using Long Division to Divide Polynomials (Zero Remainder)  (3:23 min)

▶ Play the video and work along.

Divide.

$$\frac{y^3 - 5y^2 + 3y - 15}{y^2 + 3}$$

| Review and complete | You try |
|---|---|
| 6.  Divide. | 7.  Divide. |
| $$\dfrac{x^3 - 3x^2 + 10x - 30}{x^2 + 10}$$ | $$\dfrac{x^3 - 10x^2 + 6x - 60}{x^2 + 6}$$ |

$$
\begin{array}{r}
x - 3 \\
x^2 + 0x + 10 \overline{\smash{\big)}\, x^3 - 3x^2 + 10x - 30} \\
\underline{-(x^3 + 0x^2 + 10x)\phantom{-30}} \\
-3x^2 + 0x - 30 \\
\underline{-3x^2 + 0x - 30} \\
0
\end{array}
$$

## Concept 3: Synthetic Division

**Video 5: Using Synthetic Division to Divide Polynomials  (3:54 min)**

▶ Play the video and work along.

Divide.   $(4x^3 + 7x^2 - 3) \div (x - 2)$

**Video 6: Dividing Polynomials Using Synthetic Division (Remainder Zero) (2:30 min)**

Play the video and work along.

Divide. $\dfrac{t^3 + 64}{t + 4}$

---

| Review and complete | You try |
|---|---|
| 8. Divide. | 9. Divide. |

**Review and complete**

8. Divide.

$$\frac{t^3 - 1000}{t - 10}$$

Use synthetic division.
$r = 10$

$$\begin{array}{r|rrrr} 10 & 1 & 0 & 0 & -1000 \\ & & 10 & 100 & 1000 \\ \hline & 1 & 10 & 100 & 0 \end{array}$$

Constant remainder: 0
Quotient: $t^2 + 10t + 100$

**You try**

9. Divide.

$$\frac{x^3 + 8}{x + 2}$$

---

### Video 7: Dividing Polynomials: A Summary (3:06 min)

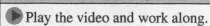 Play the video and work along.

Determine a correct procedure to use to divide the polynomials.

$$\frac{4x^3 + 16x^2 - 8x}{x^2 + 2} \qquad\qquad \frac{4x^3 + 16x^2 - 8x}{2x} \qquad\qquad \frac{4x^3 + 16x^2 - 8x}{x + 2}$$

---

| Review and complete | You try |
|---|---|

10. Determine a correct procedure to divide the polynomials.

$$\frac{x^4 + 6x^2 - 40}{x^2 + 10}$$

$$
\begin{array}{r}
x^2 \qquad\quad -4 \\
x^2 + 0x + 10 \overline{\smash{\big)}\ x^4 + 0x^3 + 6x^2 + 0x - 40} \\
-\left(x^4 + 0x^3 + 10x^2\right) \\
\hline
-4x^2 + 0x - 40 \\
-\left(-4x^2 + 0x - 40\right) \\
\hline
0
\end{array}
$$

$$\frac{x^4 + 6x^2 - 40}{x^2 + 10} = x^2 - 4$$

11. Determine a correct procedure to divide the polynomials.

$$\frac{34x^3 + 9x^2 - 45}{9x}$$

**Concept 1: Factoring Out the Greatest Common Factor**

| Video 1: Identifying the Greatest Common Factor  (3:34 min) |
| --- |
| ▶ Play the video and work along. |
| Identify the greatest common factor.   $20x^5 - 8x^4 + 12x^2$ |
| ⏸ Pause the video and try this yourself. |
| Identify the greatest common factor.   $18a^3b^2 - 24a^4bc$ |
| ▶ Play the video and check your answer. |

| Review and complete | You try |
| --- | --- |
| 1.  Identify the greatest common factor. | 2.  Identify the greatest common factor. |
| $12x^4y + 30x^2y^2 + 54xy$ | $6mn^5 + 40m^4n^4 - 14n^3$ |
| $12x^4y = 2 \cdot 2 \cdot 3 \cdot x \cdot x \cdot x \cdot x \cdot y$ $30x^2y^2 = 2 \cdot 3 \cdot 5 \cdot x \cdot x \cdot y \cdot y$ $54xy = 2 \cdot 3 \cdot 3 \cdot 3 \cdot x \cdot y$ | |
| $\text{GCF} = 2 \cdot 3 \cdot x \cdot y = 6xy$ | |

| Video 2: Factoring Out the Greatest Common Factor  (3:15 min) |
| --- |
| ▶ Play the video and work along. |
| Factor out the greatest common factor (GCF). |
| $20x^5 - 8x^4 + 12x^2$ |

| Review and complete | You try |
| --- | --- |
| 3.  Factor out the greatest common factor (GCF). | 4.  Factor out the greatest common factor (GCF). |
| $12x^4y + 30x^2y^2 + 54xy$ $= 6xy\left(2x^3 + 5xy + 9\right)$ | $6mn^5 + 40m^4n^4 - 14n^3$ |

## Video 3: Factoring Out the Greatest Common Factor (2:28 min)

▶ Play the video and work along.

Factor out the greatest common factor (GCF). $30x^4y^5 - 15x^3y^6 + 5x^2y^4$

| Review and complete | You try |
|---|---|
| 5. Factor out the greatest common factor (GCF).<br><br>$6p^2q^2 - 12p^2q^3 + 90p^2q^4$<br>$= 6p^2q^2\left(1 - 2q + 15q^2\right)$ | 6. Factor out the greatest common factor (GCF).<br><br>$x^6 - x^4 - x^2$ |

## Concept 2: Factoring Out a Negative Factor

## Video 4: Factoring Out a Negative Factor (2:45 min)

▶ Play the video and work along.

Factor out $-4$ from the polynomial.
$-8y^2 + 12y + 4$

Factor out $-3x^2$ from the polynomial.
$-12x^4 - 6x^3 + 9x^2$

| Review and complete | You try |
|---|---|
| 7. Factor $-2xy$ out of the polynomial.<br>$-18x^3y + 40x^2y^2 - 32xy^3$<br><br>$= -2xy\left(9x^2 - 20xy + \underline{\quad}\right)$ | 8. Factor $-7x$ out of the polynomial.<br>$-14x^3y + 49x^2 - 700x$ |

## Concept 3: Factoring Out a Binomial Factor

### Video 5: Factoring Out a Binomial Factor (1:10 min)

▶ Play the video and work along.

Factor out the greatest common factor.    $2x(a+b)-3y(a+b)+8(a+b)$

| Review and complete | You try |
|---|---|
| 9.  Factor out the greatest common factor.<br>    $a(x+y)+5x(x+y)+(x+y)$<br>    $=(x+y)(a+5x+1)$ | 10.  Factor out the greatest common factor.<br>    $m(n-1)+n(n-1)-8(n-1)$ |

## Concept 4: Factoring by Grouping

### Video 6: Factoring by Grouping (2:35 min)

▶ Play the video and work along.

Factor by grouping.    $2ax+3a+8bx+12b$

### Video 7: Factoring by Grouping (1:29 min)

 Play the video and work along.

Factor by grouping.    $4x^3-3x^2-20x+15$

| Review and complete | You try |
|---|---|

| | |
|---|---|
| 11. Factor by grouping.<br><br>$2x^2 - 2xy^2 - xy + y^3$<br>$= 2x(x - y^2) - y(x - y^2)$<br>$= (x - y^2)(2x - y)$ | 12. Factor by grouping.<br><br>$3y^4 - 6y^3 - xy + 2x$ |

**Video 8: Factoring Out GCF and Factoring by Grouping  (3:04 min)**

 Play the video and work along.

Factor completely.    $200m^3 - 40m^2 - 150m^2t + 30mt$

| **Review and complete** | **You try** |
|---|---|
| 13. Factor completely.<br><br>$18n^3 + 9n^2p - 72n^2 - 36np$<br>$= 9n(2n^2 + np - 8n - 4p)$<br>$= 9n((n(2n + p) - 4(2n + p))$<br>$= 9n(n - 4)(2n + p)$ | 14. Factor completely.<br>$5x^2y + 25x^2 - 5xy - 25x$ |

**Video 9: Factoring by Grouping where Rearranging Terms is Necessary (2:26 min)**

Play the video and work along.

Factor by grouping.  $3a^2 + 5c + 15a + ca$

| **Review and complete** | **You try** |
|---|---|

15. Factor by grouping.

$$9x^2 - x - x^3 + 9$$

$$x(9x - 1) - 1(x^3 - 9)$$

Binomial factors do not match.

Rearrange terms.

$$9x^2 + 9 - x^3 - x$$

$$9(x^2 + 1) - x(x^2 + 1)$$

$$(x^2 + 1)(9 - x)$$

16. Factor by grouping.

$$mn^2 + 10n + 2m + 5n^3$$

## Concept 1: Factoring Trinomials: AC Method

### Video 1: Factoring a Trinomial Using the ac-Method (3:22 min)

▶ Play the video and work along.

Factor completely.    $6x^2 + 23x + 15$

| Review and complete | You try |
|---|---|
| 1. Factor completely.  $2x^2 + 17x + 35$ | 2. Factor completely.  $15x^2 + 14x + 3$ |

Find factors of $a \cdot c = (2)(35) = 70$  whose sum is $b = 17$.

$$70 = 1(70)$$
$$\phantom{70} = 2(35)$$
$$\phantom{70} = 5(14)$$
$$\phantom{70} = 7(10)$$

Replace $17x$ with $7x+10x$ and then factor by grouping.

$$2x^2 + 17x + 35$$
$$2x^2 + 7x + 10x + 35$$
$$x(2x+7) + 5(2x+7)$$
$$(2x+7)(x+5)$$

### Video 2: Factoring a Trinomial Using the ac-Method (3:33 min)

▶ Play the video and work along.

Factor completely.    $10m^2 + 7mn - 12n^2$

| Review and complete | You try |
|---|---|
| 3.  Factor completely.  $10x^2 + xy - 3y^2$ | 4.   Factor completely.  $3x^2 + 14xy + 15y^2$ |

List all the factors of $a \cdot c = (10)(-3) = -30$ , and find the pair whose sum equals $b = 1$.

| | |
|---|---|
| $-1(30)$ | $1(-30)$ |
| $-2(15)$ | $2(-15)$ |
| $-3(10)$ | $3(-10)$ |
| $-5(6)$ | $5(-6)$ |

Replace $1xy$ with $-5xy + 6xy$. Then factor by grouping.

$$10x^2 + xy - 3y^2$$
$$10x^2 - 5xy + 6xy - 3y^2$$
$$5x(2x - y) + 3y(2x - y)$$
$$(2x - y)(5x + 3y)$$

### Video 3: Factoring a Trinomial Using the ac-Method and by Removing the GCF (1:59 min)

▶ Play the video and work along.

Factor completely.    $6x^2 y + 12y - 27xy$

| Review and complete | You try |
|---|---|
| 5. Always check to see if there is a GCF other than one before attempting other factoring methods. The GCF _____(remains/does not remain) as part of the factored form. | |

| Review and complete | You try |
|---|---|
| 6. Factor completely. $21m^3 + 245m^2 - 84m$<br><br>Factor out the GCF = 7m. Then find the factors of $a \cdot c = (3)(-12) = -36$ whose sum is $b = 35$. Those factors are 36 and -1.<br><br>$7m(3m^2 + 35m - 12)$<br>$7m(3m^2 + 36m - 1m - 12)$  Replace 35m with 36m - 1m.<br>$7m[3m(m+12) - 1(m+12)]$  Factor by grouping.<br>$7m(m+12)(3m-1)$ | 7. Factor completely. $6cd^2 + 9cd - 42c$ |

## Concept 2: Factoring Trinomials: Trial and Error Method

**Video 4: Factoring a Trinomial by the Trial-and-Error Method (Leading Coefficient Not Equal to 1) (5:34 min)**

▶ Play the video and work along.

Factor completely by using the trial and error method.    $6x^2 + 23x + 15$

$$(x + \underline{\quad})(6x + \underline{\quad}) \text{ or } (2x + \underline{\quad})(3x + \underline{\quad})$$

| 1 | 15 | 1 | 15 |
|---|----|---|----|
| 3 | 5 | 3 | 5 |
| 5 | 3 | 5 | 3 |
| 15 | 1 | 15 | 1 |

| Review and complete | You try |
|---|---|
| 8. Factor completely by using the trial and error method.<br><br>$9x^2 + 47x + 10$<br><br>$(9x + \underline{\quad})(x + \underline{\quad})$<br><br>or<br><br>$(3x + \underline{\quad})(3x + \underline{\quad})$<br><br>with 5,2 or 2,5 or 1,10 or 10,1<br><br>in the blanks.<br><br>Factored form: _____ | 9. Factor completely by using the trial and error method.<br><br>$14t^2 + 13t - 12$ |

**Video 5: Factoring a Trinomial with Leading Coefficient not Equal to 1 (Trial and Error Method) (2:26 min)**

▶ Play the video and work along.

Factor.    $-x + 2x^2 - 6$

$(2x - 1)(x + 6) = 2x^2 + 12x - 1x - 6 = 2x^2 + 11x - 6$

$(2x - 3)(x + 2) = 2x^2 + 4x - 3x - 6 = 2x^2 + x - 6$

$(2x + 3)(x - 2) = 2x^2 - 4x + 3x - 6 = 2x^2 - x - 6$

$(2x - 1)(x + 6) = 2x^2 - 12x + x - 6 = 2x^2 - 11x - 6$

| Review and complete | You try |
|---|---|
| 10. Factor.  $-7x-30+15x^2$ <br><br> $15x^2-7x-30, \text{GCF}=1$ <br><br> $(3x-\underline{\phantom{xx}})(5x+\underline{\phantom{xx}})$ <br><br> $(3x+\underline{\phantom{xx}})(5x-\underline{\phantom{xx}})$ <br><br> $(15x-\underline{\phantom{xx}})(x+\underline{\phantom{xx}})$ <br><br> $(15x+\underline{\phantom{xx}})(x-\underline{\phantom{xx}})$ <br><br> Fill blanks with factors of 30. <br><br> Factored form: _____ | 11. Factor completely by using the trial and error method. <br><br> $3m^2-5m-12$ |

**Video 6: Factoring a Trinomial Using the Trial-and-Error Method and where Rearranging Terms is Necessary (2:27 min)**

▶ Play the video and work along.

Factor completely.  $6x^2y+12y-27xy$

| Review and complete | You try |
|---|---|
| 12. Factor completely. <br><br> $14t^4-14t^2-45t^3$ <br><br> $14t^4-45t^3-14t^2$ <br><br> $t^2(14t^2-45t-14)$ <br><br> $t^2(2t-\underline{\phantom{xx}})(7t+\underline{\phantom{xx}})$ | 13. Factor completely. <br><br> $16m^2n-9n+70mn$ |

**Video 7: Recognizing a Prime Trinomial  (2:14 min)**

▶Play the video and work along.

Factor completely.  $2x^2+4x+3$

| Review and complete | You try |
|---|---|
| 14. A polynomial that cannot be factored is called a _____ polynomial. | |

| Review and complete | You try |
|---|---|
| 15. One of these polynomials is prime. Which one?<br><br>  a.  $2x^2 + 17x - 9$<br><br>  b.  $2x^2 + 17x - 8$ | 16. One of these polynomials is prime. Which one?<br><br>  a.  $3x^2 - 35x - 11$<br><br>  b.  $3x^2 - 35x - 12$ |

### Video 8: Factoring Trinomials with a Leading Coefficient 1  (3:07 min)

 Play the video and work along.

Factor completely.    $x^2 - 10x + 21$

| Review and complete | You try |
|---|---|
| 17. Multiplication is commutative. That means that $(x-3)(x-7)\_\_(x-7)(x-3)$. Fill in the blank with = or ≠ . | |

| Review and complete | You try |
|---|---|
| 18. Factor completely.    $x^2 - 3x - 10$<br><br>Think: What two integers multiply to be $-10$ and add to be $-3$?<br><br>$x^2 - 3x - 10 = ($ _____ $)($ _____ $)$ | 19. Factor completely.    $n^2 - 9n + 20$ |

### Video 9: Factoring a Trinomial with a Negative Leading Term  (1:52 min)

▶ Play the video and work along.

Factor completely.   $16m - m^2 - 15$

| Review and complete | You try |
|---|---|
| 20. Factor completely.<br>$50 - m^2 - 5m$     *Rearrange terms. Factor out -1.*<br>$-m^2 - 5m + 50$<br>$-1(m^2 + 5m - 50)$<br>$-1(m + 10)(m - 5)$ | 21. Factor completely.<br>$3t + 18 - t^2$ |

## Concept 3: Factoring Perfect Square Trinomials

| Video 10: Recognizing and Factoring a Perfect Square Trinomial (2:55 min) |
|---|

▶ Play the video and work along.

Factor completely.  $4x^2 + 12x + 9$

| Review and complete | You try |
|---|---|
| 22. Insert a middle term so that the trinomial is a perfect square.<br><br>$x^2 + \underline{\hspace{1cm}} + 81$ | 23. Insert a middle term so that the trinomial is a perfect square.<br><br>$4x^2 + \underline{\hspace{1cm}} + 25$ |
| 24. Factor completely.<br><br>$25x^2 + 30x + 9$<br>The first and third terms are perfect squares.<br>$(\underline{\hspace{1cm}})^2 = 25x^2$<br>$(\underline{\hspace{1cm}})^2 = 9$<br>So the factored form is $\underline{\hspace{3cm}}$. | 25. Factor completely.<br><br>$m^4 + 20m^2 + 100$ |

| Video 11: Identifying and Factoring Perfect Square Trinomials (3:19 min) |
|---|

▶ Play the video and work along.

Factor completely.
$81x^5 - 90x^4 y + 25x^3 y^2$    $4t^2 + 20t + 9$

---

Procedure for Factoring Trinomials
1. Factor out the GCF.
2. Check whether the trinomial is a perfect square trinomial.
   $a^2 + 2ab + b^2 = (a+b)^2$
   $a^2 - 2ab + b^2 = (a-b)^2$
3. Otherwise, factor by the trial and error method or the ac-method.

---

| Review and complete | You try |
|---|---|
| 26. Factor completely.<br><br>$-5x^4 - 90x^3 - 405x^2$<br><br>$-5x^2(x^2 + 18x + 81)$<br><br>This is a perfect square trinomial.<br><br>$-5x^2(x+9)^2$ | 27. Factor completely.<br>$-4x^3 + 32x^2 + 80x$ |

## Concept 4: Factoring by Using Substitution

**Video 12: Factoring by Using Substitution  (1:59 min)**

▶ Play the video and work along.

Factor completely.  $(3a+4)^2 - 6(3a+4) + 8$

| | |
|---|---|
| Check:<br>Factored form:<br><br>$3a(3a+2)$<br><br>$= 9a^2 + 6a$ | Original:<br><br>$(3a+4)^2 - 6(3a+4) + 8$<br><br>$= 9a^2 + 24a + 16 - 18a - 24 + 8$<br><br>$= 9a^2 + 6a$ |

| Review and complete | You try |
|---|---|
| 28. Factor completely.<br><br>$(34-t)^2 - 7(34-t) + 6$<br><br>$u = 34 - t$<br><br>$u^2 - 7u + 6$<br><br>$(u-1)(u-6)$<br><br>$((34-t)-1)((34-t)-6)$<br><br>$(33-t)(28-t)$ | 29. Factor completely.<br>$(n-2)^2 + 12(n-2) + 36$ |

**Video 13: Factoring a Higher Degree Polynomial by Using Substitution  (2:19 min)**

▶ Play the video and work along.

Factor completely.  $m^6 - 4m^3 - 5$

| Review and complete | You try |
|---|---|
| 30. Factor completely.<br><br>$16t^4 - 56t^2 + 49$<br><br>$u = t^2$<br><br>$16u^2 - 56u + 49$<br><br>$(4u - 7)(4u - 7)$<br><br>$(4u - 7)^2$<br><br>$(4t^2 - 7)^2$ | 31. Factor completely.<br>$t^6 - 16t^3 + 63$ |

## Concept 1: Difference of Squares

### Video 1: Introduction to Factoring a Difference of Squares (2:47 min)

▶ Play the video and work along.

Factor completely, if possible.

$x^2 - 25$                                                                     $x^2 + 25$

### Video 2: Factoring a Difference of Squares (2:37 min)

▶ Play the video and work along.

Factor completely.

$49m^2 - 81n^2$                    $32x^5y^2 - 8x^5$

| Perfect Squares | | Perfect Squares |
|---|---|---|
| $1^2 = 1$ | | $(x^1)^2 = x^2$ |
| $2^2 = 4$ | $8^2 = 64$ | $(x^2)^2 = x^4$ |
| $3^2 = 9$ | $9^2 = 81$ | $(x^3)^2 = x^6$ |
| $4^2 = 16$ | $10^2 = 100$ | $(x^4)^2 = x^8$ |
| $5^2 = 25$ | $11^2 = 121$ | |
| $6^2 = 36$ | $12^2 = 144$ | |
| $7^2 = 49$ | $13^2 = 169$ | |

⏸ Pause the video and try this yourself.

$x^2 - 64x$

▶ Play the video and check your answer.

| Review and complete | You try |
|---|---|
| 1. Factor completely. | 2. Factor completely. |
| a. $x^2 + 9$<br>This is the sum of squares and is prime. | a. $4x^2 - 1$ |
| b. $5x^2 - 605$<br>$= 5(x^2 - 121)$<br>$= 5((x)^2 - 11^2)$<br>$= 5(x - 11)(x + 11)$ | b. $50c^2 - 72d^2$ |

**Video 3: Factoring a Difference of Squares where Repeated Factoring is Required (1:26 min)**

▶ Play the video and work along.

Factor completely.    $16x^4 - 1$

| Perfect Squares | | Perfect Squares |
|---|---|---|
| $1^2 = 1$ | | $(x^1)^2 = x^2$ |
| $2^2 = 4$ | $8^2 = 64$ | $(x^2)^2 = x^4$ |
| $3^2 = 9$ | $9^2 = 81$ | $(x^3)^2 = x^6$ |
| $4^2 = 16$ | $10^2 = 100$ | $(x^4)^2 = x^8$ |
| $5^2 = 25$ | $11^2 = 121$ | |
| $6^2 = 36$ | $12^2 = 144$ | |
| $7^2 = 49$ | $13^2 = 169$ | |

| Review and complete | You try |
|---|---|

Before factoring by other methods, always check for a GCF other than 1. After factoring by other methods, make sure additional factoring is not possible.

| | |
|---|---|
| 3.  Factor completely. <br><br> $m^4 - 81$ <br><br> $(m^2 - 9)(m^2 + 9)$ <br><br> $(m^2 - 9)(m^2 + 9)$ <br><br> $(m - 3)(m + 3)(m^2 + 9)$ <br><br> *Examine your factored form to see if additional factoring is possible.* | 4.  Factor completely. <br><br> $n^4 - 625$ |

## Concept 2: Using a Difference of Squares in Grouping

**Video 4: Factoring by Grouping and Difference of Squares (1:39 min)**

▶ Play the video and work along.

Factor completely.    $x^3 + 2x^2 - 9x - 18$

| Review and complete | You try |
|---|---|
| 5.  Factor completely. <br><br> *Examine your factored form to see if additional factoring is possible.* | 6.  Factor completely.  $9x^3 - 5x^2 - 36x + 20$ |

Video and Study Guide to accompany Intermediate Algebra, 5th ed., Miller/O'Neill/Hyde

$$m^2 x + 2m^2 y - x - 2y$$
$$m^2(x+2y) - 1(x+2y)$$
$$(x+2y)(m^2-1)$$
$$(x+2y)(m+1)(m-1)$$

## Video 5: Factoring by Grouping One Term with Three Terms (3:08 min)

▶ Play the video and work along.

Factor completely.    $4c^2 - d^2 - 10d - 25$

| Review and complete | You try |
|---|---|
| 7. If grouping a four-term polynomial into two groups of two terms does not yield a matching binomial factor, then try grouping the terms into _____ terms and _____ term, and looking for a perfect square trinomial represented by the three term group. | |

| Review and complete | You try |
|---|---|
| 8. Factor completely. $9x^2 + 48x + 64 - 36y^2$ <br> $\left(9x^2 + 48x + 64\right) - 36y^2$ <br> $(3x+8)^2 - 36y^2$ <br> $(3x+8)^2 - (6y)^2$ <br> $((3x+8)-6y)((3x+8)+6y)$ <br> $(3x+8-6y)(3x+8+6y)$ | 9. Factor completely. $m^4 - 14m^2 + 49 - z^4$ |

## Concept 3: Sum and Difference of Cubes

## Video 6: Introduction to Factoring a Sum or Difference of Cubes (4:09 min)

▶ Play the video and work along.

### Factoring the Sum or Difference of Cubes

Difference of Cubes:     $a^3 - b^3 = (a-b)(a^2 + ab + b^2)$

Sum of Cubes          $a^3 + b^3 = (a+b)(a^2 - ab + b^2)$

SOAP = Same Opposite Always Positive

Difference of Cubes:     $a^3 - b^3 = (a-b)(a^2 + ab + b^2)$

Sum of Cubes     $a^3 + b^3 = (a+b)(a^2 - ab + b^2)$

| Review and complete | You try |
|---|---|
| 10. True or False? A sum of squares can be factored. | 11. True or False? A sum of cubes can be factored. |
| 12. Fill in the blanks.<br>The polynomial $x^3 + 8$ is a sum of _____ and _____ (can/cannot) be factored. | 13. Fill in the blanks.<br>The polynomial $27x^3 - 1$ is a difference of _____ and _____ (can/cannot) be factored. |

## Video 7: Factoring a Difference of Cubes  (2:06 min)

▶ Play the video and work along.

Factor completely.  $8x^3 - 27$

$(a \; - \; b)(a^2 \; + \; ab \; + \; b^2)$

| Perfect Cubes | Perfect Cubes |
|---|---|
| $1^3 = 1$ | $(x^1)^3 = x^3$ |
| $2^3 = 8$ | $(x^2)^3 = x^6$ |
| $3^3 = 27$ | $(x^3)^3 = x^9$ |
| $4^3 = 64$ | $(x^4)^3 = x^{12}$ |
| $5^3 = 125$ | $(x^5)^3 = x^{15}$ |

| Review and complete | You try |
|---|---|
| 14. Factor completely. $1 - 64x^6$ | 15. Factor completely.  $c^3 - 125d^3$ |

$1 - 64x^6 = (1)^3 - (4x^2)^3$

$\qquad = (1 - 4x^2)(1^2 + 1(4x) + (4x)^2)$

$\qquad = (1 - 4x)(1 + 4x + 16x^2)$

## Video 8: Factoring a Sum of Cubes (1:26 min)

▶ Play the video and work along.

Factor completely.  $t^3 + \dfrac{1}{125}$

$(a \; + \; b)(a^2 \; - \; ab \; + \; b^2)$

| Review and complete | You try |
|---|---|
| 16. Factor completely. | 17. Factor completely. |

16. Factor completely.

$$125y^3 + \frac{1}{8} = (5y)^3 + \left(\frac{1}{2}\right)^3$$

$$= \left(5y + \frac{1}{2}\right)\left(25y^2 - \frac{5y}{2} + \frac{1}{4}\right)$$

17. Factor completely.

$$\frac{x^3}{8} + 64$$

### Video 9: Factoring a Sum of Cubes (1:47 min)

▶ Play the video and work along.

Factor completely.   $64x^6 + 1$

| Review and complete | You try |
|---|---|

18. Circle the types of binomials that can be factored.

     Sum of cubes     Sum of squares     Difference of cubes     Difference of squares

| | |
|---|---|
| 19. Factor completely. $1 + c^6 d^6$ | 20. Factor completely. $8x^6 + 125$ |

19. Factor completely. $1 + c^6 d^6$

$$1 + c^6 d^6 = (1)^3 + (c^2 d^2)^3$$

$$= \left(1 + c^2 d^2\right)\left(1 - c^2 d^2 + c^4 d^4\right)$$

### Concept 4: Summary of Factoring Binomials

### Video 10: Practice Factoring Binomials (3:14 min)

▶ Play the video and work along.

Factor completely.

$$x^3 - y^6 \qquad\qquad\qquad\qquad\qquad 98 - 2t^2$$

$$p^4 + p \qquad\qquad\qquad\qquad\qquad 81y^4 - 16$$

| Review and complete | You try |
|---|---|
| 21. Factor completely.<br><br>$64x^4 + x$<br><br>$x(64x^3 + 1)$<br><br>Sum of cubes<br><br>$(a+b)(a^2 - ab + b^2)$<br><br>$a = 4x; b = 1$<br><br>$x(4x+1)(16x^2 - 4x + 1)$ | 22. Factor completely.<br>$7m^2 + 28$ |

## Concept 5: Factoring Binomials of the Form x^6 – y^6

**Video 11: Factoring a Binomial as both a Difference of Squares and as a Difference of Cubes (3:06 min)**

▶ Play the video and work along.

Factor completely. $x^6 - y^6$

**Approach 1:** Interpret this initially as a <u>difference of cubes</u>.

$a^3 - b^3 = (a - b)(a^2 + ab + b^2)$

$x^6 - y^6 = (x^2)^3 - (y^2)^3$

**Approach 2:** Interpret this initially as a difference of squares.

$a^2 - b^2 = (a - b)(a + b)$

$x^6 - y^6 = (x^3)^2 - (y^3)^2$

| Review and complete | You try |
|---|---|
|  |  |

23. Factor completely.

$$64m^6 - n^6$$
$$= (4m^2)^3 - (n^2)^3$$
$$= (4m^2 - n^2)(16m^4 + 4m^2n^2 + n^4)$$
$$= (2m - n)(2m + n)$$
$$\qquad (4m^2 - 2mn + n^2)(4m^2 + 2mn + n^2)$$

24. Factor completely.

$$c^6 - 15625$$

## Concept 1: Solving Equations by Using the Zero Product Rule

### Video 1: Definition of a Quadratic Equation (2:26 min)

▶ Play the video and work along.

> If $a$, $b$, and $c$ are real numbers such that $a \neq 0$, then a **quadratic equation** is an equation that can be written in the form
> $$ax^2 + bx + c = 0$$

For example, the following are quadratic equations.

$$3x^2 - 10x + 8 = 0 \qquad\qquad x(x - 6) = 0 \qquad\qquad -2x^2 = -18$$

| Review and complete | You try |
|---|---|
| 1. Is the equation a quadratic equation? | 2. Is the equation a quadratic equation? |
|     a. $x^2 = 9$ <br>      Yes <br>     b. $2x + 1 = 9$ <br>      No <br>     c. $x(x - 4) = 5x$ <br>      Yes |     a. $2x(x - 3) = 9$ <br><br>     b. $x^2(x - 4) = 5$ <br><br>     c. $45 - x = x - 45$ |

### Video 2: Solving a Quadratic Equation Using the Zero Product Rule  (2:28 min)

▶ Play the video and work along.

Left side is factored    Other side is zero

> **Zero Product Rule**
>
> If $ab = 0$, then $a = 0$ or $b = 0$.

Solve the equation.   $(x - 3)(x + 7) = 0$

| Review and complete | You try |
|---|---|
| 3. Factoring is an important skill because quadratic equations in factored form can be solved using the _____ rule, as long as the other side is zero. | |

4. Solve the equation.

$$(x+5)(x-4)=0$$

$$x+5=0 \text{ or } x-4=0$$

$$x=-5 \text{ or } x=4$$

Check:

| $x=-5$ | $x=4$ |
|---|---|
| $(-5+5)(x-4)=0$ | $(x+5)(4-4)=0$ |
| $0(-4)=0$ true | $5(0)=0$ true |

The solution set is $\{-5, 4\}$.

5. Solve the equation.

$$(x+2)(x+7)=0$$

## Video 3: Solving a Quadratic Using the Zero Product Rule (2:38 min)

▶ Play the video and work along.

Left side is factored    Other side is zero

$$10x(2x+5)=0$$

Left side is factored    Other side is zero

$$10(2x-1)(x+5)=0$$

## Review and complete

6. Solve the equation. $5(x-4)(2x+1)=0$

Set each factor to zero and solve.

$$5=0 \qquad x-4=0 \qquad 2x+1=0$$

No solution $\qquad x=4 \qquad x=-\dfrac{1}{2}$

The solution set is $\left\{-\dfrac{1}{2}, 4\right\}$.

## You try

7. Solve the equation. $6(x+2)(3x-4)=0$

|  |  |
| --- | --- |

### Video 4: Solving a Quadratic Using the Zero Product Rule (1:40 min)

▶ Play the video and work along.

Solve the equation.    $x^2 - x = 12$

| Review and complete | You try |
| --- | --- |

8.  To solve a quadratic equation by factoring and the zero product rule, first ensure that one side of the equation is equal to _____.

| | |
|---|---|
| 9.  Solve the equation. $2x^2 = 7x + 4$ <br><br> $2x^2 - 7x - 4 = 0$ <br><br> $(2x+1)(x-4) = 0$ <br><br> $2x+1 = 0$ or $x - 4 = 0$ <br><br> $x = -\dfrac{1}{2}$ or $x = 4$ <br><br> $\left\{ -\dfrac{1}{2}, 4 \right\}$ | 10. Solve the equation. $x^2 + 8x = 9$ |

### Video 5: Solving a Quadratic Using the Zero Product Rule (1:26 min)

▶ Play the video and work along.

Solve the equation.    $3x^2 + 21x = 0$

| Review and complete | You try |
| --- | --- |
| 11. Solve the equation. | 12. Solve the equation. <br> $4m^2 - 20m = 0$ |

$4x^2 - 28x = 0$

$4x(x-7) = 0$

$4x = 0$ or $x - 7 = 0$

$x = 0$ or $x = 7$

$\{0, 7\}$

## Video 6: Solving a Quadratic Using the Zero Product Rule (3:25 min)

▶ Play the video and work along.

Solve the equation.    $(2x+1)(x-2) = 4 - 2x$

| Review and complete | You try |
|---|---|
| 13. Solve.  $(3x+1)(x+15) = 65 + 21x$ <br><br> Clear parentheses on the left. Then set one side equal to 0 by subtracting 65 and 21x from each side. <br><br> $\qquad 3x^2 + 46x + 15 = 65 + 21x$ <br> $\qquad 3x^2 + 25x - 50 = 0$ <br> $\quad 3x^2 + 30x - 5x - 50 = 0$ <br> $3x(x+10) - 5(x+10) = 0$ <br> $\qquad (3x-5)(x+10) = 0$ <br> $\qquad\quad 3x - 5 = 0$ or $x + 10 = 0$ <br><br> $\qquad\qquad x = \dfrac{5}{3}$ or $x = -10$ <br><br> The solution set is $\left\{-10, \dfrac{5}{3}\right\}$. | 14. Solve.  $11x^2 - 3(5x+2) = 2x^2$ |

## Video 7: Identifying Equation Type and Solving a Linear Equation  (1:35 min)

▶ Play the video and work along.

Solve.  $2x(x+4) - 5 = 2x^2 + 7x + 1$

| Review and complete | You try |
|---|---|
| 15. Solve. $$x(x-4)+4x = x^2-x-1$$ $$x^2-4x+4x = x^2-x-1$$ $$x^2 = x^2-x-1$$ $$x^2-x^2 = x^2-x^2-x-1$$ $$0 = -x-1$$ $$x = -1$$ The solution set is $\{-1\}$. | 16. Solve. $$x^2-4x-2 = (x+3)(x-5)$$ |

> Simplify completely to determine the type of equation.

## Video 8: Solving Higher Degree Polynomial Equations Using the Zero Product Rule  (3:17 min)

▶ Play the video and work along.

Solve the equations.

$$3(x+5)(x-2)(4x+1) = 0 \qquad\qquad m^3+2m^2 = 4m+8$$

| Review and complete | You try |
|---|---|
| | |

**17. Solve.** $7x(x-1)(x+4)=0$

The equation is in factored form. Set each factor equal to 0 and solve.

$$7x(x-1)(x+4)=0$$

$$7x=0 \qquad x-1=0 \qquad x+4=0$$
$$x=0 \qquad x=1 \qquad x=-4$$

The solution set is $\{-4,0,1\}$.

**18. Solve.** $3m^3+m^2-48m-16=0$

## Concept 2: Applications of Quadratic Equations

**Video 9: Translating an English Phrase to a Quadratic Equation and Solving the Equation (2:28 min)**

▶ Play the video and work along.

The product of two consecutive even integers is 48. Find the integers.

| Review and complete | You try |
|---|---|
| **19.** The product of two consecutive odd integers is 9999 . Find the integers. | **20.** The product of two consecutive integers is 42 . Find the integers. |

Let $n$ represent the smaller odd integer.
Let $n+2$ represent the next odd integer.

> Both even and odd consecutive integers are separated by 2.

$$n(n+2)=9999$$
$$n^2+2n-9999=0$$
$$(n-99)(n+101)=0$$
$$n=99$$
$$n=-101$$

The integers are either 99 and 101 or
-101 and -99 .

---

### Video 10: Solving a Geometry Application Using a Quadratic Equation (Area of a Rectangle) (2:25 min)

▶ Play the video and work along.

The area of a rectangular field is 240 yd$^2$. The length is 8 yd longer than the width. Find the dimensions of the field.

---

| Review and complete | You try |
|---|---|
| 21. The area of a rectangular locker room is 375 yd$^2$. The room is 10 yd longer than it is wide. Find the dimensions of the locker room.<br><br>Let w represent the width.<br>Then $w + 10$ represents the length.<br><br>$A = L \cdot W$<br><br>$375 = (w + 10) \cdot w$<br><br>$375 = w^2 + 10w$<br><br>$w^2 + 10w - 375 = 0$<br><br>$(w + 25)(w - 15) = 0$<br><br>~~$w = -25$~~<br><br>$w = 15$<br>The width is 15 yd and the length is 25 yd. | 22. The top of a large coffee table is twice as long as it is wide. Find its dimensions if its area is 18 ft$^2$. |

---

### Video 11: Solving a Geometry Application Using a Quadratic Equation (Area of a Triangle) (2:37 min)

▶ Play the video and work along.

Winston's hobby is flying stunt kites. One of his kites is triangular in shape, and the height is 2 ft less than the base. If the area of the kite is 7.5 ft$^2$, find the base and height of the kite.

---

| Review and complete | You try |
|---|---|
| 23. The base of a triangle is four units less than twice its height. If the area of the triangle is $35 \text{ cm}^2$, find the dimensions of the triangle. | 24. The height of a triangle is $1$ in more than twice the base. If the area is $18 \text{ in}^2$, find the base and height of the triangle. |

$$A = \frac{1}{2}bh$$

Let n represent the height of the triangle.
Then 2n – 4 is its base.

$$\frac{1}{2}(2n-4)n = 35$$

$$(n-2)(n) = 35$$

$$n^2 - 2n - 35 = 0$$

$$(n+5)(n-7) = 0$$

$$\cancel{n = -5}; n = 7$$

The height of the triangle is 7 cm. The base is 10 cm.

---

## Video 12: Using a Quadratic Equation in an Application with the Pythagorean Theorem (3:39 min)

▶ Play the video and work along.

A ramp is built so that the horizontal leg is 1 ft less than the length of the ramp, and the vertical leg is 32 ft less than the length of the ramp. Find the length of the ramp.

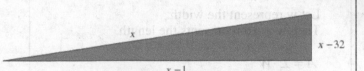

$x$

$x - 32$

$x - 1$

---

| Review and complete | You try |
|---|---|

25. A 20-ft extension ladder is leaning on a wall of an apartment building. If the top of the ladder is touching the wall at a point 8 ft less than twice the distance that the bottom of the ladder is from the base of the wall, find the height of the point where the top of the ladder is touching the wall.

Let x be the base of the right triangle that represents this situation.

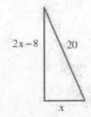

Then $2x - 8$ is the longer leg.

$$a^2 + b^2 = c^2$$
$$x^2 + (2x-8)^2 = 20^2$$
$$x^2 + 4x^2 - 32x + 64 = 400$$
$$5x^2 - 32x - 336 = 0$$
$$(x-12)(5x+28) = 0$$
$$x = 12; \quad x = \frac{-28}{5}$$

> Recall the procedure to square a binomial.

The bottom of the ladder is 12 ft from the wall. The ladder reaches a height of 16 ft on the wall.

26. A right triangle has sides that are represented by three consecutive even integers. Find the side lengths.

## Concept 3: Definition of a Quadratic Function

### Video 13: Definition of a Quadratic Function and Discussion on x- and y-intercepts (2:22 min)

▶ Play the video and work along.

**Definition of a Quadratic Function**
Let *a*, *b*, and *c* be real numbers such that $a \neq 0$. Then a function defined by $f(x) = ax^2 + bx + c$ is a **quadratic function**.

| Review and complete | You try |
|---|---|
| 27. The graph of a quadratic function can have ___, ___, or ___ x-intercepts. The graph of a quadratic function has only one y-intercept. | |

## Video 14: Determining x- and y-intercepts of the Graph of a Quadratic Function (2:06 min)

▶ Play the video and work along.

Determine the x- and y-intercepts of the function defined by $f(x) = x^2 - 2x - 3$

$$f(x) = x^2 - 2x - 3$$

x-intercepts

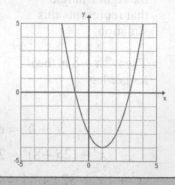

y-intercept:

| Review and complete | You try |
|---|---|
| 28. Determine the x- and y-intercepts of the function defined by $f(x) = x^2 + 7x - 18$.  <br><br> x-intercepts        y-intercept <br> $0 = x^2 + 7x - 18$ <br> $0 = (x+9)(x-2)$      $f(0) = 0^2 + 7(0) - 18$ <br> $x = -9, x = 2$         $(0, -18)$ <br> $(-9, 0)$ and $(2, 0)$ | 29. Determine the x- and y-intercepts of the function defined by $f(x) = x^2 - 4x - 12$. <br><br> x-intercept(s)        y-intercept |

## Concept 4: Applications of Quadratic Functions

## Video 15: An Application Involving Quadratic Equations (Object in Free-Fall) (4:08 min)

▶ Play the video and work along.

A rocket is shot straight upward from ground level with an initial velocity of 128 ft/sec. The function defined by $h(t) = -16t^2 + 128t$ relates the rocket's height, $h(t)$ (in feet) to the time $t$ (in seconds) after launch.

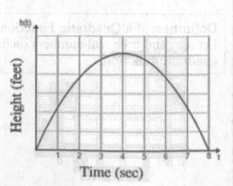

**Height of Rocket vs Time**

Determine the values of $h(3)$ and $h(6)$. Interpret their meaning in the context of the problem.

Determine the t-intercepts and interpret their meaning in the context of the problem.

Determine when the rocket will reach a height of 240 ft.

| Review and complete | You try |
|---|---|
| 30. A rocket is shot straight upward from ground level with an initial velocity of 245 m/sec. The function defined by $h(t) = -4.9t^2 + 245t$ relates the rocket's height, $h(t)$, (in meters) to the time $t$ (in seconds) after launch. | 31. A rocket is shot straight upward from ground level with an initial velocity of 490 m/sec. The function defined by $h(t) = -4.9t^2 + 490t$ relates the rocket's height, $h(t)$, (in meters) to the time $t$ (in seconds) after launch. |

**30. (continued)**

a. Determine the value of $h(20)$ and interpret its meaning in the context of the problem.

$$h(20) = -4.9(20)^2 + 245(20)$$

$$= 2940$$

At 20 sec after launch, the rocket is 2940 ft above the ground.

b. Determine the $t$-intercepts and interpret their meaning in the context of the problem.

$$0 = -4.9x^2 + 245x$$

$$0 = -4.9x(x - 50)$$

The intercepts are $(0, 0)$ and $(50, 0)$.

$$x = 0; x = 50$$

The rocket is at ground level at 0 sec and 50 sec.

**31. (continued)**

a. Determine the value of $h(10)$ and interpret its meaning in the context of the problem.

b. Determine the $t$-intercepts and interpret their meaning in the context of the problem.

**Answers: Section 4.1**

1. opposite    3a. 16    3b. −16    3c. −64    4. 1    6a. 1

6b. 5    8a. $\dfrac{1}{k^4}$    8b. $\dfrac{1}{27}$    10. $m^5$    12. $\dfrac{20}{9}$    14. $\dfrac{189x^9}{y^{25}}$

16. $8 \times 10^6$    18. $3.189 \times 10^8$    19. left; right    21. 0.0000000012    23. $1.872 \times 10^{10}$

25. $3.158 \times 10^{-8}$    27. 12,000,000    29. $2 \times 10^{24}$

**Answers: Section 4.2**

1. terms    3a. True    3b. False    3c. True    5a. $-x^3$

5b. 3    7. $7x^3 - 2x^2 + 5x - 2$    9. $5.84x^2 - 7.5x + 6.28$

11. $x^2 - 7xy + 2y^2$    13. $10mn^2 + 6m^2n - 9mn$    15. $3x^2 + 16xy$

16. $\neq$    18. $\dfrac{1}{3}x^2 + \dfrac{7}{2}y^2$    19. $f(x)$ and $m(x)$

21. −1    23. $P(x) = 3x + 12$    25. 18%

**Answers: Section 4.3**

1. commutative, associative    3. $2x^2 y^7$    5. $21m^2 n^2 - 3m^3 n$

7. $x^2 + 7x - 8$    8. $cd$    9. $6x^2 - 5xy + y^2$

10. $-7m$    11. $2n^3 + 17n^2 - 31n + 11$    12. binomial

13. trinomial    15. $9t^2 - 25$    16. binomial

18. $81x^2 - 36x + 4$    20. $a^2 - c^2 - 18c - 81$    22. $x^3 + 3x^2 + 3x + 1$

23. $y + 7$    24. $y^2 + \dfrac{1}{16}$    26a. $A(x) = \dfrac{3}{2}x^2 + \dfrac{5}{2}x$

26b. $2500 \text{ in}^2$

**Answers**: Section 4.4

2. $6x^3 y - \dfrac{2x}{y} - \dfrac{2}{5x}$    3. zero    5. $4x^2 + 15x + 30 + \dfrac{57}{x-2}$    7. $x - 10$

9. $x^2 - 2x + 4$    11. $\dfrac{34x^2}{9} + x - \dfrac{5}{x}$

**Answers**: Section 4.5

2. $2n^3$    4. $2n^3(3mn^2 + 20m^4 n - 7)$    6. $x^2(x^4 - x^2 - 1)$

7. $16y^2$    8. $-7x(2x^2 y - 7x + 100)$    10. $(n-1)(m+n-8)$

12. $(3y^3 - x)(y-2)$    14. $5x(x-1)(y+5)$    16. $(m+5n)(n^2+2)$

**Answers**: Section 4.6

2. $(3x+1)(5x+3)$    4. $(3x+5y)(x+3y)$    5. remains

7. $3c(2d+7)(d-2)$    8  $(9x+2)(x+5)$    9. $(7t-4)(2t+3)$

10. $(3x-5)(5x+6)$    11. $(3m+4)(m-3)$    12. $t^2(2t-7)(7t+2)$

13. $n(8m-1)(2m+9)$    14. prime    15. b. $2x^2 + 17x - 8$

16. a. $3x^2 - 35x - 11$    17. $=$    18. $(x-5)(x+2)$

19. $(n-4)(n-5)$    21. $-1(t-6)(t+3)$    22. $18x$

23. $20x$    24. $5x, 3, (5x+3)^2$    25. $(m^2+10)^2$

27. $-4x(x-10)(x+2)$

29. $(n+4)^2$

31. $(t^3-9)(t^3-7)$

**Answers:** Section 4.7

2a. $(2x-1)(2x+1)$

2b. $2(5c-6d)(5c+6d)$

4. $(n^2+25)(n-5)(n+5)$

6. $(9x-5)(x+2)(x-2)$

7. three; one

9. $(m^2-7-z^2)(m^2-7+z^2)$

10. False

11. True

12. cubes; can

13. cubes; can

15. $(c-5d)(c^2+5cd+25d^2)$

17. $\left(\dfrac{x}{2}+4\right)\left(\dfrac{x^2}{4}-2x+16\right)$

18. All but the sum of squares

20. $(2x^2+5)(4x^4-10x^2+25)$

22. $7(m^2+4)$

24. $(c-5)(c+5)(c^2+5c+25)(c^2-5c+25)$

**Answers:** Section 4.8

2a. Yes

2b. No

2c. No

3. zero product

5. $\{-7,-2\}$

7. $\left\{-2,\dfrac{4}{3}\right\}$

8. zero

10. $\{-9,1\}$

12. $\{0,5\}$

14. $\left\{-\dfrac{1}{3},2\right\}$

16. $\left\{\dfrac{13}{2}\right\}$

18. $\left\{-4,-\dfrac{1}{3},4\right\}$

20. $-7$ and $-6$ or $6$ and $7$

22. 3 ft by 6 ft

24. Height 9 in, base 4 in

26. 6, 8, 10

27. 0, 1, or 2

29. x-int: $(6,0),(-2,0)$ ; y-int: $(0,-12)$

31a. $h(10) = 4410$ ; The rocket is 4410 m above the ground in 10 sec after it is launched.

31b. $(0,0),(100,0)$ ; The rocket is at ground level 0 sec and 100 sec after launch.

**Concept 1**: Rational Functions

| Video 1: Introduction to Rational Expressions and Functions (3:07 min) |
| :--- |

▶ Play the video and work along.

**Definition of a Rational Expression**

A **rational expression** is a ratio of two polynomials, $\dfrac{p}{q}$ , where $q \neq 0$ .

Rational numbers:

Rational Expressions:

⏸ Pause the video and try this yourself

Find the restricted values of $m$ for $\dfrac{m^3 - 5m}{4}$ .

▶ Play the video and check your answer.

**Definition of a Rational Function**

A **rational function** is defined by $f(x) = \dfrac{p(x)}{q(x)}$ , where $p(x)$ and $q(x)$ are polynomials and $q(x) \neq 0$ .

Rational functions: $\qquad$ $f(x) = \dfrac{x-9}{x^2-4}$ ; $x \neq \underline{\quad}, x \neq \underline{\quad}$

❚❚ Pause the video and try this yourself

Find the restriction on $x$ for $g(x) = \dfrac{6x-5}{x}$ ; $x \neq \underline{\quad}$

▶ Play the video and check your answer.

| Review and complete | You try |
|---|---|
| 1. Rational functions are different than polynomial functions because they can have _____ values on the independent variable. | |

| | |
|---|---|
| 2. Find the restricted value(s) of the rational expression or function. | 3. Find the restricted value(s) of the rational expression or function. |
| a. $\dfrac{5x}{x-8}$ ; $x \neq 8$ <br><br> Restricted value(s) of the independent variable make the denominator equal to zero. | a. $\dfrac{6-y}{5y}$ |
| b. $h(x) = \dfrac{4x}{x^2-9}$ ; $x \neq 3$ , $x \neq -3$ | b. $f(x) = \dfrac{x-3}{x^2-9x}$ |

**Video 2: Evaluating Rational Expressions and Functions for Selected Values in the Domain** (2:48 min)

▶ Play the video and work along.

Evaluate the function for the given values of $x$.    $f(x) = \dfrac{x+1}{x-3}$

$f(0) = \dfrac{0+1}{0-3} =$                                  $f(2) =$

$f(-2) =$                                          $f(3) =$

Write the domain of the function in set-builder notation and in interval notation.

$f(x) = \dfrac{x+1}{x-3}$                    $\longrightarrow$

| Review and complete | You try |
|---|---|
| 4.  Exclude from the domain of a rational function any input values for which the denominator is _____. |  |

5. Write the domain of the function in set-builder notation and in interval notation.

$$g(x) = \frac{x^2 - 1}{x - 5}$$

The value of $x - 5$ cannot be zero, so $x \neq 5$.

$$\{x \mid x \neq 5\}$$

$$(-\infty, 5) \cup (5, \infty)$$

Exclude 5 from the domain.

6. Write the domain of the function in set-builder notation and in interval notation.

$$h(x) = \frac{x + 5}{x + 4}$$

## Video 3: Determining Domain of a Rational Function (4:32 min)

▶ Play the video and work along.

Determine the domain of each function.

$$f(x) = \frac{x + 1}{3x - 2}$$

⏸ Pause the video and try this yourself.

$$g(t) = \frac{t - 5}{t^2 - 7t + 12}$$

$$h(a) = \frac{a}{a^2 + 9}$$

▶ Play the video and check your answer.

| Review and complete | You try |
|---|---|

7.  If the denominator of a rational function is always positive, the domain is _____.

8.  Determine the domain of the function. Write in interval notation.

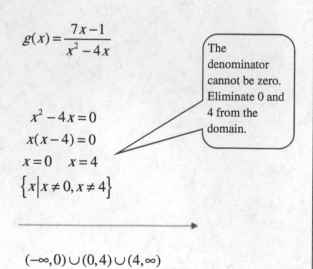

$$g(x) = \frac{7x-1}{x^2-4x}$$

> The denominator cannot be zero. Eliminate 0 and 4 from the domain.

$$x^2 - 4x = 0$$
$$x(x-4) = 0$$
$$x = 0 \quad x = 4$$
$$\{x \mid x \neq 0, x \neq 4\}$$

$$(-\infty, 0) \cup (0, 4) \cup (4, \infty)$$

9.  Determine the domain of the function. Write in interval notation.

$$f(t) = \frac{t+1}{t^2-t-6}$$

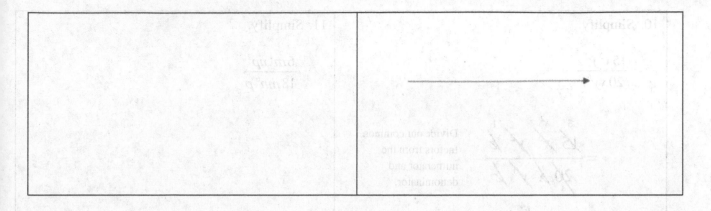

**Concept 2**: Simplifying Rational Expressions

| Video 4: Simplifying a Rational Expression with Monomials (3:18 min) |
|---|

▶ Play the video and work along.

Simplify the rational expressions.

$$\frac{35}{75}$$

$$\frac{18x^2y^5z}{12xy^2z^7}$$

| Review and complete | You try |
|---|---|

Fundamental Principle of Rational Expressions

Let $p$, $q$, and $r$ represent polynomials such that $q \neq 0$ and $r \neq 0$. Then

$$\frac{pr}{qr} = \frac{p}{q} \cdot \frac{r}{r} = \frac{p}{q} \cdot 1 = \frac{p}{q}$$

**10. Simplify.**

$$\frac{15x^3y^2z}{20xy^6z}$$

$$= \frac{\overset{3}{\cancel{15}} \ \overset{x^2}{\cancel{x^3}} \ \cancel{y^2} \ \overset{1}{\cancel{z}}}{\underset{4}{\cancel{20}} \ \cancel{x} \ \cancel{y^6} \ \cancel{z}}$$  Divide out common factors from the numerator and denominator.

$$\qquad\qquad y^4$$

$$= \frac{3x^2}{4y^4}$$

**11. Simplify.**

$$\frac{6m^4np^2}{18mn^6p}$$

---

## Video 5: Simplifying a Rational Expression (3:39 min)

▶ Play the video and work along.

Simplify the expression.  $\dfrac{2x-6}{x^2-8x+15}$

---

| Review and complete | You try |
|---|---|

12. Restricted values of a rational expression _____ (are/are not) eliminated when the expression is simplified.

13. It is incorrect to divide out **factors** from the numerator and the denominator before _____.

| | |
|---|---|
| 14. Simplify the expression. Identify any restricted values. $$\frac{x^2-4}{5x+10}$$ $$= \frac{(x-2)(x+2)}{5(x+2)}$$ Factor the numerator and denominator. Divide out the common factor. $$= \frac{x-2}{5}$$ Restrictions on x: $x \neq -2$ | 15. Simplify the expression. Identify any restricted values. $$\frac{3x-3}{x^2+9x-10}$$ |

**Video 6: Identifying the Restricted Values for a Rational Expression and Simplifying the Expression (3:20 min)**

▶ Play the video and work along.

Factor the numerator and denominator.

$$\frac{x^3 + x^2 - x - 9}{x^2 + 4x + 3}$$

❚❚ Pause the video and try this yourself.

Identify the restricted value(s) for $x$ in this expression. Simply the expression.

Write the domain of the function $f(x) = \dfrac{x^3 + x^2 - x - 9}{x^2 + 4x + 3}$ in set-builder notation.

▶ Play the video and check your answer.

| Review and complete | You try |
|---|---|

16. Identify restricted values of a rational function or expression _____ (before/after) simplifying.

| | |
|---|---|
| 17. Simplify the expression. Identify any restricted values. $$\frac{n^2 + 12n + 27}{n^3 + 9n^2 + 5n + 45}$$ $$= \frac{(n+3)(n+9)}{n^2(n+9) + 5(n+9)}$$ $$= \frac{(n+3)(n+9)}{(n+9)(n^2 + 5)}$$ The value $n = -9$ makes the denominator zero (and no value of $n$ makes $n^2 + 5 = 0$, so there is only one restriction: $n \neq -9$. $$\frac{(n+3)\,\overset{1}{\cancel{(n+9)}}}{\cancel{(n+9)}\,(n^2 + 5)} = \frac{n+3}{n^2 + 5} \text{ provided } x \neq -9.$$ | 18. Simplify the expression. Identify any restricted values. $$\frac{2x^2 + 19x + 35}{x^2 - 49}$$ |

**Concept 3**: Simplifying Ratios of -1

| Video 7: Recognizing a Ratio of -1 (2:59 min) |
| --- |

▶ Play the video and work along.

Recognizing a ratio of 1

$$\frac{5}{5} =$$    $$\frac{x-4}{x-4} =$$

Recognizing a ratio of -1

$$\frac{5}{-5} =$$    $$\frac{x-4}{4-x} =$$

| Review and complete | You try |
| --- | --- |

Subtraction is not commutative, so $a - b \neq b - a$. However, we can factor out -1 from $b - a$ to form $a - b = -1(-a + b) = -1(b - a)$.

| 19. Simplify. | 20. Simplify. |
| --- | --- |
| a. $\dfrac{-4}{4} = -1$ | a. $\dfrac{5-x}{x-5}$ |
| b. $\dfrac{1-n}{n-1} = -1$ | b. $\dfrac{c+6}{c-6}$ |
| c. $\dfrac{6x-1}{6x+1}$ This cannot be simplified. | c. $\dfrac{9x+1}{9x+1}$ |

| Video 8: Simplifying a Rational Expression with Common Factor -1 (1:47 min) |
| --- |

► Play the video and work along.

Simplify the expression.    $\dfrac{x^2-36}{30-5x}$

Three ways to write the same fraction: $-\dfrac{1}{3}=\dfrac{-1}{3}=\dfrac{1}{-3}$

⏸ Pause the video and try this yourself.

Write the three possible ways to express the simplified form of $\dfrac{x^2-36}{30-5x}$ .

► Play the video and check your answer.

| Review and complete | You try |
|---|---|
| 21. Simplify. <br><br> $\dfrac{3x-12}{16-x^2}$ | 22. Simplify. <br><br> $\dfrac{6-6x}{5x^2+5x-10}$ |

$$= \frac{3 \, \cancel{(x-4)}^{-1}}{(4+x) \cancel{(4-x)}}$$

$$= -\frac{3}{4+x}$$

$$= -\frac{3}{4+x} = \frac{3}{-(4+x)}$$

## Concept 1: Multiplication of Rational Expressions

| **Video 1: Multiplying Rational Expressions (2:37 min)** |
|---|

▶ Play the video and work along.

Multiply the rational expressions.

$$\frac{3}{5} \cdot \frac{10}{9}$$

⏸ Pause the video and try this yourself.

$$\frac{8a^2 b}{7a^5} \cdot \frac{21}{3b^3}$$

▶ Play the video and check your answer.

| **Review and complete** | **You try** |
|---|---|
| To **multiply rational expressions**, multiply the numerators and multiply the denominators. Then simplify if possible. | |

1.  Common factors can be _____ out of both the numerator and denominator in a rational expression.

| 2. Multiply. | 3. Multiply. |
|---|---|
| $$\frac{4}{9a} \cdot \frac{3a^2}{8b}$$ | $$\frac{6x^2}{5y} \cdot \frac{25y^2}{2x}$$ |
| $$= \frac{12a^2}{72ab}$$    Divide out common factors. | |
| $$= \frac{\overset{1}{\cancel{2}} \cdot \overset{1}{\cancel{2}} \cdot \overset{1}{\cancel{3}} \cdot \overset{1}{\cancel{a}} \cdot a}{\overset{1}{\cancel{2}} \cdot \overset{1}{\cancel{2}} \cdot 2 \cdot \overset{1}{\cancel{3}} \cdot 3 \cdot \overset{1}{\cancel{a}} \cdot b}$$ | |
| $$= \frac{a}{6b}$$ | |

| **Video 2: Multiplying Rational Expressions (2:07 min)** |
|---|

▶ Play the video and work along.

Multiply and simplify.  $$\frac{3m+3n}{12m-36n} \cdot \frac{m^2 - mn - 6n^2}{m^2 + 2mn + n^2}$$

| Review and complete | You try |
|---|---|
| **4.** Write the _____ form of each numerator and denominator before multiplying rational expressions. | |
| **5.** Multiply and simplify. $$\frac{x^2+5x+6}{2x+4} \cdot \frac{2x+2}{x^2-9}$$ $$=\frac{(x+2)(x+3)}{2(x+2)} \cdot \frac{2(x+1)}{(x-3)(x+3)}$$ $$=\frac{\overset{1}{(\cancel{x+2})}\ \overset{1}{(\cancel{x+3})}}{\underset{1}{\cancel{2}\,(\cancel{x+2})}} \cdot \frac{\overset{1}{\cancel{2}}(x+1)}{(x-3)(\cancel{x+3})}$$ $$=\frac{x+1}{x-3}$$ Divide out common factors. | **6.** Multiply and simplify. $$\frac{x+5}{x^2+4x+4} \cdot \frac{x^2-x-6}{2x+10}$$ |

## Concept 2: Division of Rational Expressions

**Video 3: Dividing Rational Expressions (1:38 min)**

▶ Play the video and work along.

Divide and simplify. $\dfrac{3c^2}{4d} \div \dfrac{15c^3}{8}$

| Review and complete | You try |
|---|---|

7.  To divide rational expressions, multiply the first fraction (the dividend) by the _____ of the second (the divisor). Then simplify if possible.

8.  Divide and simplify.

$$\frac{12a^2}{5b} \div \frac{10a^2}{3b}$$

$$= \frac{12a^2}{5b} \cdot \frac{3b}{10a^2}$$

$$= \frac{12a^2 \cdot 3b}{5b \cdot 10a^2}$$

Divide out common factors to finish to get

$$\frac{\boxed{\phantom{xxxx}}}{\boxed{\phantom{xxxx}}}$$

9.  Divide and simplify.

$$\frac{16m^2}{3n} \div \frac{4m}{n^2}$$

**Video 4: Dividing Rational Expressions  (1:55 min)**

▶ Play the video and work along.

Divide and simplify.    $\dfrac{y^2 - 8y + 16}{5y + 35} \div \dfrac{y^2 - 16}{20}$

| Review and complete | You try |
|---|---|
| 10. Divide and simplify. | 11.  Divide and simplify. |
| $\dfrac{x^2 - 4}{4x + 4} \div \dfrac{x^2 - 2x}{x^2 + 8x + 7}$ | $\dfrac{x^2 + 7x + 10}{4x - 12} \div \dfrac{x^2 + 10x + 25}{16x^2 - 48x}$ |
| $= \dfrac{x^2 - 4}{4x + 4} \cdot \dfrac{x^2 + 8x + 7}{x^2 - 2x}$ | |
| $= \dfrac{\cancel{(x-2)}(x+2)}{4\cancel{(x+1)}} \cdot \dfrac{\cancel{(x+1)}(x+7)}{x\cancel{(x-2)}}$ | |
| $= \dfrac{(x+2)(x+7)}{4x}$ | |

**Video 5: Dividing Rational Expressions and Using the Order of Operations (4:00 min)**

▶ Play the video and work along.

Divide and simplify.    $(8 - 2x) \div \dfrac{3x^2 - 11x - 4}{6x^2 + 23x + 7} \div (4x^2 - 49)$

Video and Study Guide to accompany Intermediate Algebra, 5th ed., Miller/O'Neill/Hyde

| Review and complete | You try |
|---|---|

12. When there are two divisions in an expression, perform the divisions in order from _____ to _____.

13. The first and third expressions in the example illustrated in Video 5 were rewritten with denominators of _____ to keep track of which polynomials are in the numerator and which are in the denominator.

| | |
|---|---|
| 14. Divide and simplify. | 15. Divide and simplify. |

14. Divide and simplify.

$$(6x+2) \div \frac{3x^2 - 11x - 4}{9 - x} \div (x^2 - 81)$$

$$= \left( \frac{6x+2}{1} \cdot \frac{9-x}{3x^2 - 11x - 4} \right) \div \frac{(x^2 - 81)}{1}$$

$$= \left( \frac{6x+2}{1} \cdot \frac{9-x}{3x^2 - 11x - 4} \right) \cdot \frac{1}{(x^2 - 81)}$$

$$= \frac{2(3x+1) \cdot \overset{-1}{(9-x)} \cdot 1}{1 \cdot (3x+1)(x-4) \cdot (x-9)(x+9)}$$

$$= \frac{-2}{(x-4)(x+9)} \quad \text{or} \quad -\frac{2}{(x-4)(x-9)}$$

15. Divide and simplify.

$$(x-6) \div \frac{2x^2 + 14x + 24}{x^2 + 6x + 9} \div (36 - x^2)$$

**Concept 1: Addition and Subtraction of Rational Expressions with Like Denominators**

| Video 1: Adding and Subtracting Rational Expressions having the Same Denominator (2:52 min) |
| --- |

▶ Play the video and work along.

Add or subtract as indicated. Then simplify the result.

$$\frac{1}{9}+\frac{5}{9}$$

$$\frac{6x}{3x-1}+\frac{5x}{3x-1}$$

⏸ Pause the video and try this yourself.

$$\frac{x^2}{x-4}-\frac{x+12}{x-4}$$

▶ Play the video and check your answer.

| Review and complete | You try |
| --- | --- |

To **add or subtract rational expressions** with like denominators, add or subtract the numerators Then write the result over the common denominator. Simplify the result if possible.

$$\frac{p}{q}+\frac{r}{q}=\frac{p+r}{q} \quad \text{and} \quad \frac{p}{q}-\frac{r}{q}=\frac{p-r}{q}$$

Video and Study Guide to accompany Intermediate Algebra, 5th ed., Miller/O'Neill/Hyde
Copyright © 2018 McGraw-Hill Education

1. Add or subtract as indicated. Then simplify.

$$\frac{x^2}{x+1} - \frac{x+2}{x+1}$$

$$= \frac{x^2 - (x+2)}{x+1}$$

$$= \frac{x^2 - x - 2}{x+1}$$

$$= \frac{(x+1)(x-2)}{x+1}$$

> Factor to determine if there are common factors.

$$= \frac{(x+1)(x-2)}{x+1}$$

$$= x - 2$$

2. Add or subtract as indicated. Then simplify.

$$\frac{2x+1}{x+2} + \frac{3}{x+2}$$

## Concept 2: Least Common Denominator

**Video 2: Adding Numerical Fractions  (3:23 min)**

▶ Play the video and work along.

Add.    $\frac{5}{12} + \frac{7}{8}$

| Review and complete | You try |
|---|---|
|  |  |

3. Add.

$$\frac{5}{6}+\frac{7}{10}$$

$$=\frac{5}{6}\cdot\frac{5}{5}+\frac{7}{10}\cdot\frac{3}{3}$$

Convert each fraction to an equivalent fraction with the LCD, 30, as its denominator.

$$=\frac{25}{30}+\frac{21}{30}$$

$$=\frac{46}{30}$$   Add.

$$=\frac{23}{15}$$   Simplify.

4. Add.

$$\frac{3}{20}+\frac{1}{12}$$

---

## Video 3: Determining Least Common Denominator Between Two Rational Expressions (2:36 min)

▶ Play the video and work along.

Determine the LCD for each group of fractions.

$$\frac{5}{9x^2y^7},\frac{7}{3x^4y^2z}$$

$$\frac{x+4}{5x-10},\frac{3}{x^2-4x+4}$$

| Review and complete | You try |
| --- | --- |

5. Determine the LCD for each group of fractions.

a. $\dfrac{6}{5x^5 y^2}, \dfrac{3}{10x^2 y^3 z}$

$\dfrac{6}{5x^5 y^2}$

$\dfrac{3}{10x^2 y^3 z} = \dfrac{3}{2 \cdot 5 x^2 y^3 z}$

$LCD = 2 \cdot 5 \cdot x^5 y^3 z$

b. $\dfrac{5}{x^2 + 3x}, \dfrac{6}{x^2 + 12x + 27}$

$\dfrac{5}{x^2 + 3x} = \dfrac{5}{x(x+3)}$

$\dfrac{6}{x^2 + 12x + 27} = \dfrac{6}{(x+3)(x+9)}$

$LCD = x(x+3)(x+9)$

6. Determine the LCD for each group of fractions.

a. $\dfrac{19}{3x^2 yz}, \dfrac{4}{15xy^4 z^3}$

b. $\dfrac{10}{x^2 + 4x + 4}, \dfrac{3}{5x + 10}$

## Concept 3: Equivalent Rational Expressions

**Video 4: Writing Equivalent Fractions (2:27 min)**

▶ Play the video and work along.

Convert each expression to an equivalent expression with the indicated denominator.

$$\frac{5}{4m^2} = \frac{}{8m^5}$$

$$\frac{t}{t+2} = \frac{}{t^2+3t+2}$$

| Review and complete | You try |
|---|---|

7. To convert a fraction to an equivalent fraction with a different denominator, multiply the fraction by a convenient form of _____.

| | |
|---|---|
| 8. Convert each expression to an equivalent expression with the indicated denominator. | 9. Convert each expression to an equivalent expression with the indicated denominator. |

8.
  a.  $\dfrac{6}{5a^2b^3} = \dfrac{}{40a^5b^3}$

   Multiply the numerator and the denominator by the missing factor, $8a^3$.

   $$\dfrac{6}{5a^2b^3} \cdot \dfrac{8a^3}{8a^3} = \dfrac{48a^3}{40a^5b^3}$$

  b.  $\dfrac{6}{x+m} = \dfrac{}{x^2-m^2}$

   Multiply the numerator and the denominator by the missing factor, $(x-m)$.

   $$\dfrac{6}{x+m} = \dfrac{}{(x-m)(x+m)}$$
   $$\dfrac{6}{(x+m)} \cdot \dfrac{(x-m)}{(x-m)} = \dfrac{6(x-m)}{(x-m)(x+m)}$$

9.
  a.  $\dfrac{5}{6x^3y} = \dfrac{}{60x^4y^5}$

  b.  $\dfrac{7}{x+2} = \dfrac{}{x^2-12x-28}$

**Concept 4: Addition and Subtraction of Rational Expressions with Unlike Denominators**

| Video 5: Adding Rational Expressions (1:32 min) |
| --- |

▶ Play the video and work along.

Add.    $\dfrac{5}{11m}+\dfrac{3}{4m^4}$

| Review and complete | You try |
| --- | --- |

10. To convert both fractions to new fractions with the common denominator, multiply by a convenient form of _____.

| 11. Add. | 12. Add. |
| --- | --- |

11. Add.

$$\dfrac{12}{5x^2}+\dfrac{1}{3x^4}$$

Convert each expression to an equivalent expression with a denominator of the $LCD=15x^4$.

$$\dfrac{12}{5x^2}+\dfrac{1}{3x^4}=\dfrac{12}{5x^2}\cdot\dfrac{3x^2}{3x^2}+\dfrac{1}{3x^4}\cdot\dfrac{5}{5}$$

$$=\dfrac{36x^2}{15x^4}+\dfrac{5}{15x^4}$$

$$=\dfrac{36x^2+5}{15x^4}$$

12. Add.

$$\dfrac{4}{13x^2}+\dfrac{7}{2x}$$

| Video 6: Adding Rational Expressions (3:38 min) |
| --- |

▶ Play the video and work along.

Add.    $\dfrac{5}{y+3}+\dfrac{y^2+y+24}{y^2-9}$

| Review and complete | You try |
|---|---|
| **Avoiding Mistakes:** Do not try to reduce or "cancel" terms in the numerator and denominator, such as the $y$ – variables or the 3's in $\dfrac{y+3}{y-3}$ . | |

| Review and complete | You try |
|---|---|
| 13. Add. $$\frac{m-2}{m-4}+\frac{2m^2-15m+12}{m^2-16}$$ Rewrite the expression so that each denominator is factored. $$\frac{m-2}{m-4}+\frac{2m^2-15m+12}{(m-4)(m+4)}$$ Convert each expression to an equivalent expression with a denominator of the $\text{LCD}=(m-4)(m+4)$. $$=\frac{m-2}{m-4}\cdot\frac{(m+4)}{(m+4)}+\frac{2m^2-15m+12}{(m-4)(m+4)}$$ $$=\frac{m^2+2m-8+2m^2-15m+12}{(m-4)(m+4)}$$ $$=\frac{3m^2-13m+4}{(m-4)(m+4)}$$ $$=\frac{(3m-1)\,\cancel{(m-4)}}{\cancel{(m-4)}\,(m+4)}$$ $$=\frac{3m-1}{m+4}$$ | 14. Add. $$\frac{x}{x-3}+\frac{5}{x-6}$$ |

**Video 7: Subtracting Rational Expressions (2:28 min)**

▶ Play the video and work along.

Subtract. $\dfrac{3}{m-2} - \dfrac{m}{m-5}$

| Review and complete | You try |
|---|---|
| 15. Subtract.<br><br>$\dfrac{3x}{x+1} - \dfrac{1}{2x-1}$<br><br>Convert each expression to an equivalent expression with a denominator of the $LCD = (x+1)(2x-1)$.<br><br>$= \dfrac{3x}{x+1} \cdot \dfrac{(2x-1)}{(2x-1)} - \dfrac{1}{2x-1} \cdot \dfrac{(x+1)}{(x+1)}$<br><br>$= \dfrac{3x(2x-1)-1(x+1)}{(2x-1)(x+1)}$<br><br>$= \dfrac{6x^2-3x-x-1}{(2x-1)(x+1)}$<br><br>$= \dfrac{6x^2-4x-1}{(2x-1)(x+1)}$ | 16. Subtract.<br><br>$\dfrac{6x}{x-4} - \dfrac{x}{x+2}$ |
| **Video 8: Subtracting Rational Expressions in Which the Denominators are Opposites (1:57 min)** | |

▶ Play the video and work along.

Subtract. $\dfrac{t^2}{t-8} - \dfrac{-8t}{8-t}$

| Review and complete | You try |
| --- | --- |
| 17. When two denominators are opposites, they differ by a factor of _____. | |

**18. Subtract.**

$$\frac{x-2}{x-7}-\frac{x+2}{7-x}$$

The denominators $x-7$ and $7-x$ are opposites and differ by a factor of -1. Note that $7-x=-(x-7)=-1(x-7)$. Multiply the numerator and denominator of the 2nd fraction by -1 to get an LCD of $x-7$.

$$=\frac{x-2}{x-7}-\frac{x+2}{7-x}\cdot\frac{-1}{-1}$$

$$=\frac{x-2}{x-7}-\frac{-x-2}{x-7}$$

Carefully subtract the terms of the numerator of the second rational expression.

$$=\frac{x-2-(-x)-(-2)}{x-7}$$

$$=\frac{x-2+x+2}{x-7}$$

$$=\frac{2x}{x-7}$$

**19. Subtract.**

$$\frac{7}{x-2}-\frac{5x-1}{2-x}$$

## Concept 1: Simplifying Complex Fractions by Method 1

### Video 1: Introduction to Complex Fractions (1:56 min)

▶ Play the video and work along.

Simplify the complex fraction.

$$\dfrac{\dfrac{4p^3}{9y}}{\dfrac{10p}{3y^4}}$$

| Review and complete | You try |
|---|---|

1. The middle fraction bar in a complex fraction represents _____.

| | |
|---|---|
| 2. Simplify the complex fraction. $$\dfrac{\dfrac{30x}{7y^4}}{\dfrac{6x^2}{5y^3}}$$ | 3. Simplify the complex fraction. $$\dfrac{\dfrac{3x^2}{4yz}}{\dfrac{15wx}{z^2}}$$ |

$= \dfrac{\overset{5}{\cancel{30}}\,\cancel{x}}{7\,\cancel{y^4}\big/_y} \cdot \dfrac{5\,\cancel{y^3}}{\cancel{6}\,\cancel{x^2}\big/_x}$

$= \dfrac{25}{7xy}$

Multiply the numerator of the complex fraction by the reciprocal of the denominator of the complex fraction. Divide out common factors. Simplify.

## Video 2: Simplifying a Complex Fraction Using Method I (2:33 min)

▶ Play the video and work along.

Simplify the complex fraction using Method I.

$$\dfrac{\dfrac{1}{5} - \dfrac{3}{10} + 1}{\dfrac{1}{4} + \dfrac{7}{20}}$$

| Review and complete | You try |
|---|---|
| 4.  Simplify the complex fraction using Method I.  $$\dfrac{\dfrac{1}{6} + \dfrac{1}{12} - \dfrac{1}{2}}{1 - \dfrac{2}{3}}$$ | 5. Simplify the complex fraction using Method I.  $$\dfrac{1 + \dfrac{1}{3}}{\dfrac{5}{6} - 1}$$ |

$$= \dfrac{\dfrac{2}{12} + \dfrac{1}{12} - \dfrac{6}{12}}{\dfrac{3}{3} - \dfrac{2}{3}}$$

The LCD in the numerator is 12.
The LCD in the denominator is 3.

$$= \dfrac{\dfrac{-3}{12}}{\dfrac{1}{3}}$$

Simplify the numerator and denominator.

$$= \dfrac{-3}{12} \cdot \dfrac{3}{1}$$

Multiply the numerator of the complex fraction by the reciprocal of the fraction in the denominator of the complex fraction.

$$= \dfrac{-3}{4}$$

Simplify.

---

## Video 3: Simplifying a Complex Fraction Using Method I (2:10 min)

▶ Play the video and work along.

Simplify the complex fraction by using Method I.    $\dfrac{1 - \dfrac{4}{x^2}}{\dfrac{1}{x} + \dfrac{2}{x^2}}$

| Review and complete | You try |
|---|---|

6. Method I follows the order of operations to simplify the _____ and _____ of the complex fraction before dividing.

| | |
|---|---|
| 7. Simplify by using Method I. $\dfrac{2 - \dfrac{1}{2y}}{\dfrac{4}{y^2} + y}$ | 8. Simplify by using Method I. $\dfrac{\dfrac{3}{x^2} + \dfrac{2}{x}}{\dfrac{4}{x^2} - \dfrac{9}{x}}$ |

$$= \frac{\dfrac{2y}{2y} \cdot \dfrac{2}{1} - \dfrac{1}{2y}}{\dfrac{4}{y^2} + \dfrac{y^2}{y^2} \cdot \dfrac{y}{1}}$$

Numerator LCD $= 2y$

Denominator LCD $= y^2$

$$= \frac{\dfrac{4y-1}{2y}}{\dfrac{4+y^3}{y^2}}$$

Simplify the numerator and denominator.

$$= \frac{4y-1}{2\cancel{y}} \cdot \frac{\cancel{y^2}^{\,y}}{4+y^3}$$

Multiply the numerator of the complex fraction by the reciprocal of the fraction in the denominator of the complex fraction.

$$= \frac{y(4y-1)}{2(4+y^3)}$$

Simplify.

## Concept 2: Simplifying Complex Fractions by Method 2

### Video 4: Simplifying Complex Fractions Using Method II (2:32 min)

▶ Play the video and work along.

Simplify the complex fraction by using Method II.

$$\frac{\dfrac{1}{5} - \dfrac{3}{10} + 1}{\dfrac{1}{4} + \dfrac{7}{20}}$$

| Review and complete | You try |
|---|---|
| 9. The principle behind Method II is to multiply the numerator and denominator of the complex fraction by the _____ of all the individual fractions. | |

**10. Simplify by using Method II.**

$$\frac{\dfrac{6}{11}-\dfrac{5}{2}}{\dfrac{1}{4}+1}$$

$$=\frac{44\cdot\left(\dfrac{6}{11}-\dfrac{5}{2}\right)}{44\cdot\left(\dfrac{1}{4}+\dfrac{1}{1}\right)}$$

The LCD of all the fractions is 44. Multiply the numerator and denominator of the complex fraction by 44.

$$=\frac{\overset{4}{\cancel{44}}\cdot\dfrac{6}{\cancel{11}}-\overset{22}{\cancel{44}}\cdot\dfrac{5}{\cancel{2}}}{\overset{11}{\cancel{44}}\cdot\dfrac{1}{\cancel{4}}+44\cdot\dfrac{1}{1}}$$

Apply the distributive property.

$$=\frac{24-110}{11+44}$$

Simplify each product.

$$=-\frac{86}{55}$$

Simplify.

**11. Simplify by using Method II.**

$$\frac{\dfrac{7}{8}+\dfrac{3}{4}}{\dfrac{1}{3}-\dfrac{5}{6}}$$

---

## Video 5: Simplifying a Complex Fraction Using Method II (4:20 min)

▶ Play the video and work along.

Simplify the complex fraction by using Method II. $\dfrac{p^{-2}t-pt^{-2}}{t^{-1}-p^{-1}}$

---

| Review and complete | You try |
|---|---|

**12.** Negative exponents tell us to take the _____ of the base, and change the sign of the exponent.

**13.** Simplify by using Method II.

$$\frac{a^{-2}-9}{a+a^{-1}}$$

$$=\frac{\dfrac{1}{a^2}-9}{a+\dfrac{1}{a}}$$
The LCD of all the fraction is $a^2$.

$$=\frac{a^2\cdot\left(\dfrac{1}{a^2}-\dfrac{9}{1}\right)}{a^2\cdot\left(\dfrac{a}{1}+\dfrac{1}{a}\right)}$$
Multiply the numerator and denominator of the complex fraction by the LCD.

$$=\frac{a^2\cdot\dfrac{1}{a^2}-a^2\cdot\dfrac{9}{1}}{a^2\cdot\dfrac{a}{1}+a^2\cdot\dfrac{1}{a}}$$
Apply the distributive property.

$$=\frac{1-9a^2}{a^3+a}$$
Simplify each product.

$$=\frac{(1-3a)(1+3a)}{a^2(a^2+1)}$$
Factor to determine if there are common factors.

**14.** Simplify by using Method II.

$$\frac{x^{-1}+x^{-2}}{5x^{-2}}$$

---

**Video 6: Simplifying a Complex Fraction Using Method II (2:47 min)**

▶ Play the video and work along.

Simplify the complex fraction by using Method II.
$$\frac{\dfrac{2}{c-3}+\dfrac{1}{c+4}}{\dfrac{6}{c+4}+\dfrac{4}{c-3}}$$

Factor to see if there are common factors.

**Review and complete**                     **You try**

**15. Simplify by using Method II.**

$$\frac{\dfrac{5}{c-5}+\dfrac{1}{c+5}}{\dfrac{5}{c+5}-5}$$

Multiply the numerator and denominator of the complex fraction by the $LCD=(c+5)(c-5)$. Apply the distributive property. Simplify.

$$=\frac{(c+5)(c-5)\cdot\dfrac{5}{c-5}+(c+5)(c-5)\cdot\dfrac{1}{c+5}}{(c+5)(c-5)\cdot\dfrac{5}{c+5}-5\cdot(c+5)(c-5)}$$

$$=\frac{5(c+5)+1(c-5)}{5(c-5)-5(c+5)(c-5)}$$

$$=\frac{5c+25+c-5}{5c-25-5(c^2-25)}$$

$$=\frac{6c+20}{5c-25-5c^2+125}$$

$$=\frac{6c+20}{-5c^2+5c-150}$$

$$=\frac{2(3c+10)}{-5(c^2-c+6)}$$

$$=\frac{2(3c+10)}{-5(c-3)(c+2)}$$

**16. Simplify by using Method II.**

$$\frac{\dfrac{2}{x}-\dfrac{3}{x+1}}{\dfrac{2}{x+1}-\dfrac{3}{x}}$$

**Concept 1:** Solving Rational Equations

| **Video 1: Introduction to Rational Equations: Review of Clearing Fractions (2:34 min)** |
| --- |

▶ Play the video and work along.

Solve the equation.    $-\dfrac{1}{2}+\dfrac{3}{4}x=\dfrac{1}{8}x+2$

| **Review and complete** | **You try** |
| --- | --- |

1. Fractions can be cleared from an equation by multiplying both sides by the _____ of all terms.

| 2. Solve the equation. | 3. Solve the equation. |
| --- | --- |
| $\dfrac{2}{3}x+\dfrac{1}{2}=\dfrac{1}{6}x$ | $\dfrac{2}{5}x-8=\dfrac{4}{15}x$ |

2.
$$6\left(\dfrac{2}{3}x\right)+6\left(\dfrac{1}{2}\right)=6\left(\dfrac{1}{6}x\right)$$  Multiply both sides by the LCD, 6. Apply the distributive property.

$$4x+3=x$$

$$3x=-3$$   Solve the resulting equation.

$$x=-1$$

Check:

$$\dfrac{2}{3}(-1)+\dfrac{1}{2}\overset{?}{=}\dfrac{1}{6}(-1)$$

$$-\dfrac{2}{3}+\dfrac{1}{2}\overset{?}{=}\dfrac{-1}{6}$$

$$\dfrac{-4}{6}+\dfrac{3}{6}\overset{?}{=}\dfrac{-1}{6}$$

$$\dfrac{-1}{6}=\dfrac{-1}{6}$$

The solution $x=-1$ makes the equation true. The solution set is $\{-1\}$.

## Video 2: Solving a Rational Equation (2:28 min)

▶ Play the video and work along.

Solve the equation. $\dfrac{1}{3} + \dfrac{5}{x} = \dfrac{3}{4}$

| Review and complete | You try |
|---|---|
| 4. Solve the equation. $$\dfrac{2}{x} + \dfrac{1}{10} = \dfrac{1}{2}$$ Multiply each side by the LCD, $10x$. Solve the resulting equation. | 5. Solve the equation. $$\dfrac{1}{2} - \dfrac{3}{2x} = \dfrac{4}{x} - \dfrac{5}{12}$$ |

$$\frac{10x}{1}\left(\frac{2}{x}+\frac{1}{10}\right)=\frac{10x}{1}\left(\frac{1}{2}\right)$$

$$\frac{10\cancel{x}}{1}\left(\frac{2}{\cancel{x}}\right)+\frac{\cancel{10}x}{1}\left(\frac{1}{\cancel{10}}\right)=\frac{\overset{5}{\cancel{10}}x}{1}\left(\frac{1}{\cancel{2}}\right)$$

$$20+x=5x$$

$$20=4x$$

$$5=x$$

Check:

$$\frac{2}{5}+\frac{1}{10}\overset{?}{=}\frac{1}{2}$$

$$\frac{4}{10}+\frac{1}{10}\overset{?}{=}\frac{1}{2}$$

$$\frac{5}{10}\overset{?}{=}\frac{1}{2}$$

$$\frac{1}{2}=\frac{1}{2}$$

The value $x=5$ makes the equation true. So, the solution set is {5}.

---

## Video 3: Solving a Rational Equation with No Solution (4:13 min)

▶ Play the video and work along.

Solve the equation.    $\dfrac{5}{y-4}-\dfrac{2}{y-3}=\dfrac{2y-3}{y^2-7y+12}$

---

| Review and complete | You try |

6. Rational equations cannot have solutions that are restricted values because substituting them into the original equation will cause division by _____.

7. Solve the equation.

$$\frac{p}{p-2}+\frac{2}{3}=\frac{2}{p-2}$$

The LCD of all terms in the equation is $3(p-2)$. Note that there is a restriction that $p \neq 2$. Multiply each side by the LCD and apply the distributive property.

$$\frac{p}{p-2}+\frac{2}{3}=\frac{2}{p-2}$$

$$3(p-2)\left(\frac{p}{(p-2)}\right)+3(p-2)\left(\frac{2}{3}\right)=\left(\frac{2}{(p-2)}\right)3(p-2)$$

$$3p+2(p-2)=6$$

$$3p+2p-4=6$$

$$5p=10$$

$$p=2$$

The value 2 is a restriction on p. The value $p=2$ does not check since it makes a denominator 0. Since no other potential solution exists, the equation has no solution. The solution set is the empty set, $\{\ \}$.

8. Solve the equation.

$$5-\frac{8}{x+2}=\frac{4x}{x+2}$$

## Video 4: Solving a Rational Equation (2:16 min)

▶ Play the video and work along.

Solve the equation.    $1+\frac{5}{x}=\frac{14}{x^2}$

| Review and complete | You try |
|---|---|
| 9. Solve the equation. | 10. Solve the equation. |

**Review and complete**

9. Solve the equation.

$$\frac{7}{2}+\frac{1}{x}=\frac{3}{x^2}+\frac{3}{2}$$

The LCD of all the terms in the equation is $2x^2$. Note that there is a restriction that $x \neq 0$. Multiply each side by the LCD and apply the distributive property.

$$2x^2\left(\frac{7}{2}\right)+2x^2\left(\frac{1}{x}\right)=\left(\frac{3}{x^2}\right)2x^2+\left(\frac{3}{2}\right)2x^2$$

$$7x^2+2x=6+3x^2$$

$$4x^2+2x-6=0$$

$$2(2x^2+x-3)=0$$

$$2(2x+3)(x-1)=0$$

$$2x+3=0 \text{ or } x-1=0$$

$$x=-\frac{3}{2} \qquad x=1$$

Neither value is restricted; both values check.

The solution set is $\left\{-\frac{3}{2},1\right\}$.

**You try**

10. Solve the equation.

$$1+\frac{3}{x}=\frac{28}{x^2}$$

**Video 5: Solving a Rational Equation (2:58 min)**

▶ Play the video and work along.

Solve the equation.   $\dfrac{m}{m-4} = \dfrac{m}{2} + \dfrac{4}{m-4}$

| Review and complete | You try |
|---|---|

When fractions are cleared from a rational equation, a quadratic equation can result. Quadratic equations can result in two solutions. If one of the solutions is a restricted value of the rational equation, we **eliminate that solution from the solution set.**

11. Solve the equation.

$$\dfrac{4}{x-2} - \dfrac{1}{x+2} = \dfrac{x^2}{x^2-4}$$

The LCD is $(x-2)(x+2)$. The restrictions are that $x \neq 2, x \neq -2$. Multiply each side by the LCD.

$$(x-2)(x+2)\cdot\dfrac{4}{x-2} - (x-2)(x+2)\cdot\dfrac{1}{x+2} = \dfrac{x^2}{(x-2)(x+2)}\cdot(x-2)(x+2)$$

$$4(x+2)-1(x-2) = x^2$$

$$4x+8-x+2 = x^2$$

$$3x+10 = x^2$$

$$0 = x^2-3x-10$$

$$0 = (x-5)(x+2)$$

$$x = 5; \quad \cancel{x = -2}$$

One potential solution is a restricted value. So, the solution set is $\{5\}$.

12. Solve the equation.

$$\dfrac{1}{m} = \dfrac{1}{2m-1} - \dfrac{2m}{2m-1}$$

**Concept 2:** Formulas Involving Rational Equations

| Video 6: Solving a Formula for a Different Variable (1:13 min) |
| --- |

▶ Play the video and work along.

Solve for V.    $\dfrac{PV}{n} = rt$

| Video 7: Solving a Literal Rational Equation Where Factoring is Required to Isolate the Variable (4:27 min) |
| --- |

▶ Play the video and work along.

Solve for $y$.    $c = \dfrac{a+y}{ay}$

| Review and complete | You try |
| --- | --- |

13. Solving for a specific variable means to get that variable by itself on _____.

| 14. Solve for $a$. | 15. Solve for $b$. |
| --- | --- |
| $\dfrac{a+b}{ab} = \dfrac{1}{m}$ | $\dfrac{b+a}{ab} = \dfrac{1}{m}$ |

**14.** Solve for $a$.

$$\dfrac{a+b}{ab} = \dfrac{1}{m}$$

$\cancel{ab}\,m\left(\dfrac{a+b}{\cancel{ab}}\right) = \left(\dfrac{1}{\cancel{m}}\right)ab\,\cancel{m}$    Multiply both sides by the LCD, $abm$.

$m(a+b) = ab$    Clear fractions.

$am + bm = ab$    Apply the distributive property.

$am - ab = -bm$    Collect terms with $a$ on one side.

$a(m-b) = -bm$    Factor out $a$.

$\dfrac{a\cancel{(m-b)}}{\cancel{(m-b)}} = \dfrac{-bm}{(m-b)}$    Divide by $(m-b)$.

$a = \dfrac{-bm}{m-b}$ or $\dfrac{bm}{b-m}$

## Concept 1: Solving Proportions

### Video 1: Introduction to Proportions (2:07 min)

▶ Play the video and work along.

Solve the proportion.        $\dfrac{6}{11} = \dfrac{42}{y}$

| Review and complete | You try |
|---|---|
| 1. The ratio of a to b can be written as _____, where $b \neq 0$. An equation that equates two ratios is called a _____. To solve a proportion, multiply each side of the equation by the _____. | |

| | |
|---|---|
| 2. Solve the proportion.<br><br>$\dfrac{5}{x} = \dfrac{8}{56}$<br><br>$56x\left(\dfrac{5}{x}\right) = 56x\left(\dfrac{8}{56}\right)$  The LCD is $56x$. Note that $x \neq 0$. Multiply both sides by the LCD.<br><br>$\quad 280 = 8x$  Clear fractions.<br><br>$\quad\quad x = 35$  Solve the resulting equation.<br><br>The solution set is $\{35\}$. | 3. Solve the proportion.<br><br>$\dfrac{14}{21} = \dfrac{8}{x}$ |

## Concept 2: Applications of Proportions

### Video 2: An Application of Rational Equations: Solving a Proportion (2:06 min)

▶ Play the video and work along.

Franco drove 301 miles on 7 gal of gas in his Honda hybrid. How many gallons will he need for a 2021 mi trip cross country?

| Review and complete | You try |
|---|---|
| 4. Mari can get to work and back all week, a total of 230 mi, on 8 gal of gas. How many gallons will she need for a 1035 mi one-way trip to Chicago?<br><br>Let $x$ represent the amount of gas for the Chicago trip.<br><br>$$\frac{230}{8} = \frac{1035}{x} \longleftarrow \begin{array}{l}\text{Number of miles}\\ \text{Gallons of gas}\end{array}$$<br><br>$$8x\left(\frac{230}{8}\right) = 8x\left(\frac{1035}{x}\right) \quad \begin{array}{l}\text{Multiply both sides}\\ \text{by the LCD, } 8x.\end{array}$$<br><br>$$230x = 8280 \qquad \begin{array}{l}\text{Solve the resulting}\\ \text{equation.}\end{array}$$<br>$$x = 36$$<br><br>Mari will need 36 gal of gas for the trip to Chicago. | 5. Abdul drives a truck and notices that he used 68 gal of gas on a 714-mi trip. How much gas should he use on a 441-mi trip? |

## Video 3: An Application of Rational Equations: Solving a Proportions (2:25 min)

▶ Play the video and work along.

The ratio of female students to male students taking algebra class is 6 to 5. If the total number of students taking the algebra class is 1232, how many are male?

| Review and complete | You try |
|---|---|
| 6. If the total number of students is 1232, and there are $m$ males, the expression that represents the number of females is _____. | |

7. The ratio of students who are math majors to students who are not math majors in a certain class is 2 to 5. If the class has 35 students, how many are not math majors?

Let $n =$ the number of nonmath majors.
Then $35 - n$ is the number of majors.

$$\frac{\text{math}}{\text{non-math}} \qquad \frac{2}{5} = \frac{35-n}{n}$$

$$\frac{2}{5} = \frac{35-n}{n}$$

$$\cancel{5}n\left(\frac{2}{\cancel{5}}\right) = 5\cancel{n}\left(\frac{35-n}{\cancel{n}}\right)$$

$$2n = 175 - 5n$$

$$7n = 175$$

$$n = 25$$

There are 25 nonmath majors.

8. The ratio of welders with a college degree in welding to those without a degree at a certain shop is 2 to 9. If there are 242 welders in the shop, how many do not have the degree?

## Concept 3: Similar Triangles

### Video 4: An Application of Rational Equations: Similar Triangles  (2:21 min)

▶ Play the video and work along.

The triangles $ABC$ and $DEF$ are similar triangles. Solve for $x$ and $y$.

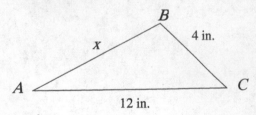

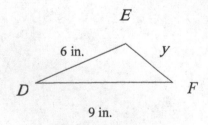

| Review and complete | You try |
|---|---|
| 9. Triangle $ABC$ is similar to triangle $DEF$. Solve for $x$ and $y$. | 10. Triangle $ABC$ is similar to triangle $DEF$. Solve for $x$ and $y$. |

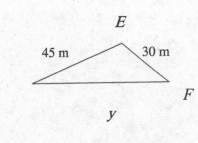

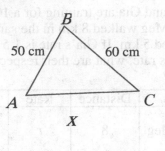

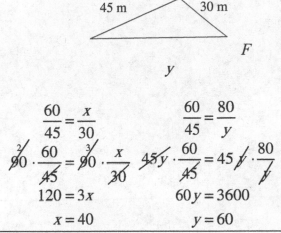

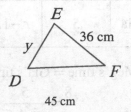

$$\frac{60}{45} = \frac{x}{30} \qquad \frac{60}{45} = \frac{80}{y}$$

$$\overset{2}{\cancel{90}} \cdot \frac{60}{\cancel{45}} = \overset{3}{\cancel{90}} \cdot \frac{x}{30} \qquad 45y \cdot \frac{60}{\cancel{45}} = 45\cancel{y} \cdot \frac{80}{\cancel{y}}$$

$$120 = 3x \qquad\qquad 60y = 3600$$

$$x = 40 \qquad\qquad y = 60$$

**Concept 4:** Applications of Rational Equations

---

## Video 5: An Application of Rational Equations: A Uniform Motion Application (3:13 min)

▶ Play the video and work along.

Michelle can run 6 mi in the same amount of time that she can walk 2.25 mi. If she walks 5 mph slower than she runs, find her speed running, and her speed walking.

| | Distance | Rate | Time |
|---|---|---|---|
| Running | | | |
| Walking | | | |

| Review and complete | You try |
|---|---|

**11.** In the distance/time/rate relationship, time is equal to _____ over _____.

---

**12.** Meg and Gia are training for a 10 km race. One day, Meg walked 8 km in the same time that Gia walked 5 km. If Gia's rate is 1.5 km/h slower than Meg's rate, what are their respective walking rates?

|  | Distance | Rate | Time |
|---|---|---|---|
| Meg | 8 | $r$ | $\dfrac{8}{r}$ |
| Gia | 5 | $r-1.5$ | $\dfrac{5}{r-1.5}$ |

Meg's time = Gia's time

$$\frac{8}{r} = \frac{5}{r-1.5}$$

$$r(r-1.5)\left(\frac{8}{r}\right) = r(r-1.5)\cdot\left(\frac{5}{r-1.5}\right)$$

$$8(r-1.5) = 5r$$

$$8r-12 = 5r$$

$$3r = 12$$

$$r = 4$$

Meg walks at a rate of 4 km/h and Gia walks at a rate of 2.5 km/h.

**13.** Wendy and LaRon both bike in a Wednesday night group. On a recent Wednesday, Wendy and LaRon had rides that took exactly the same time. Wendy rode 30 mi and LaRon rode 50 mi. If LaRon's rate is 3 mph faster than Wendy's, what is Wendy's average speed? What is LaRon's average speed?

## Video 6: An Application of Rational Equations: A Uniform Motion Application (3:51 min)

▶ Play the video and work along.

Tony and Alina are distance cyclists and often ride their bikes in fundraising events. One event was a 90-mi ride and Alina started the race at 7:30 am. Tony slept late and left 1 hr later. Suppose Tony averaged 3 mph faster than Alina and they both reached the finish line at the same time. Find the average speed of each rider.

|       | Distance | Rate | Time |
|-------|----------|------|------|
| Alina |          |      |      |
| Tony  |          |      |      |

| | **Review and complete** | **You try** |
|---|---|---|

**Review and complete**

14. Kris does triathlons. She runs 2 mph faster than her friend Joe, a weekend athlete. They each run 12 mi but Kris finishes one half hour ahead of Joe. Determine each person's running speed.

| | Distance | Rate | Time |
|---|---|---|---|
| Kris | 12 | $r+2$ | $\dfrac{12}{r+2}$ |
| Joe | 12 | $r$ | $\dfrac{12}{r}$ |

Time for Kris = Time for Joe, minus $\dfrac{1}{2}$ hr

$$\frac{12}{r+2} = \frac{12}{r} - \frac{1}{2}$$

$$2r(r+2)\left(\frac{12}{r+2}\right) = 2r(r+2)\left(\frac{12}{r}\right) - 2r(r+2)\left(\frac{1}{2}\right)$$

$$24r = 12 \cdot 2 \cdot (r+2) - r(r+2)$$

$$24r = 24r + 48 - r^2 - 2r$$

$$24r = -r^2 + 22r + 48$$

$$r^2 + 2r - 48 = 0$$

$$(r+8)(r-6) = 0$$

$$r = 6; \; r = -8$$

The negative answer is not reasonable since speed cannot be negative. Joe runs 6 mph. Kris runs 8 mph.

**You try**

15. On a snowy day, a car averages 21 mph faster than a bus on a 196-mi trip. If the car takes 3 hr less time to complete the trip than the bus, what are the speeds of the bus and car?

| Distance | Rate | Time |
|---|---|---|
| | | |
| | | |

---

### Video 7: An Application of Rational Equations: A "Work" Application (2:26 min)

▶ Play the video and work along.

Stan can paint a room in 8 hr, and Ollie can paint a room of similar size in 10 hr. How long will it take then if they worked together?

Let $x$ represent the time required to do the job together.

$$\left(\text{Speed that Stan Works}\right) + \left(\text{Speed that Ollie Works}\right) = \left(\text{Speed working together}\right)$$

| Review and complete | You try |
|---|---|

16. If two workers together paint a room in $x$ hours, then their speed of painting together is $\dfrac{\underline{\quad} \text{ room}}{\underline{\quad} \text{ hr}}$.

---

17. One carpenter can complete a kitchen in 8 days. Another can finish a similar job alone in 12 days. How long will it take to complete a kitchen of similar size if they work together?

$$
\begin{array}{ccc}
\text{Rate of} & \text{Rate of} & \text{Rate} \\
\text{Carpenter 1} + & \text{Carpenter 2} = & \text{Together}
\end{array}
$$

$$\frac{1 \text{ kitchen}}{8 \text{ days}} + \frac{1 \text{ kitchen}}{12 \text{ days}} = \frac{1 \text{ kitchen}}{x \text{ days}}$$

$$\frac{1}{8} + \frac{1}{12} = \frac{1}{x}$$

$$\overset{3}{\cancel{24}} x \left( \frac{1}{\cancel{8}} \right) + \overset{2}{\cancel{24}} x \left( \frac{1}{\cancel{12}} \right) = 24 \cancel{x} \left( \frac{1}{\cancel{x}} \right)$$

$$3x + 2x = 24$$

$$5x = 24$$

$$x = 4.8$$

The kitchen will take 4.8 days with the carpenters working together.

18. Marj can change the advertisement on a billboard in 4 hr. Bea can do the same job in 5 hr. How long will it take them working together?

---

## Concept 1: Definition of Direct and Inverse Variation

### Video 1: Introduction to Direct and Inverse Variation (2:08 min)

▶ Play the video and work along.

**Definition of Direct and Inverse Variation**
Let $k$ be a nonzero constant real number. Then the following statements are equivalent.

1. $\left.\begin{array}{l} y \text{ varies directly as } x. \\ y \text{ is directly proportional to } x. \end{array}\right\}$ .

2. $\left.\begin{array}{l} y \text{ varies inversely as } x. \\ y \text{ is inversely proportional to } x. \end{array}\right\}$

Note: The value of $k$ is called the **constant of variation**.

**Distance vs. Time**

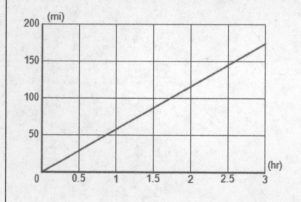

**Time of Travel vs. Speed**

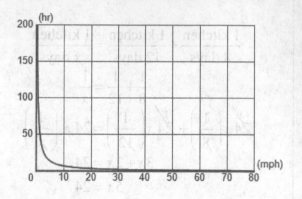

| Review and complete | You try |
| --- | --- |

1. As time passes, the distance traveled by a car goes up. This is direct variation. The equation for direct variation is $y = kx$, where $k$ is the constant of variation.

2. As speed increases, the time a trip takes goes _____. This is inverse variation. The equation for inverse variation is _____, where $k$ is the constant of variation.

## Concept 2: Translations Involving Variation

| Video 2: Writing an English Statement as a Variation Model (1:17 min) |
|---|

▶ Play the video and work along.

Write each statement as an equivalent mathematical model. Use $k$ as the constant of variation.

   a.   The variable $p$ varies directly as $w$.

⏸ Pause the video and try this yourself.

   b.   The variable $x$ varies inversely as $v$.

▶ Play the video and check your answer.

▶ Play the video and work along.

   c.   The variable $t$ varies inversely as the square of $y$.

⏸ Pause the video and try this yourself.

   d.   The variable $c$ is directly proportional to the square root of $z$.

▶ Play the video and check your answer.

| Review and complete | You try |
|---|---|
| Write each statement as an equivalent variation model. Use $k$ as the constant of variation. | Write each statement as an equivalent variation model. Use $k$ as the constant of variation. |
| 3a. The variable $p$ varies directly as $m$.<br><br>   Solution:   $p = km$ | 4a. The variable $y$ is inversely proportional to the square of $m$. |
| 3b. The variable $y$ varies inversely as the cube of $n$.<br><br>   Solution:   $y = \dfrac{k}{n^3}$ | 4b. The variable $q$ is directly proportional to the cube of $n$. |

## Video 3: Introduction to Joint Variation (1:30 min)

▶ Play the video and work along.

DEFINITION: Joint Variation
Let $k$ be a nonzero constant real number. Then the following statements are equivalent.

$\left.\begin{array}{l} y \text{ varies jointly as } w \text{ and } z. \\ y \text{ is jointly proportional to } w \text{ and } z. \end{array}\right\} y = \underline{\hspace{1.5cm}} .$

Write each statement as an equivalent mathematical model. Use $k$ as the constant of variation.

    a.   $z$ varies jointly as $x$ and the square of $y$.

⏸ Pause the video and try this yourself.

    b.   $c$ varies jointly as $m$ and $n$, and inversely as the square root of $t$.

▶ Play the video and check your answer.

| Review and complete | You try |
|---|---|
| 5. Write the statement as an equivalent mathematical model. Use $k$ as the constant of variation.<br><br>    The variable $I$ varies jointly as $t$ and $r$.<br><br>    Solution:   $I = ktr$ | 6. Write the statement as an equivalent mathematical model. Use $k$ as the constant of variation.<br><br>    The variable $p$ varies jointly as $t$ and $m$ and inversely as $n$. |

## Concept 3: Applications of Variation

## Video 4: Writing a Variation Model Based on Given Information (1:42 min)

▶ Play the video and work along.

$m$ varies directly as $t$. When $t$ is 6, $m$ is 48.

    a.   Write a variation model for this situation. Use $k$ as the constant of variation.

    b.   Solve for the constant of variation.

    c.   Find the value of $m$ when $t$ is 7.5.

| Review and complete | You try |
|---|---|
| 7. $y$ varies inversely as $x$. When $x$ is 6, $y$ is 2. | 8. $y$ varies directly as the square of $x$. When $x$ is 4, $y$ is 12. |
|    a.  Write a variation model for this situation. Use $k$ as the constant of variation.<br>   b.  Solve for the constant of variation.<br>   c.  Find the value of $y$ when $x$ is 1.5. |    a.  Write a variation model for this situation. Use $k$ as the constant of variation.<br>   b.  Solve for the constant of variation.<br>   c.  Find the value of $y$ when $x$ is 10. |

Solution:

a.  $y = \dfrac{k}{x}$

     $2 = \dfrac{k}{6}$

b.  $6 \cdot \dfrac{2}{1} = \dfrac{k}{6} \cdot 6$

     $12 = k$

The constant of variation is 12.

    $y = \dfrac{12}{x}$

c.  $y = \dfrac{12}{1.5}$

    $y = 8$

When $x = 1.5$, $y = 8$.

## Video 5: Solving a Variation Application (1:56 min)

▶ Play the video and work along.

The water pressure exerted on a scuba diver varies directly as the depth. If the pressure at a depth of 22 ft is 9.8 lb/in.$^2$, determine the pressure at a depth of 50 ft.

| Review and complete | You try |
|---|---|

9. The first step in solving an application of variation is to declare a variable to represent the unknown quantity. Then write the _____ model.

10. The amount of medicine a nurse practitioner prescribes for a patient varies directly with the patient's weight. She prescribes 55 mg for a patient who weighs 150 lb. How much should be prescribed for a 270-lb patient?

Solution:

Let $p$ be the amount prescribed and $w$ be the patient's weight.

$$p = kw$$

$$55 = k \cdot 150$$

$$\frac{55}{150} = \frac{k \cdot 150}{150}$$

$$k = \frac{11}{30}$$

Knowing the value of $k$ gives us the variation model.

$$p = \frac{11}{30}w$$

Replace $w$ with 270 to get

$$p = \frac{11}{30} \cdot 270$$

$$p = 99$$

She should prescribe 99 mg for the 270-lb patient.

11. The amount of medicine a nurse practitioner prescribes for a patient varies directly with the patient's weight. She prescribes 14 mg for a child who weighs 60 lb. How much should be prescribed for a 150-lb patient?

## Video 6: Solving an Application Involving Inverse Variation   (2:17 min)

▶ Play the video and work along.

The electrical resistance of wire varies inversely as the square of the diameter of the wire. The resistance of a wire 3 mm in diameter is 8 ohms. Find the resistance of a wire of equal length and the same material, but 4 mm in diameter.

| Review and complete | You try |
|---|---|
| 12. The electrical resistance of wire varies inversely as the square of the diameter of the wire. The resistance of a wire 3 mm in diameter is 5 ohms. Find the resistance of a wire of equal length and the same material, but 12 mm in diameter. | 13. An author self publishes a craft manual and finds that the number of manuals she can sell varies inversely with the price. When she charges $12, she sells 250 copies per month. How many could she sell if she drops the price to $10? |

Solution:
R = resistance (ohms) , d = diameter (mm)

First, find $k$ given that $R = 5$ and $d = 3$.

$$R = \frac{k}{d^2}$$

$$5 = \frac{k}{(3)^2}$$

$$5 = \frac{k}{9}$$

$$9 \cdot 5 = 9 \cdot \frac{k}{9}$$

$$45 = k$$

Now find the value of $R$ given that $d = 12$ and $k = 45$.

$$R = \frac{45}{12^2}$$

$$R = \frac{45}{144}$$

$$R = 0.3125$$

## Video 7: Solving an Application Involving Joint Variation  (2:04 min)

▶ Play the video and work along.

The cost, C, of copper pipe varies jointly as the length of the pipe, L, and the diameter of the pipe, d. If 10 ft of pipe that is 1 in. in diameter costs $45, how much would 64 ft of pipe cost if the pipe is 1 ½ in. in diameter?

**Answers:** Section 5.1

1. restricted

3a. $y \neq 0$

3b. $x \neq 0; x \neq 9$

4. zero

6. $\{x \mid x \neq -4\}$; $(-\infty,-4)\cup(-4,\infty)$

7. all real numbers

8. $(-\infty,0)\cup(0,4)\cup(4,\infty)$

9. $(-\infty,-2)\cup(-2,3)\cup(3,\infty)$

11. $\dfrac{m^3 p}{3n^5}$

12. are not

13. factoring

15. $\dfrac{3}{x+10}$ provided $x \neq 1, -10$

16. before

18. $\dfrac{2x+5}{x-7}$ provided $x \neq 7, x \neq -7$

20a. -1

20b. Cannot be simplified

20c. 1

22. $\dfrac{-6}{5(x+2)}$ or $-\dfrac{6}{5(x+2)}$ or $\dfrac{6}{-5(x+2)}$

**Answers:** Section 5.2

2. divided

3. $15xy$

4. factored

6. $\dfrac{x-3}{2(x+2)}$

7. reciprocal

8. $\dfrac{18}{25}$

9. $\dfrac{4mn}{3}$

11. $\dfrac{4x(x+2)}{x+5}$

12. left; right

13. 1

15. $\dfrac{-(x+3)}{2(x+4)(x+6)}$ or $-\dfrac{x+3}{2(x+4)(x+6)}$

**Answers:** Section 5.3

2. 2

4. $\dfrac{7}{30}$

6a. LCD $=15x^2 y^4 z^3$

6b. LCD $=5(x+2)^2$

7. 1

9a. $50xy^4$

9b. $7(x-14)$

10. 1

12. $\dfrac{8+91x}{26x^2}$

14. $\dfrac{x^2-x-15}{(x-3)(x-6)}$

16. $\dfrac{x(5x+16)}{(x-4)(x+2)}$

17. $-1$

19. $\dfrac{5x+6}{x-2}$

**Answers:** Section 5.4

1. division

3. $\dfrac{xz}{20wy}$

5. $-8$

6. numerator; denominator

8. $\dfrac{3+2x}{4-9x}$

9. LCD

11. $\dfrac{-13}{4}$

12. reciprocal

14. $\dfrac{x+1}{5}$

16. $\dfrac{-x+2}{-x-3}$

**Answers**: Section 5.5

1. LCD

3. $\{60\}$

5. $\{6\}$

6. zero

8. $\{\,\}$, no solution

10. $\{-7, 4\}$

12. $\{-1\}$

13. one side of the equation

15. $b = \dfrac{ma}{m-a}$ or $b = \dfrac{-ma}{a-m}$

**Answers**: Section 5.6

1. $\dfrac{a}{b}$; proportion; LCD

3. $\{12\}$

5. 42 gal

6. $1232 - m$

8. 198 welders do not have a degree

10. $x = 75$ cm; $y = 30$ cm

11. distance; rate

13. Wendy: 4.5 mph; LaRon: 7.5 mph

15. Car 49 mph; bus 28 mph

16. $\dfrac{1 \text{ room}}{x \text{ hr}}$

18. $2\dfrac{2}{9}$ hr

**Answers**: Section 5.7

2a. down

2b. $y = \dfrac{k}{x}$

4a. $y = \dfrac{k}{m^2}$

4b. $q = kn^3$

7. $p = \dfrac{ktm}{n}$

8a. $y = kx^2$

8b. $k = \dfrac{3}{4}$

8c. $y = 75$

9. variation

11. 35 mg

13. 300 manuals

## Concept 1: Definition of a Square Root

| Video 1: Definition of a Square Root (1:43 min) |
|---|

▶ Play the video and work along.

Determine the square roots of the given real number.

64

⏸ Pause the video and try these yourself.

Determine the square roots of the given real number.

81                                                        −81

▶ Play the video and check your answer.

| Review and complete | You try |
|---|---|
| 1.  A number *b* is a square root of *a* if _____. All positive real numbers have _____ square roots. | |
| 2.  Determine the square roots of each number.<br><br>  a.  100<br>   The square roots of 100 are 10 and −10.<br><br>  b.  9<br>   The square roots of 9 are 3 and −3.<br><br>  c.  −9<br>   There are no real number square roots of −9. | 3.  Determine the square roots of each number.<br><br>  a.  49<br><br>  b.  −49<br><br>  c.  16 |

| Video 2: Defining Square Roots Using Radical Notation (2:27 min) |
|---|

▶ Play the video and work along.

Simplify the expression.

$\sqrt{64}$ $\qquad\qquad\sqrt{\dfrac{25}{49}}$ $\qquad\qquad\sqrt{-4}$

⏸ Pause the video and try these yourself.

$-\sqrt{4}$ $\qquad\qquad\sqrt{0.36}$

▶ Play the video and check your answers.

| Review and complete | You try |
|---|---|
| 4. Simplify the expressions. | 5. Simplify the expressions. |
| a. $\sqrt{36}=6$ | a. $\sqrt{10000}$ |
| | b. $-\sqrt{\dfrac{1}{4}}$ |
| b. $-\sqrt{144}=-12$ | |
| | c. $\sqrt{0.25}$ |
| c. $\sqrt{0}=0$ | |

**Concept 2: Definition of an nth Root**

| Video 3: Definition of an nth Root (2:44 min) |
|---|

▶ Play the video and work along.

Find the 3$^{rd}$ root of 8. $\qquad\qquad\qquad$ Find the 4$^{th}$ root(s) of 81.

Writing an nth root using radical notation.

---

**SUMMARY**

a.  If $n$ is a positive even integer, and $a > 0$, then $\sqrt[n]{a}$ is the principal (positive) nth root of $a$.

b.  If $n > 1$ is an odd integer, then $\sqrt[n]{a}$ is the principal (positive) nth root of $a$.

c.  If $n > 1$ is an integer, then $\sqrt[n]{0} = 0$.

---

| Review and complete | You try |
|---|---|
| 6.  Negative real numbers do not have real number square roots, but they can have an $n$th roots as long as $n$ is _____. | |

| | |
|---|---|
| 7.  Find the fourth root(s) of 16.<br><br>    $2$ and $-2$ , since $(2)^4 = 16$ and $(-2)^4 = 16$ . | 8.  Find the third root(s) of $-125$ . |

**Video 4: Evaluating nth Roots  (3:48 min)**

▶ Play the video and work along.

Simplify each expression.

a.  $\sqrt{64}$

      b.  $\sqrt[3]{64}$

| Perfect Squares | Perfect Cubes |
|---|---|
| $1^2 = 1$ | $1^3 = 1$ |
| $2^2 = 4$ | $2^3 = 8$ |
| $3^2 = 9$ | $3^3 = 27$ |
| $4^2 = 16$ | $4^3 = 64$ |
| $5^2 = 25$ | $5^3 = 125$ |
| $6^2 = 36$ | |
| $7^2 = 49$ | **Perfect 4th Powers** |
| $8^2 = 64$ | $1^4 = 1$ |
| $9^2 = 81$ | $2^4 = 16$ |
| $10^2 = 100$ | $3^4 = 81$ |
| $11^2 = 121$ | $4^4 = 256$ |
| $12^2 = 144$ | $5^4 = 625$ |
| $13^2 = 169$ | |

c.  $\sqrt[6]{64}$

      d.  $\sqrt{-1}$

e.  $\sqrt[3]{-1}$       f.  $\sqrt[4]{-1}$

---

Video and Study Guide to accompany Intermediate Algebra, 5th ed., Miller/O'Neill/Hyde

g. $\sqrt[4]{-16}$

h. $-\sqrt[4]{16}$

i. $\sqrt[3]{-\dfrac{1}{27}}$

j. $\sqrt[3]{0.008}$

| Review and complete | You try |
|---|---|
| 9. Simplify. | 10. Simplify. |
| a. $\sqrt[3]{-8} = -2$ | a. $\sqrt[3]{-1}$ |
| b. $\sqrt[4]{16} = 2$ | b. $\sqrt[4]{10000}$ |
| c. $\sqrt[5]{1} = 1$ | c. $\sqrt[3]{-0.008}$ |

|  |  |
|--|--|
|  |  |

## Concept 3: Roots of Variable Expressions

### Video 5: Determine the Principal nth Root of an nth Power(2:29 min)

▶ Play the video and work along.

Simplify the expressions.

$\sqrt[3]{x^3}$

$\sqrt{x^2}$

$\sqrt[3]{(-7)^3}$

$\sqrt{(-7)^2}$

$\sqrt[10]{(2y-z)^{10}}$

$\sqrt[5]{(c+d)^5}$

| Review and complete | You try |
|---|---|
| 11. For roots with an even index, we require the answer to be _____. | |
| 12. Simplify the expressions. | 13. Simplify the expressions. |
| a. $\sqrt[5]{y^5} = y$ | a. $\sqrt[4]{y^4}$ |
| b. $\sqrt{(9x-2)^2} = \lvert 9x-2 \rvert$ | b. $\sqrt[7]{(-4)^7}$ |
| c. $\sqrt{(m^6)^2} = \lvert m^6 \rvert = m^6$ | c. $\sqrt{(m+n)^2}$ |

### Video 6:  Simplifying the nth Root of Perfect nth Powers (2:51 min)

▶ Play the video and work along.

Simplify the expressions. Assume all variables represent positive real numbers.

$\sqrt{t^6}$

| **Perfect Squares** | **Perfect Cubes** |
|---|---|
| $(x^1)^2 = x^2$ | $(x^1)^3 = x^3$ |
| $(x^2)^2 = x^4$ | $(x^2)^3 = x^6$ |
| $(x^3)^2 = x^6$ | $(x^3)^3 = x^9$ |
| $(x^4)^2 = x^8$ | $(x^4)^3 = x^{12}$ |
| exponents | exponents |
| are multiples | are multiples |
| of two | of three |

$\sqrt{25w^2}$

$-\sqrt[3]{64m^6}$

$\sqrt[4]{\dfrac{a^4}{b^8}}$

| **Review and complete** | **You try** |
|---|---|

14. Simplify the expressions. Assume all variables represent positive real numbers.

    a.  $\sqrt[4]{16a^8} = 2a^2$

    b.  $-\sqrt[3]{-8x^6} = -\left(-2x^2\right) = 2x^2$

15. Simplify the expressions. Assume all variables represent positive real numbers.

    a.  $\sqrt[5]{32a^{15}}$

    b.  $\sqrt[4]{y^{16}}$

### Concept 4: Pythagorean Theorem

**Video 7: Reviewing the Pythagorean Theorem (1:52 min)**

▶ Play the video and work along.

**Pythagorean Theorem**

If $a$ and $b$ represent the lengths of the legs of a right triangle, and $c$ is the length of the hypotenuse, then

$$a^2 + b^2 = c^2$$

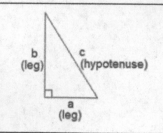

The triangle shown here is a right triangle. Determine the length of the missing side.

?

5 in.

13 in.

| Review and complete | You try |
|---|---|
| 16. Determine the length of the missing side. | 17. A right triangle has legs measuring 10 cm and 24 cm. Find the length of the hypotenuse. |

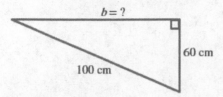

$$a^2 + b^2 = c^2$$
$$60^2 + b^2 = 100^2$$
$$3600 + b^2 = 10{,}000$$
$$b^2 = 6400$$
$$b = \sqrt{6400}$$
$$b = 80$$

The missing side is 80 cm.

## Video 8: An Application of the Pythagorean Theorem (1:54 min)

▶ Play the video and work along.

Jarven and Travis leave the college parking lot at 3:00 pm. Jarven drives south 8 mi, and Travis drives west 6 mi. How far apart are they?

| Review and complete | You try |
|---|---|
| **18.** Finding lengths of sides in a right triangle using the Pythagorean Theorem uses the concept of principal or _____ square root, since length cannot be negative. . | |

| | |
|---|---|
| **19.** Anneke drives north 21 mi from her home in Clare, Michigan. Brayden drives 28 mi east from the exact same spot. How far apart are they? | **20.** Paul drives west 16 mi from his home in Topeka, Kansas. Queenie drives 12 mi north from the same spot. How far apart are they? |

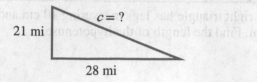

$$a^2 + b^2 = c^2$$
$$21^2 + 28^2 = c^2$$
$$441 + 784 = c^2$$
$$1225 = c^2$$
$$c = \sqrt{1225}$$
$$c = 35$$

Anneke and Brayden are 35 mi apart.

## Concept 5: Radical Functions

### Video 9: Introduction to Radical Functions (2:55 min)

▶ Play the video and work along.

$f(x) = \sqrt{x}$

| x | y |
|---|---|
| 0 | 0 |
| 1 | 1 |
| 4 | 2 |
| 9 | 3 |

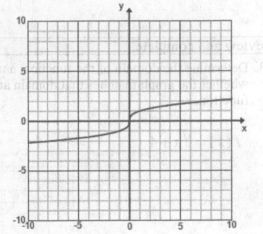

| x | y |
|---|---|
| -8 |  |
| -1 |  |
| 0 |  |
| 1 |  |
| 8 |  |

$f(x) = \sqrt[3]{x}$

| **Review and complete** | **You try** |
|---|---|
| 21. Make a table of values and sketch the graph. | 22. Make a table of values and sketch the graph. |

$g(x) = \sqrt{x+1}$

| x | y |
|---|---|
| -2 | Undefined |
| -1 | 0 |
| 0 | 1 |
|  | 1.73 |

$h(x) = 2 + \sqrt[3]{x}$

| x | y |
|---|---|
|  |  |
|  |  |
|  |  |
|  |  |

## Video 10: Determining the Domain of Radical Functions (2:36 min)

▶ Play the video and work along.

Determine the domain of each function and match the function with the graph.

$f(x) = \sqrt{x-2}$ _____    $g(x) = \sqrt[3]{x-2}$ _____    $h(x) = \sqrt[4]{2-x}$ _____

a.

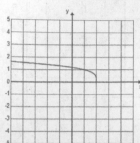

b.

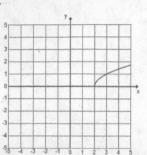

c.

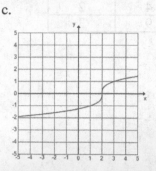

| Review and complete | You try |
|---|---|

**23.** Determine the domain of the function, and decide whether the graph matches that domain and the function.

$$f(x) = \sqrt[4]{x+4}$$

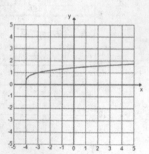

**24.** Determine the domain of the function, and decide whether the graph matches that domain and the function.

$$f(x) = -\sqrt[3]{x+1}$$

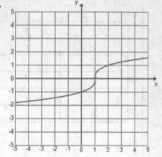

The radicand must be nonnegative, so

$$x + 4 \geq 0$$
$$x \geq -4$$

The domain is, therefore, $\left[-4, \infty\right)$.

The graph demonstrates this domain, and plotting points such as (0, 1.19) confirms the graph is correct.

**Video 11: Graphing a Radical Function (3:32 min)**

▶ Play the video and work along.

Given $f(x) = \sqrt{4-x}$ ,

    Write the domain of $f$ in interval notation.

_____→

    Graph $f$ by making a table of points.

| $x$ | $f(x)$ |
|-----|--------|
|     |        |
|     |        |
|     |        |
|     |        |
|     |        |

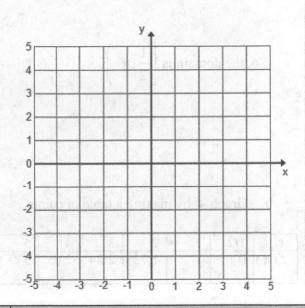

| Review and complete | You try |
|---------------------|---------|

25. In a radical function, we require the radicand to be _____.

26. Given $f(x) = \sqrt{2x-1}$,

    a. Write the domain of $f$ in interval notation.

$$2x - 1 \geq 0$$
$$2x \geq 1$$
$$x \geq \frac{1}{2}$$

    So, the domain is $\left[\frac{1}{2}, \infty\right)$.

    b. Graph $f$ by making a table of points.

| $x$ | 1/2 | 1 | 2 | 3 | 4 | 5 |
|------|-----|---|------|------|------|---|
| $f(x)$ | 0 | 1 | 1.73 | 2.24 | 2.65 | 3 |

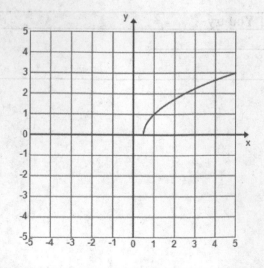

27. Given $f(x) = -\sqrt{x-3}$,

    a. Write the domain of $f$ in interval notation.

    b. Graph $f$ by making a table of points.

## Concept 1: Definition of $a^{1/n}$ and $a^{m/n}$

| Video 1: Definition of "a" to the 1/n Power  (3:02 min) |
| --- |

▶ Play the video and work along.

Write each expression in radical notation and simplify.

$49^{1/2}$                                         $64^{1/3}$

$(-16)^{1/4}$                        $-16^{1/4}$

$25^{-1/2}$                          $\left(\dfrac{1}{9}\right)^{-1/2}$

| Review and complete | You try |
| --- | --- |

1. For a real number $a$ and an integer, $n$, greater than 1, the expression $a^{1/n} = $ _____.

| 2. Write each expression in radical notation and simplify. | 3. Write each expression in radical notation and simplify. |
| --- | --- |
| a.  $100^{1/2} = \sqrt{100} = 10$ | a.  $8^{1/3}$ |
| b.  $81^{-1/2} = \left(\dfrac{1}{81}\right)^{1/2} = \sqrt{\dfrac{1}{81}} = \dfrac{1}{9}$ | b.  $8^{-1/3}$ |

| Video 2: Definition of "a" to the m/n Power  (3:49 min) |
| --- |

▶ Play the video and work along.

Write the expression in radical notation and simplify.

$8^{2/3}$

⏸ Pause the video and try these yourself.

Video and Study Guide to accompany Intermediate Algebra, 5th ed., Miller/O'Neill/Hyde
Copyright © 2018 McGraw-Hill Education

$100^{3/2}$                                                    $16^{-3/4}$

▶ Play the video and check your answers. Then work along.

$(-4)^{3/2}$                                                    $-4^{3/2}$

| Review and complete | You try |
|---|---|
| 4.  The difference in the last two expressions of Video 2 is that the negative sign is part of the _____ in the first example, and not part of it in the second. | |
| 5.  Write the expression in radical notation and simplify.<br><br>    $36^{3/2}$<br><br><br><br><br><br>    $=\left(\sqrt{36}\right)^3 = 6^3 = 216$ | 6.  Write the expression in radical notation and simplify.<br><br>    $-9^{5/2}$ |

**Concept 2: Converting Between Rational Exponents and Radical Notation**

| Video 3: Converting Between Radical Notation and Rational Exponents (2:44 min) |
|---|
| ▶ Play the video and work along.<br><br>Convert each expression to radical notation. Assume all variables represent positive real numbers.<br><br>$y^{2/3}$                      $5z^{1/4}$                      $(7z^3)^{1/4}$<br><br><br><br>Convert each expression to an expression with rational exponents. Assume all variables represent positive real numbers.<br><br>$\sqrt[3]{t}$                                      $\sqrt[5]{c^2}$<br><br><br><br>❚❚ Pause the video and try these yourself.<br><br>                              $11\sqrt{k}$ |

Video and Study Guide to accompany Intermediate Algebra, 3ʳᵈ ed., Miller/O'Neill/Hyde
Copyright © 2018 McGraw-Hill Education

$\sqrt{11k}$

▶ Play the video and check your answers.

| Review and complete | You try |
|---|---|
| 7. Convert the expression to an equivalent form (from rational exponents to radical notation or from radical notation to rational exponents). Assume all variables represent positive real numbers. | 8. Convert the expression to an equivalent form (from rational exponents to radical notation or from radical notation to rational exponents). Assume all variables represent positive real numbers |
| a. $(3z)^{1/3} = \sqrt[3]{3z}$ | a. $5a^{1/2}$ |
| b. $\sqrt{x^3} = x^{3/2}$ | b. $4\sqrt{x^7}$ |

## Concept 3: Properties of Rational Exponents

### Video 4: Properties of Rational Exponents  (3:21 min)

> ▶ Play the video and work along.

**Properties of Rational Exponents:** Let $a$ and $b$ be nonzero real numbers. Let $m$ and $n$ be rational numbers such that $a^m$, $a^n$, and $b^m$ are real numbers.

| Description | Property | Example |
|---|---|---|
| Multiplying like bases | $a^m a^n = a^{mn}$ | $x^{2/3} x^{4/3}$ |
| Dividing like bases | $\dfrac{a^m}{a^n} = a^{m-n}$ | $\dfrac{5^{3/4}}{5^{1/2}}$ |
| Power rule | $(a^m)^n = a^{mn}$ | $(p^{3/5})^{1/3}$ |
| Power of a product | $(ab)^m = a^m b^n$ | $(c^{1/3} d^{1/2})^6$ |
| Power of a quotient | $\left(\dfrac{a}{b}\right)^m = \dfrac{a^m}{b^n}$ | $\left(\dfrac{25}{b^2}\right)^{1/2}$ |

| Review and complete | You try |
|---|---|
| 9. Use the properties of exponents to simplify each expression. Assume all variables represent positive real numbers.<br><br>a. $y^{1/3} y^{2/3} = y^{1/3+2/3} = y^{3/3} = y^1 = y$ | 10. Use the properties of exponents to simplify each expression. Assume all variables represent positive real numbers.<br><br>a. $y^2 y^{1/2}$ |

Video and Study Guide to accompany Intermediate Algebra, 5$^{th}$ ed., Miller/O'Neill/Hyde
Copyright © 2018 McGraw-Hill Education

b. $(x^{1/2}x^2)^2$
$= (x^{1/2})^2(x^2)^2$
$= (x^{(1/2)(2)})(x^{2\cdot2})$
$= x^1 \cdot x^4$
$= x^5$

b. $(x^{14}x^{1/2})^4$

c. $\left(\dfrac{x^{2/3}}{y^{1/2}}\right)^6 = \dfrac{(x^{2/3})^6}{(y^{1/2})^6} = \dfrac{x^{(2/3)(6)}}{y^{(1/2)(6)}} = \dfrac{x^4}{y^3}$

c. $\left(\dfrac{a^{2/5}}{b^{1/5}}\right)^{10}$

## Video 5: Evaluating Exponential Expressions (1:53 min)

▶ Play the video and work along.

### Definition of Negative and Zero Exponents

Let $a$ be a nonzero real number. Let $m$ be a rational number such that $a^m$ is a real number.

| Definition | Example | |
|---|---|---|
| $a^{-m} = \left(\dfrac{1}{a}\right)^m = \dfrac{1}{a^m}$ | $16^{-1/2}$ | |
| $a^0 = 1$ | $(-1)^0$ | $5x^0$ |
|  | $-1^0$ | $(5x)^0$ |

| Review and complete | You try |
|---|---|
| 11. Any nonzero quantity raised to the zero power is equal to _____. | |

| Review and complete | You try |
|---|---|
| 12. Simplify the expressions. | 13. Simplify the expressions. |
| a. $64^{-1/3}$ | a. $16^{-1/4}$ |
| $= \left(\dfrac{1}{64}\right)^{1/3} = \sqrt[3]{\dfrac{1}{64}} = \dfrac{1}{4}$ | |
| b. $(4x^2)^0 = 1$ | b. $-6y^0$ |

## Video 6: Simplifying Expressions with Rational Exponents (1:18 min)

▶ Play the video and work along.

Simplify the expression. Assume that $a$ is a positive real number.

$$\dfrac{a^{2/7}a^{3/7}}{a^{1/2}}$$

| Review and complete | You try |
|---|---|
| 14. Simplify the expression. Assume that $x$ is a positive real number. | 15. Simplify the expression. Assume that $x$ is a positive real number. |
| $\dfrac{x^{1/8}x^{1/2}}{x^{1/4}}$ | $\dfrac{x^{9/10}x^{1/5}}{x^{1/10}}$ |
| $= \dfrac{x^{1/8+1/2}}{x^{1/4}} = \dfrac{x^{5/8}}{x^{2/8}} = x^{5/8-2/8} = x^{3/8}$ | |

## Video 7: Simplifying Expressions with Rational Exponents (2:04 min)

▶

Play the video and work along.

Simplify the expression, and write the answer with positive exponents. Assume that $m$ and $n$ represent positive real numbers.

$$\left( \frac{36mn^{-1/2}}{m^{-4/3}} \right)^{1/2}$$

| Review and complete | You try |
|---|---|
| 16. Simplify the expression, and write the answer with positive exponents. Assume all variables represent positive real numbers. $$\left( \frac{a^{-1/3}}{16b^{-2}} \right)^{-3/4}$$ $$= \left( 16^{-1} a^{-1/3} b^{2} \right)^{-3/4}$$ $$= 16^{3/4} a^{1/4} b^{-3/2}$$ $$= \frac{(\sqrt[4]{16})^{3} a^{1/4}}{b^{3/2}}$$ $$= \frac{8a^{1/4}}{b^{3/2}}$$ | 17. Simplify the expression, and write the answer with positive exponents. Assume all variables represent positive real numbers. $$\left( \frac{x^{4}}{81y^{-2}} \right)^{3/4}$$ |

## Video 8: Simplifying Expressions with Rational Exponents (1:35 min)

▶ Play the video and work along.

Simplify the expression, and write the answer with positive exponents. Assume $c$ and $d$ represent positive real numbers.

$$(c^{-3/5} d^{1/3})^{-1/5} (c^{1/2} d^{0})^{4}$$

| Review and complete | You try |
|---|---|
| 18. Simplify the expression, and write the answer with positive exponents. Assume all variables represent positive real numbers. | 19. Simplify the expression, and write the answer with positive exponents. Assume all variables represent positive real numbers. |

$$\left(x^{2/3}y^{-3/4}\right)^{-1/2}\left(x^0 y^{-3}\right)^{-1/3}$$
$$=\left(x^{2/3}\right)^{-1/2}\left(y^{-3/4}\right)^{-1/2}\left(1\cdot y^{-3}\right)^{-1/3}$$
$$=x^{-1/3}\,y^{3/8}\,y^{1}$$
$$=x^{-1/3}\,y^{3/8+8/8}$$
$$=\frac{y^{11/8}}{x^{1/3}}$$

$$\left(a^{-1/3}b^{1/2}\right)^{4}\left(a^{-1/2}b^{3/5}\right)^{10}$$

## Video 9: Recognizing and Simplifying Similar Looking Expressions (4:14 min)

▶ Play the video and work along.

Simplify the expressions.

⏸ Pause the video and try this yourself.

$81^{3/4}$

$-81^{3/4}$

$(-81)^{3/4}$

$81^{-3/4}$

$-81^{-3/4}$

$(-81)^{-3/4}$

$$\left(\frac{1}{81}\right)^{-3/4} \qquad\qquad \left(-\frac{1}{81}\right)^{0}$$

▶ Play the video (Press Continue) and check your answers.

| Review and complete | You try |
|---|---|
| 20. Simplify the expressions. | 21. Simplify the expressions. |
| a. $25^{3/2}$    b. $25^{-3/2}$    c. $125^{-2/3}$ | a. $64^{3/2}$    b. $64^{-2/3}$    c. $\left(\dfrac{1}{64}\right)^{-2/3}$ |
| $=(\sqrt{25})^3$ $=125$   $=\left(\sqrt{\dfrac{1}{25}}\right)^3$   $=\left(\sqrt[3]{\dfrac{1}{125}}\right)^2$ | |
| $=\dfrac{1}{125}$   $=\dfrac{1}{5^2}=\dfrac{1}{25}$ | |

## Concept 1: Multiplication Property of Radicals

### Video 1: Introduction to the Multiplication Property of Radicals (0:56 min)

▶ Play the video and work along.

**Multiplication Property of Radicals**

Let $a$ and $b$ represent real numbers such that $\sqrt[n]{a}$ and $\sqrt[n]{b}$ are real numbers.

$$\sqrt[n]{a} \cdot \sqrt[n]{b} = \sqrt[n]{ab}$$

$$\sqrt[n]{a} \cdot \sqrt[n]{b} = \sqrt[n]{ab}$$

| Review and complete | You try |
|---|---|
| 1. Radicals with the same _____ can be multiplied according to the multiplication property of radicals. | |
| 2. Can the radicals be multiplied with the multiplication property of radicals?<br><br>$\sqrt{10}$ and $\sqrt{15}$<br><br>Yes, these can be multiplied. Both radicals have an index of 2. | 3. Can the radicals be multiplied with the multiplication property of radicals?<br><br>$\sqrt[3]{10}$ and $\sqrt{10}$ |

## Concept 2: Simplifying Radicals Using the Multiplication Property of Radicals

### Video 2: Criteria for a Radical to be in Simplified Form (1:40 min)

▶ Play the video and work along.

**Properties of a Simplified Radical** Consider a radical expression where the radicand is written as a product of prime factors. The radical is in *simplified form* if all of the conditions are met.

1. The radicand has no factor raised to a power greater than or equal to the index.

2. The radicand does not contain a fraction.

3. There are no radicals in the denominator of a fraction.

| Review and complete | You try |
|---|---|
| **4.** Why are the radicals *not* simplified? | **5.** Why are the radicals *not* simplified? |
| a. $\sqrt{360}$ <br> A factor of $6^2 = 36$ is in the radicand. | a. $\sqrt{\dfrac{6}{13}}$ |
| b. $-\sqrt{\dfrac{144}{49}}$ <br> A fraction is in the radicand. | b. $\sqrt{400}$ |
| c. $\dfrac{1}{\sqrt{7}}$ <br> There is a radical in the denominator. | c. $\dfrac{37}{\sqrt{37}}$ |

## Video 3: Simplifying Radicals (3:58 min)

▶ Play the video and work along.

Simplify the expressions. Assume that $x$ and $y$ represent positive real numbers.

$\sqrt{x^2}$          $\sqrt{x^4}$          $\sqrt{x^{20}}$

$\sqrt{x^7}$

❚❚ Pause the video and try this yourself.

$\sqrt[3]{x^{17}}$

▶ Play the video and check your answer.

**Recall:** All factors within the radicand must be raised to a power less than the index.

Video and Study Guide to accompany Intermediate Algebra, 5ᵗʰ ed., Miller/O'Neill/Hyde
Copyright © 2018 McGraw-Hill Education

| Review and complete | You try |
|---|---|
| 6. The absolute value bars on the even root of an even power can be left off if we assume that variables represent _____ numbers. | |

| | |
|---|---|
| 7. Simplify the expressions. Assume that variables represent positive real numbers. <br><br> a. $\sqrt{a^6} = a^3$ <br><br><br><br> b. $\sqrt[4]{x^9} = \sqrt[4]{x^8 \cdot x} = \sqrt[4]{x^8} \cdot \sqrt[4]{x} = x^2 \sqrt[4]{x}$ | 8. Simplify the expressions. Assume that variables represent positive real numbers. <br><br> a. $\sqrt[3]{c^9}$ <br><br><br><br> b. $\sqrt{m^{11}}$ |

## Video 4: Simplifying Radicals (2:33 min)

▶ Play the video and work along.

Simplify the expressions. Assume that $a$, $b$, and $c$ are positive real numbers.

$$\sqrt{a^2 b^9 c}$$

⏸ Pause the video and try this yourself.

$$\sqrt[3]{a^6 b^5 c^{22}}$$

▶ Play the video and check your answer.

> **Recall:** All factors within the radicand must be raised to a power less than the index.

| Review and complete | You try |
|---|---|
| 9. Simplify the expression. Assume that a, b, and c are positive real numbers. <br><br> $\sqrt[3]{a^8 b^2 c^9}$ | 10. Simplify the expression. Assume that variables represent positive real numbers. <br><br> $\sqrt{m^3 n^4 p}$ |

$$= \sqrt[3]{(a^6 c^9) a^2 b^2}$$

$$= \sqrt[3]{a^6 c^9} \cdot \sqrt[3]{a^2 b^2}$$

$$= a^2 c^3 \sqrt[3]{a^2 b^2}$$

## Video 5: Simplifying a Radical with a Numerical Factor in the Radicand (4:36 min)

▶ Play the video and work along.

Simplify the expression.

$$\sqrt{72}$$

| Review and complete | You try |
|---|---|
| Always factor the radicand **completely** to simplify a radical expression. | |

11. Simplify the expression.

$$\sqrt{120}$$

$$= \sqrt{2 \cdot 2 \cdot 2 \cdot 3 \cdot 5}$$  [Simplify.]

$$= \sqrt{2^2 \cdot 2 \cdot 3 \cdot 5}$$

$$= 2\sqrt{30}$$

12. Simplify the expression.

$$\sqrt{250}$$

## Video 6: Simplifying a Radical (2:09 min)

▶ Play the video and work along.

Simplify the expression. Assume that $k$ and $p$ are positive real numbers.

$$\sqrt[3]{135 k^{13} p^2}$$

| Review and complete | You try |
|---|---|

13. Simplify the expression. Assume variables represent positive real numbers.

$$\sqrt[3]{80a^4b^5}$$

$$= \sqrt[3]{2^4 \cdot 5a^4b^5}$$

$$= \sqrt[3]{\left(2^3\,a^3b^3\right)} \cdot \sqrt[3]{5 \cdot 2ab^2}$$

$$= 2ab\sqrt[3]{5 \cdot 2ab^2}$$

$$= 2ab\sqrt[3]{10ab^2}$$

14. Simplify the expression. Assume variables represent positive real numbers.

$$\sqrt{45x^6y^9}$$

---

### Video 7: Simplifying Radicals by Using an Alternative Method (3:44 min)

▶ Play the video and work along.

Simplify the expressions. Assume $x$, $y$, and $z$ represent positive real numbers.

$$\sqrt[3]{x^{14}}$$

⏸ Pause the video and try this yourself.

$$\sqrt{8y^9z^4}$$

▶ Play the video and check your answer.

| Review and complete | You try |

| 15. Simplify the expression. Assume $a$ represents a positive real number. | 16. Simplify the expression. Assume $x$ and $y$ represent positive real numbers. |
|---|---|
| $\sqrt[4]{32a^{11}} = \sqrt[4]{2^5 a^{11}}$  $4\overline{)5} = 1 \text{ r } 1 \; ; \; 4\overline{)11} = 2 \text{ r } 3$  $\sqrt[4]{2^5 a^{11}} = 2^1 a^2 \sqrt[4]{2^1 a^3} = 2a\sqrt[4]{2a^3}$ | $\sqrt[3]{10000x^7 y^{10}}$ |

## Concept 3: Simplifying Radicals Using the Order of Operations

**Video 8: Simplifying a Radical in which a Fraction Appears in the Radicand  (1:27 min)**

▶ Play the video and work along.

Simplify the expressions. Assume that $x$ is a positive real number.

$$\sqrt{\frac{4x^7}{x}}$$

$$\sqrt[3]{\frac{2}{128}}$$

| **Review and complete** | **You try** |
|---|---|
| 17. Simplify the expression. Assume that $x$ is a positive real number.  $\sqrt{\dfrac{36x^4}{9x^2}}$  $= \sqrt{\dfrac{\overset{4}{\cancel{36}}x^{\overset{2}{\cancel{4}}}}{\cancel{9}\,x^{\cancel{2}}}}$  $= \sqrt{4x^2}$  $= 2x$ | 18. Simplify the expression. Assume that $c$ is a positive real number.  $\sqrt{\dfrac{200c^{14}}{8c^6}}$ |

**Video 9: Simplifying a fraction with Radicals to Lowest Terms (1:53 min)**

▶ Play the video and work along.

Simplify the expression.

$$\frac{5\sqrt{63}}{9}$$

| Review and complete | You try |
|---|---|
| 19. Simplify the expression. | 20. Simplify the expression. |
| $$\frac{6\sqrt{50}}{5}$$ | $$\frac{11\sqrt{44}}{12}$$ |
| $$= \frac{6\sqrt{5^2 \cdot 2}}{5}$$ | |
| $$= \frac{6 \cdot \cancel{5}\sqrt{2}}{\cancel{5}}$$ | |
| $$= 6\sqrt{2}$$ | |

## Concept 1: Addition and Subtraction of Radicals

| Video 1: Introduction to Addition and Subtraction of Radicals (2:21 min) |

▶ Play the video and work along.

Add or subtract as indicated.

$3x + 7x - 4x$ $\qquad\qquad$ $3\sqrt{x} + 7\sqrt{x} - 4\sqrt{x}$

$3x + 7x - 4x$ $\qquad\qquad$ $3\sqrt{x} + 7\sqrt{x} - 4\sqrt{x}$

| Review and complete | You try |
|---|---|
| 1. Like radicals have the same _____ and the same _____. | |
| 2. Add or subtract as indicated.<br><br>$3\sqrt[3]{y} + 9\sqrt[3]{y} - \sqrt[3]{y}$<br>$\quad = (3 + 9 - 1)\sqrt[3]{y}$<br>$\quad = 11\sqrt[3]{y}$ | 3. Add or subtract as indicated.<br><br>$5\sqrt{7} + \sqrt{7} - 7\sqrt{7}$ |

## Video 2: Adding and Subtracting Like Radicals (2:40 min)

▶ Play the video and work along.

Add or subtract as indicated.

$$12z\sqrt[3]{2z} - 5z\sqrt[3]{2z} - z\sqrt[3]{2z}$$

$$\sqrt{5} + \sqrt{5}$$

$$4\sqrt{y} + 2\sqrt{x} - 5\sqrt[3]{x}$$

| Review and complete | You try |
|---|---|
| 4. Add or subtract as indicated. | 5. Add or subtract as indicated. |
| $3\sqrt{x} + \sqrt{x} - \sqrt{y}$ <br> $= (3+1)\sqrt{x} - \sqrt{y}$ <br> $= 4\sqrt{x} - \sqrt{y}$ | $\sqrt[3]{x} + 5\sqrt{x} + 7\sqrt[3]{x}$ |

**Video 3: Adding and Subtracting Radicals (2:50 min)**

▶ Play the video and work along.

Add or subtract as indicated.

$$\sqrt{98x^3} - 3x\sqrt{8x} + \sqrt{2x^3}$$

| Review and complete | You try |
|---|---|

6.  If the *indices* of two radical terms match, simplify to see if the _____ will match to determine if the radicals can be added.

| | |
|---|---|
| 7.  Add or subtract as indicated.<br><br>$6\sqrt{a^2b} - 4a\sqrt{b}$<br>$= 6a\sqrt{b} - 4a\sqrt{b}$  ← Simplifying the first radical means the radicands now match.<br>$= 2a\sqrt{b}$ | 8.  Add or subtract as indicated.<br><br>$\sqrt{12t} - \sqrt{27t} + 2\sqrt{3t}$ |

**Video 4: Add and Subtract Radicals (1:50 min)**

▶ Play the video and work along.

Add or subtract as indicated.

$\sqrt[3]{81} + 4\sqrt[3]{24} - 7\sqrt[3]{3}$

| Review and complete | You try |
|---|---|
| 9. Add or subtract as indicated. | 10. Add or subtract as indicated. |

9. Add or subtract as indicated.

$2\sqrt{40} - \sqrt{8} + 3\sqrt{50}$

$= 4\sqrt{10} - 2\sqrt{2} + 15\sqrt{2}$ — Only terms 2 and 3 are like terms.

$= 4\sqrt{10} + 13\sqrt{2}$

10. Add or subtract as indicated.

$\sqrt{45} + 2\sqrt{20} + \sqrt{810}$

## Video 5: Adding Radicals (1:53 min)

▶ Play the video and work along.

Add. $2x\sqrt{3y^3} + 5\sqrt{3x^2 y}$

| Review and complete | You try |
|---|---|

11. Are the parentheses in $(2y + 5)x\sqrt{3y}$ required for mathematical accuracy?

Video and Study Guide to accompany Intermediate Algebra, 5th ed., Miller/O'Neill/Hyde
Copyright © 2018 McGraw-Hill Education

**12. Add.**

$$3\sqrt{3x^3} + \sqrt{75x}$$
$$= 3x\sqrt{3x} + 5\sqrt{3x}$$
$$= (3x+5)\sqrt{3x}$$

**13. Add.**

$$3\sqrt{45y^3} + 4xy\sqrt{20y}$$

## Concept 1: Multiplication Property of Radicals

| Video 1: Introduction to the Multiplication of Radicals (2:36 min) |
|---|

▶ Play the video and work along.

Multiply. Assume that $x$ represents a positive real number.

$$\sqrt{7} \cdot \sqrt{5}$$

⏸ Pause the video and try this yourself.

$$\sqrt[3]{2} \cdot \sqrt[3]{x}$$
$$\sqrt{98x} \cdot \sqrt{6x}$$

▶ Play the video and check your answer.

| Review and complete | You try |
|---|---|
| 1. Radicals can be multiplied if they have the same _____ and the individual radicals are real numbers. | |
| 2. Multiply. Assume that $x$ represents a positive real number. $$\sqrt{5} \cdot \sqrt{20x}$$ $$= \sqrt{5 \cdot 20x}$$ $$= \sqrt{100x}$$ $$= 10\sqrt{x}$$ | 3. Multiply. Assume that $x$ represents a positive real number. $$\sqrt{7} \cdot \sqrt{14x}$$ |

| Video 2: Multiplying Radical Expressions (2:32 min) |
|---|

Video and Study Guide to accompany Intermediate Algebra, 5th ed., Miller/O'Neill/Hyde

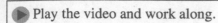 Play the video and work along.

Multiply and simplify the result. Assume that all variables represent positive real numbers.

$$\left(4\sqrt{6}\right)\left(2\sqrt{15}\right)$$

$$\left(\frac{1}{4}\sqrt[3]{xy^2}\right)\left(\frac{1}{2}\sqrt[3]{xy}\right)$$

| Review and complete | You try |
|---|---|
| 4. Multiply and simplify the result. Assume that all variables represent positive real numbers.<br><br>$\left(3x\sqrt{y}\right)\left(\sqrt{3xy}\right)$<br><br>$=3x\sqrt{y^2 \cdot 3x}$<br><br>$=3xy\sqrt{3x}$ | 5. Multiply and simplify the result. Assume that all variables represent positive real numbers.<br><br>$\left(2x\sqrt{10x}\right)\left(3\sqrt{15x^3}\right)$ |

## Video 3: Multiply Radicals (2:17 min)

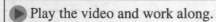 Play the video and work along.

Multiply and simplify the result. Assume that all variables represent positive real numbers.

$$5x(3x+4y-7)$$

$$5\sqrt{x}\left(3\sqrt{x}+4\sqrt{y}-7\right)$$

| Review and complete | You try |
|---|---|
| 6. Multiply and simplify the result. Assume that all variables represent positive real numbers.<br><br>$\sqrt{2}\left(4\sqrt{10x}+\sqrt{6x}+9\right)$ | 7. Multiply and simplify the result. Assume that all variables represent positive real numbers.<br><br>$\sqrt{3t}\left(\sqrt{12t}-\sqrt{t}+2\sqrt{3}\right)$ |

$$= \sqrt{2}\left(4\sqrt{10x}\right)+\sqrt{2}\left(\sqrt{6x}\right)+\sqrt{2}(9)$$

$$= 4\sqrt{20x}+\sqrt{12x}+9\sqrt{2}$$

$$= 4\sqrt{4\cdot 5x}+\sqrt{4\cdot 3x}+9\sqrt{2}$$

$$= 4(2)\sqrt{5x}+2\sqrt{3x}+9\sqrt{2}$$

$$= 8\sqrt{5x}+2\sqrt{3x}+9\sqrt{2}$$

## Video 4: Multiplying Two-Term Radical Expressions (3:10 min)

▶ Play the video and work along.

Multiply.   $(-2\sqrt{3}+3\sqrt{7})(4\sqrt{3}-\sqrt{7})$

| Review and complete | You try |
|---|---|
| 8.  Multiply. | 9.  Multiply. |
| $(2\sqrt{2}-5\sqrt{3})(6\sqrt{2}+\sqrt{3})$ | $(\sqrt{3}-\sqrt{10})(10\sqrt{3}+3\sqrt{10})$ |
| $=(2\sqrt{2})(6\sqrt{2})+(2\sqrt{2})(\sqrt{3})$ | |
| $\quad+(-5\sqrt{3})(6\sqrt{2})+(-5\sqrt{3})(\sqrt{3})$ | |
| $=12\sqrt{2^2}+2\sqrt{6}-30\sqrt{6}-5\sqrt{3^2}$ | |
| $=12\cdot 2-28\sqrt{6}-5\cdot 3$ | |
| $=24-28\sqrt{6}-15$ | |
| $=9-28\sqrt{6}$ | |

## Video 5: Multiplying Radical Expressions with Several Terms (3:02 min)

▶ Play the video and work along.

Multiply and simplify the result. Assume that all variables represent positive real numbers.

$(9\sqrt{x}-5\sqrt{z})(\sqrt{x}-2\sqrt{z}+4)$

| Review and complete | You try |
|---|---|
| 10. Multiplying a radical expression with two terms by another radical expression with three terms, as in the previous example, requires the use of the _____ property. | |

| Review and complete | You try |
|---|---|
| 11. Multiply and simplify the result. Assume that all variables represent positive real numbers.<br><br>$\left(4+2\sqrt{x}\right)\left(2\sqrt{x}-\sqrt{y}+10\right)$<br><br>$=(4)\left(2\sqrt{x}\right)+(4)\left(-\sqrt{y}\right)+(4)(10)$<br>$+\left(2\sqrt{x}\right)\left(2\sqrt{x}\right)+\left(2\sqrt{x}\right)\left(-\sqrt{y}\right)+\left(2\sqrt{x}\right)(10)$<br>$=8\sqrt{x}-4\sqrt{y}+40+4\sqrt{x^2}-2\sqrt{xy}+20\sqrt{x}$<br>$=28\sqrt{x}-2\sqrt{xy}+4\sqrt{y}+4x+40$ | 12. Multiply and simplify the result. Assume that all variables represent positive real numbers.<br><br>$\left(2+\sqrt{x}\right)\left(3-\sqrt{x}-\sqrt{y}\right)$ |

## Concept 2: Expressions of the Form $\left(\sqrt[n]{a}\right)^n$

**Video 6: Raising an nth Root to an nth Power (1:33 min)**

▶ Play the video and work along.

Perform the indicated operation. Assume that all variables represent positive real numbers.

$$\left(\sqrt{15}\right)^2 \qquad\qquad\qquad \left(\sqrt[4]{mn}\right)^4$$

❚❚ Pause the video and try this yourself.

$$\left(\sqrt{a+8}\right)^2$$

▶ Play the video and check your answer.

| Review and complete | You try |
|---|---|

| 13. Perform the indicated operation. Assume that all variables represent positive real numbers. | 14. Perform the indicated operation. Assume that all variables represent positive real numbers. |
|---|---|
| $$\left(\sqrt[4]{x+y^2}\right)^4$$ $$= x + y^2$$ | $$\left(\sqrt{3x+67}\right)^2$$ |

## Concept 3: Special Case Products

### Video 7: Squaring a Two-Term Radical Expression (3:51 min)

▶ Play the video and work along.

Square the radical expression. Assume that the variables represent positive real numbers.

$$\left(\sqrt{z}+7\right)^2$$

$$\left(3\sqrt{x}-5\sqrt{7}\right)^2$$

| Review and complete | You try |
|---|---|
| 15. True or false? $\left(\sqrt{m}+3\right)^2 = m+9$ | |

| 16. Square the radical expression. Assume that the variables represent positive real numbers. | 17. Square the radical expression. Assume that the variables represent positive real numbers. |
|---|---|
| $$\left(8-5\sqrt{x}\right)^2$$ $$= (8)^2 - 2(8)(5\sqrt{x}) + (5\sqrt{x})^2$$ $$= 64 - 80\sqrt{x} + 25x$$ | $$\left(3\sqrt{m}-4\right)^2$$ |

Video and Study Guide to accompany Intermediate Algebra, 5th ed., Miller/O'Neill/Hyde
Copyright © 2018 McGraw-Hill Education

| Video 8: Squaring a Two-Term Radical Expression (1:39 min) |

▶ Play the video and work along.

Square the radical expression. $\left(\sqrt{3m-2}-5\right)^2$

| Review and complete | You try |

18. True or False? $\left(\sqrt{3m-2}-5\right)^2 = (3m-2)+(25) = 3m+23$

| 19. Square the radical expression. | 20. Square the radical expression. |

**19.** Square the radical expression.

$$\left(\sqrt{2x+5}+7\right)^2$$

$$a^2+2ab+b^2$$

$$=\left(\sqrt{2x+5}\right)^2+2\left(\sqrt{2x+5}\right)(7)+7^2$$

$$=2x+5+14\sqrt{2x+5}+49$$

$$=2x+54+14\sqrt{2x+5}$$

**20.** Square the radical expression.

$$\left(\sqrt{x+1}-3\right)^2$$

| Video 9: Multiplying Conjugate Radical Expressions (2:17 min) |

▶ Play the video and work along.

Multiply the expressions. Assume all variables represent positive real numbers

$$\left(2\sqrt{5}-7\right)\left(2\sqrt{5}+7\right)$$

$$\left(\frac{1}{3}\sqrt{y}+\frac{2}{5}\sqrt{z}\right)\left(\frac{1}{3}\sqrt{y}-\frac{2}{5}\sqrt{z}\right)$$

| Review and complete | You try |

| 21. Multiply the expressions. | 22. Multiply the expressions. |
|---|---|
| $\left(5\sqrt{7}-1\right)\left(5\sqrt{7}+1\right)$ $=\left(5\sqrt{7}\right)^2-1^2$ $=25\cdot 7-1$ $=174$ | $\left(12-3\sqrt{5}\right)\left(12+3\sqrt{5}\right)$ |

**Concept 4: Multiplying Radicals with Different Indices**

| Video 10: Multiplying Radicals with Different Indices (3:04 min) |
|---|
| ▶ Play the video and work along. |

Multiply the expressions. Write the answer in radical form.

$\sqrt[4]{7}\cdot\sqrt{7}$                                           $\sqrt[3]{5}\cdot\sqrt[4]{2}$

▶ Play the video and check your answer.

| Review and complete | You try |
|---|---|
| 23. Multiply the expressions. Write the answer in radical form. $\sqrt[3]{6}\cdot\sqrt{5}$ $=6^{1/3}\cdot 5^{1/2}$ $=6^{2/6}\cdot 5^{3/6}$ $=\sqrt[6]{6^2}\cdot\sqrt[6]{5^3}$ $=\sqrt[6]{6^2\cdot 5^3}$ $=\sqrt[6]{4500}$ | 24. Multiply the expressions. Write the answer in radical form. $\sqrt[4]{5}\cdot\sqrt{2}$ |

**Concepts 1 and 2: Simplified Form of a Radical/Division Property of Radicals**

| Video 1: Introduction to Division Radicals (2:39 min) |
|---|

▶ Play the video and work along.

Simplify completely.

$$\sqrt{\frac{49\,y^3}{81}}$$                                              $$\frac{\sqrt[3]{x^4}}{\sqrt[3]{x}}$$

| Review and complete | You try |
|---|---|
| 1. When we remove a radical from the denominator of a fraction, this is called _____ the denominator. | |
| 2. Simplify completely. $$\frac{\sqrt{40x^3}}{\sqrt{10x}} = \sqrt{\frac{40x^3}{10x}} = \sqrt{4x^2} = 2x$$ | 3. Simplify completely. $$\frac{\sqrt{75\,m^3}}{\sqrt{3\,m^2}}$$ |

| Video 2: Applying the Division Property of Radicals (1:30 min) |
|---|

▶ Play the video and work along.

Simplify. Assume $x$ and $y$ represent positive real numbers.

$$\sqrt{\frac{x^6}{y^{12}}}$$

⏸ Pause the video and try this yourself.

$$\sqrt{\frac{18\,y^5}{49}}$$

▶ Play the video and check your answer.

| Review and complete | You try |
|---|---|
| 4. Multiply and simplify the result. Assume that all variables represent positive real numbers. | 5. Multiply and simplify the result. Assume that all variables represent positive real numbers. |

$$\sqrt{\frac{200x^3}{81y^2}} = \frac{\sqrt{200x^3}}{\sqrt{81y^2}} = \frac{10x\sqrt{2x}}{9y}$$  Simplify both radicals, numerator and denominator.

$$\sqrt{\frac{16y}{25x^4}}$$

## Video 3: Applying the Division Property of Radicals  (1:24 min)

▶ Play the video and work along.

Simplify. Assume $z$ represents a positive real number.

$$\frac{\sqrt{11z^3}}{\sqrt{z}}$$

$$\frac{\sqrt[3]{24}}{\sqrt[3]{3}}$$

| Review and complete | You try |
|---|---|
| 6. The division property can be applied to two radicals as long as they have the same _____. | |

| 7. Simplify. Assume that all variables represent positive real numbers. | 8. Simplify. Assume that all variables represent positive real numbers. |
|---|---|

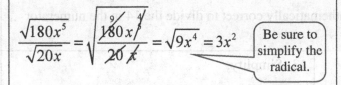

$$\frac{\sqrt{180x^5}}{\sqrt{20x}} = \sqrt{\frac{180x^5}{20x}} = \sqrt{9x^4} = 3x^2$$  Be sure to simplify the radical.

$$\frac{\sqrt{98y^4}}{\sqrt{2y}}$$

## Concept 3: Rationalizing the Denominator--One Term

## Video 4: Rationalizing the Denominator (1 Term) (3:09 min)

▶ Play the video and work along.

Simplify the radicals. Assume that $x$ represents a positive real number.

$$\frac{7}{\sqrt[3]{x}}$$

$$\frac{y}{\sqrt{5}}$$

Video and Study Guide to accompany Intermediate Algebra, 5th ed., Miller/O'Neill/Hyde
Copyright © 2018 McGraw-Hill Education

| Review and complete | You try |
|---|---|
| 9.  Simplify the radical.<br><br>$$\frac{2x}{\sqrt[3]{5}}$$<br><br>This fraction is equivalent to 1.<br><br>$$\frac{2x}{\sqrt[3]{5^1}} \cdot \frac{\sqrt[3]{5^2}}{\sqrt[3]{5^2}} = \frac{2x\sqrt[3]{5^2}}{\sqrt[3]{5^3}} = \frac{2x\sqrt[3]{25}}{5}$$ | 10.  Simplify the radical.<br><br>$$\frac{x}{\sqrt[3]{7}}$$ |

**Video 5: Rationalizing the Denominator (1 term) (1:41 min)**

▶ Play the video and work along.

Simplify.

$$\frac{14}{\sqrt{7}}$$

| Review and complete | You try |
|---|---|
| 11.  True or False: In the previous example, it is mathematically correct to divide the 14 in the numerator with the 7 in the denominator. | |
| 12.  Simplify.<br><br>$$\frac{5}{\sqrt{10}}$$<br><br>$$= \frac{5}{\sqrt{10}} \cdot \frac{\sqrt{10}}{\sqrt{10}}$$<br><br>$$= \frac{5\sqrt{10}}{\sqrt{10^2}}$$<br><br>$$= \frac{\overset{1}{\cancel{5}}\sqrt{10}}{\underset{2}{\cancel{10}}}$$<br><br>$$= \frac{\sqrt{10}}{2}$$ | 13.  Simplify.<br><br>$$\frac{8}{\sqrt{2}}$$ |

## Video 6: Removing a Fraction from a Radicand (1:35 min)

▶ Play the video and work along.

Simplify. Assume that $d$ represents a positive real number.

$$\sqrt{\dfrac{d}{18}}$$

| Review and complete | You try |
|---|---|
| 14. Simplify. Assume that $x$ represents a positive real number. | 15. Simplify. Assume that $y$ represents a positive real number. |
| $$\sqrt{\dfrac{3x}{98}}$$ | $$\sqrt{\dfrac{5y}{24}}$$ |

For the Review and complete:

$$= \frac{\sqrt{3x}}{\sqrt{7^2 \cdot 2}}$$

$$= \frac{\sqrt{3x}}{7\sqrt{2}}$$

$$= \frac{\sqrt{3x}}{7\sqrt{2}} \cdot \frac{\sqrt{2}}{\sqrt{2}}$$

$$= \frac{\sqrt{3x \cdot 2}}{7 \cdot 2}$$

Do not divide these factors of 2.

$$= \frac{\sqrt{6x}}{14}$$

## Video 7: Simplifying a Radical That Contains a Fraction (2:40 min)

▶ Play the video and work along.

Simplify the radical. Assume that $a$ and $b$ represent positive real numbers.

$$\sqrt[3]{\dfrac{16a^7}{b^4}}$$

| Review and complete | You try |
|---|---|
| 16. Simplify the radical. Assume that variables represent positive real numbers. | 17. Simplify the radical. Assume that variables represent positive real numbers. |

16. Simplify the radical. Assume that variables represent positive real numbers.

$$\sqrt[3]{\frac{81y^5}{x}}$$

$$= \frac{\sqrt[3]{81y^5}}{\sqrt[3]{x}} \cdot \frac{\sqrt[3]{x^2}}{\sqrt[3]{x^2}}$$

$$= \frac{\sqrt[3]{81x^2y^5}}{\sqrt[3]{x^3}}$$

$$= \frac{\sqrt[3]{3^4 x^2 y^5}}{x}$$

$$= \frac{3y\sqrt[3]{3x^2y^2}}{x}$$

17. Simplify the radical. Assume that variables represent positive real numbers.

$$\sqrt[3]{\frac{24y}{x^8}}$$

## Video 8: Rationalizing the Denominator (1 Term) (3:01 min)

▶ Play the video and work along.

Simplify the radical. Assume that $x$, $y$, and $z$ represent positive real numbers.

$$\frac{6}{\sqrt[4]{4xy^2z^3}}$$

| Review and complete | You try |
|---|---|
| 18. Simplify the radical. Assume that $x$, $y$, and $z$ represent positive real numbers. | 19. Simplify the radical. Assume that $x$, $y$, and $z$ represent positive real numbers. |

$$\frac{2}{\sqrt[3]{4xy^4z^2}}$$

$$= \frac{2}{\sqrt[3]{4x^1y^4z^2}} \cdot \frac{\sqrt[3]{2x^2y^2z^1}}{\sqrt[3]{2x^2y^2z^1}}$$

$$= \frac{2\sqrt[3]{2x^2y^2z}}{\sqrt[3]{8x^3y^6z^3}}$$

$$= \frac{\cancel{2}\sqrt[3]{2x^2y^2z}}{\cancel{2}xy^2z}$$

$$= \frac{\sqrt[3]{2x^2y^2z}}{xy^2z}$$

Notice there are no powers 3 or greater in the radicands.

$$\frac{3}{\sqrt[5]{9xyz^2}}$$

## Concept 4: Rationalizing the Denominator--Two terms

**Video 9: Rationalizing the Denominator (2 Terms)  (3:58 min)**

▶ Play the video and work along.

Simplify the radical.    $\dfrac{2}{\sqrt{5}-3}$

| Review and complete | You try |
|---|---|

20. When a denominator of an expression has two terms in which one or both has a square root, we can rationalize the denominator by multiplying the numerator and denominator by the _____ of the denominator.

**21. Simplify the radical.**

$$\frac{5}{5-\sqrt{17}}$$

$$=\frac{5}{\left(5-\sqrt{17}\right)}\cdot\frac{\left(5+\sqrt{17}\right)}{\left(5+\sqrt{17}\right)}$$

$$=\frac{5\left(5+\sqrt{17}\right)}{5^2-\left(\sqrt{17}\right)^2}$$

$$=\frac{25+5\sqrt{17}}{25-17}$$

$$=\frac{25+5\sqrt{17}}{8}$$

Notice there are no common factors in numerator and denominator.

**22. Simplify the radical.**

$$\frac{1}{\sqrt{15}-4}$$

---

## Video 10: Rationalizing the Denominator (2 Terms) (3:26 min)

▶ Play the video and work along.

Simplify the radical. Assume that $t$ represents a positive real number,

$$\frac{\sqrt{t}-3\sqrt{7}}{\sqrt{7}+t}$$

| Review and complete | You try |
|---|---|
| 23. Simplify the radical. Assume that $y$ represents a positive real number. | 24. Simplify the radical. |

23. Simplify the radical. Assume that $y$ represents a positive real number.

$$\frac{1-\sqrt{y}}{\sqrt{2y}+\sqrt{5}}$$

$$=\frac{\left(1-\sqrt{y}\right)}{\left(\sqrt{2y}+\sqrt{5}\right)}\cdot\frac{\left(\sqrt{2y}-\sqrt{5}\right)}{\left(\sqrt{2y}-\sqrt{5}\right)}$$

$$=\frac{\sqrt{2y}-\sqrt{5}-\sqrt{2y^2}+\sqrt{5y}}{2y-5}$$

$$=\frac{\sqrt{2y}-\sqrt{5}-y\sqrt{2}+\sqrt{5y}}{2y-5}$$

24. Simplify the radical.

$$\frac{5\sqrt{2}-\sqrt{5}}{5\sqrt{2}+\sqrt{5}}$$

## Concept 1: Solutions to Radical Equations

| Video 1: Introduction to Radical Equations (3:56 min) |
|---|

▶ Play the video and work along.

Solve the equations.

$2x - 3 = 7$                                        $\sqrt{x} - 3 = 7$

| Review and complete | You try |
|---|---|

1. The inverse operation of taking the square root of an expression is to _____ the expression.

2. Solve the equation.

$$\sqrt{m} + 4 = 9$$
$$\sqrt{m} + 4 - 4 = 9 - 4$$
$$\sqrt{m} = 5$$
$$\left(\sqrt{m}\right)^2 = (5)^2$$
$$m = 25$$

The solution checks, so the solution set is {25}.

3. Solve the equation.

$$1 + \sqrt{y} = 4$$

## Concept 2: Solving Radical Equations Involving One Radical

| Video 2: Procedure to Solve a Radical Equations (2:20 min) |
|---|

▶ Play the video and work along.

Solve the equation.        $\sqrt{p - 4} + 7 = 10$

Video and Study Guide to accompany Intermediate Algebra, 5ᵗʰ ed., Miller/O'Neill/Hyde
Copyright © 2018 McGraw-Hill Education

| Review and complete | You try |
|---|---|
| 4.  Solve the equation. | 5.  Solve the equation. |

4.  Solve the equation.

$$\sqrt{x+5}-4=1 \qquad \text{Add 4 to each side.}$$
$$\left(\sqrt{x+5}\right)^2 = \left(5\right)^2 \qquad \text{Square each side of the equation.}$$
$$x+5=25 \qquad \text{Simplify.}$$
$$x=20 \qquad \text{Subtract 5 from each side.}$$

The solution checks so the solution set is {20}.

5.  Solve the equation.

$$\sqrt{5y+1}=4$$

## Video 3: Solving an Equation where the Variable is Raised to a Rational Exponent (1:49 min)

▶ Play the video and work along.

Solve.   $(n-5)^{1/3}+9=5$

| Review and complete | You try |
|---|---|

6.  In equations where a cube root is isolated, raise each side to the power of _____ to solve the equation.

7.  Solve.

$$5=1+(x+2)^{1/3}$$
$$5=1+\sqrt[3]{x+2}$$
$$4=\sqrt[3]{x+2}$$
$$\left(4\right)^3 = \left(\sqrt[3]{x+2}\right)^3$$
$$64=x+2$$
$$x=62$$

> Be sure to check solutions to all radical equations.

Check:
$$5=1+(62+2)^{1/3}$$
$$5=1+\sqrt[3]{64}$$
$$4=\sqrt[3]{64}$$
$$4=4$$

The solution checks, so the solution set is {62}.

8.  Solve.

$$(8+3a)^{1/2}-1=4$$

Video and Study Guide to accompany Intermediate Algebra, 5<sup>th</sup> ed., Miller/O'Neill/Hyde
Copyright © 2018 McGraw-Hill Education

## Video 4: Solving a Radical Equation with No Solution (2:01 min)

▶ Play the video and work along.

Solve.    $8 + \sqrt{v+3} = 6$

| Review and complete | You try |
|---|---|
| 9.  Solve.<br><br>$$6 + \sqrt{9+x} = 3$$<br>$$\sqrt{9+x} = -3$$<br>$$\left(\sqrt{9+x}\right)^2 = \left(-3\right)^2$$<br>$$9+x = 9$$<br>$$x = 0$$<br>Check:<br>$$6 + \sqrt{9+0} = 3$$<br>$$6 + 3 \neq 3$$<br><br>The solution does not check. So, the solution set is the empty set, or { }. | 10. Solve<br><br>$$5 = \sqrt{2x+1} + 10$$ |

## Video 5: Solving a Radical Equation (2:37 min)

▶ Play the video and work along.

Solve the equation.    $2 = 5 + \sqrt[3]{2m+1}$

| Review and complete | You try |
|---|---|

**11.** True or False: Potential solutions that are negative *never* "check" in radical equations.

| | |
|---|---|
| **12.** Solve.<br><br>$$\sqrt[3]{y+18}+6=1$$<br>$$\sqrt[3]{y+18}=-5$$<br>$$\left(\sqrt[3]{y+18}\right)^3=(-5)^3$$<br>$$y+18=-125$$<br>$$y=-143$$<br><br>The solution checks. The solution set is $\{-143\}$. | **13.** Solve.<br><br>$$\sqrt[3]{3x+2}+9=7$$ |

**Video 6: Solving a Radical Equation in which One Potential Solution Does Not Check (3:31 min)**

▶ Play the video and work along.

Solve.   $\sqrt{2x+6}+3=x+2$

Recall
$$(a-b)^2=a^2-2ab+b^2$$

| Review and complete | You try |
|---|---|

| | |
|---|---|
| **14.** Solve.<br><br>$$\sqrt{m+14}-m=2$$<br>$$\sqrt{m+14}=m+2$$<br>$$\left(\sqrt{m+14}\right)^2=(m+2)^2$$<br>$$m+14=m^2+4m+4$$<br>$$0=m^2+3m-10$$<br>$$0=(m+5)(m-2)$$<br>$$m+5=0 \text{ or } m-2=0$$<br>$$m=-5 \text{ or } m=2$$ | **15.** Solve.<br><br>$$\sqrt{4-5a}-a=4$$ |

Check: $-5$

$\sqrt{(-5)+14}-(-5)=2$

$\sqrt{9}+5=2$

$8 \neq 2$

Check: 2

$\sqrt{(2)+14}-(2)=2$

$\sqrt{16}-2=2$

$2=2$

Only 2 checks in the original equation, so the solution set is $\{2\}$.

## Concept 3: Solving Radical Equations Involving More than One Radical

### Video 7: Solving a Radical Equation Containing Two Radicals (2:49 min)

▶ Play the video and work along.

Solve the equation. $(3x+1)^{1/4}-(2x+5)^{1/4}=0$

### Review and complete

16. Solve the equation.

$$\left(x+4\right)^{1/4}-\left(2x-5\right)^{1/4}=0$$

$$\sqrt[4]{x+4}-\sqrt[4]{2x-5}=0$$

$$\left(\sqrt[4]{x+4}\right)^{4}=\left(\sqrt[4]{2x-5}\right)^{4}$$

$$x+4=2x-5$$

$$9=x$$

The solution checks, so the solution set is $\{9\}$.

### You try

17. Solve the equation.

$$(3v+6)^{1/4}-(7v-6)^{1/4}=0$$

## Video 8: Solving a Radical Equation that Requires Squaring Both Sides Twice (4:50 min)

▶ Play the video and work along.

Solve.  $\sqrt{3c+1}-\sqrt{c-1}=2$

| Review and complete | You try |
|---|---|
| 18. Solve. | 19. Solve. |

**Review and complete**

18. Solve.

$\sqrt{6m+7}-\sqrt{3m+3}=1$

> Carefully square this binomial.

$\left(\sqrt{6m+7}\right)^2 = \left(\sqrt{3m+3}+1\right)^2$

$6m+7 = \left(\sqrt{3m+3}\right)^2 + 2\left(\sqrt{3m+3}\right)(1)+(1)^2$

$6m+7 = 3m+3+2\sqrt{3m+3}+1$

$6m+7 = 3m+4+2\sqrt{3m+3}$

$3m+3 = 2\sqrt{3m+3}$

$(3m+3)^2 = \left(2\sqrt{3m+3}\right)^2$

$9m^2+18m+9 = 4(3m+3)$

$9m^2+18m+9 = 12m+12$

$9m^2+6m-3 = 0$

$3(3m^2+2m-1)=0$

$3(3m-1)(m+1)=0$

$m=\dfrac{1}{3}, m=-1$

The solution set is $\left\{\dfrac{1}{3},-1\right\}$.

**You try**

19. Solve.

$\sqrt{2y+6}=\sqrt{7-2y}+1$

## Concept 4: Applications of Radical Equations and Functions

| Video 9: Solving a Formula for a Different Variable (1:02 min) |
|---|

▶ Play the video and work along.

Solve the equation for $a^2$.     $c = \sqrt{25 - a^2 - b^2}$

| Review and complete | You try |
|---|---|
| 20. Solve the equation for $a^2$. | 21. Solve the equation for $h^2$. |
| $$c = \sqrt{100 + a^2 b^2}$$ $$(c)^2 = \left(\sqrt{100 + a^2 b^2}\right)^2$$ $$c^2 = 100 + a^2 b^2$$ $$\frac{c^2 - 100}{b^2} = \frac{a^2 b^2}{b^2}$$ $$a^2 = \frac{c^2 - 100}{b^2}$$ | $$r = \pi\sqrt{r^2 + h^2}$$ |

| Video 10: An Application Involving a Radical Function (3:39 min) |
|---|

▶ Play the video and work along.

A storm off the coast of South Carolina reached tropical storm intensity at 6 a.m. and grew in intensity for 16 hr. Ricardo lives in Charleston and recorded the average wind speed at his house starting at 6:00 a.m. He modeled the wind speed as a function of time according to the function:

$w(t) = 40 + 6\sqrt{t}$ where $w(t)$ is the wind speed in mph and $t$ is the time in hours since 6:00 a.m. and $0 \le t \le 16$.

Evaluate $w(4)$ and interpret its meaning in context.

Evaluate $w(16)$ and interpret its meaning in context.

Determine the time at which the storm will reach a wind speed of 58 mph.

| Review and complete | You try |
| --- | --- |
| 22. On a certain surface, the speed $s(x)$ in mph of a car before the brakes are applied can be approximated from the length of its skid marks $x$ in feet by the function $$s(x) = 4.5\sqrt{x} \text{ where } x \ge 0.$$ | 23. On a certain surface, the speed $s(x)$ in mph of a car before the brakes are applied can be approximated from the length of its skid marks $x$ in feet by the function $$s(x) = 3.2\sqrt{x} \text{ where } x \ge 0.$$ |
| a. Find $s(49)$ and interpret its meaning in context. $$s(49) = 4.5\sqrt{49} = 4.5 \cdot 7 = 31.5$$ The car was traveling 31.5 mph. | a. Find $s(100)$ and interpret its meaning in context. |
| b. Find the length of skid mark that would indicate a speed of 90 mph. $$s(x) = 4.5\sqrt{x}$$ $$90 = 4.5\sqrt{x}$$ $$20 = \sqrt{x}$$ $$(20)^2 = \left(\sqrt{x}\right)^2$$ $$400 = x$$ The skid mark would be 400 ft long. | b. Find the length of skid mark that would indicate a speed of 80 mph. |

## Concept 1: Definition of $i$

| Video 1: Definition of the Imaginary Number, $i$ (2:31 min) |
|---|

▶ Play the video and work along.

> **DEFINITION The Imaginary Number, $i$**
>
> 1.  $i = \sqrt{-1}$
> 2.  $\sqrt{-b} = i\sqrt{b}$ provided that $b$ is a real number and $b > 0$ .

Simplify the expressions in terms of $i$.

$\sqrt{-4}$ $\qquad\qquad$ $\sqrt{-9}$ $\qquad\qquad$ $\sqrt{-12}$

⏸ Pause the video and try this yourself.

$\sqrt{-13}$

▶ Play the video and check your answer.

| Review and complete | You try |
|---|---|
| 1.  The number $i$ is equal to the square root of _____ . | |
| 2.  Simplify the expressions in terms of $i$. <br><br> a. $\sqrt{-5} = i\sqrt{5}$ <br><br><br> b. $\sqrt{-50} = i\sqrt{50} = i\sqrt{2 \cdot 5^2} = 5i\sqrt{2}$ | 3.  Simplify the expressions in terms of $i$. <br><br> a. $\sqrt{-7}$ <br><br><br> b. $\sqrt{-48}$ |

| Video 2: Simplifying a Product of Imaginary Numbers (2:09 min) |
|---|

▶ Play the video and work along.

Simplify the expression.

$\sqrt{-16} \cdot \sqrt{-4}$

**Video 3: Simplifying a Product of Radicals (2:32 min)**

▶ Play the video and work along.

Simplify the expressions.

$$\sqrt{-6} \cdot \sqrt{-15}$$

⏸ Pause the video and try these yourself.

$$\frac{\sqrt{-121}}{\sqrt{-81}}$$

$$\sqrt[3]{-1} \cdot \sqrt[3]{-8}$$

▶ Play the video and check your answer.

| Review and complete | You try |
|---|---|
| 4. When working with radicals with negative radicands, do not apply the multiplication or division property of radicals until you have rewritten each radical in terms of _____. | |

5.  Simplify the expressions.

> Rewrite in terms of $i$.

a. $\sqrt{-35} \cdot \sqrt{-5}$

$$\sqrt{-35} \cdot \sqrt{-5} = i\sqrt{35} \cdot i\sqrt{5}$$
$$= i\sqrt{35} \cdot i\sqrt{5}$$
$$= i^2\sqrt{175}$$
$$= i^2 \cdot 5\sqrt{7}$$
$$= -5\sqrt{7}$$

b. $\dfrac{\sqrt{-100}}{\sqrt{-4}} = \dfrac{\cancel{i}\sqrt{100}}{\cancel{i}\sqrt{4}} = \dfrac{10}{2} = 5$

6.  Simplify the expressions.

a. $\sqrt{-4} \cdot \sqrt{-8}$

b. $\dfrac{\sqrt{-8}}{\sqrt{-4}}$

## Concept 2: Powers of $i$

### Video 4: Introduction to the Powers of $i$ (2:59 min)

▶ Play the video and work along.

| Observations: | |
|---|---|
| $i = i$ | $i^9 = i$ |
| $i^2 = -1$ | $i^{10} = -1$ |
| $i^3 = -i$ | $i^{11} = -i$ |
| $i^4 = 1$ | $i^{12} = 1$ |
| $i^5 = i$ | $i^{13} = i$ |
| $i^6 = -1$ | $i^{14} = -1$ |
| $i^7 = -i$ | $i^{15} = -i$ |
| $i^8 = 1$ | $i^{16} = 1$ |

### Video 5: Simplifying Powers of $i$ (2:55 min)

▶

Play the video and work along.

Simplify the powers of $i$.

$i^{44}$                                          $i^{22}$

⏸ Pause the video and try this yourself.

$i^{17}$                                          $i^{23}$

▶ Play the video and check your answers.

| Review and complete | You try |
|---|---|
| • $i^{even} = 1$ or $-1$     * **If the exponent is a multiple of 4, the result is 1.** *  **If the exponent is even but not a multiple of 4, the result is -1.** | |
| • $i^{odd} = i$ or $-i$ | |
| 7.  Simplify. Show work.     $i^5 = i^4 i = 1 \cdot i = i$ | 8.  Simplify. Show work.     $i^9$ |
| 9.  True or False: All even powers of $i$ are equal to 1. ||
| 10. Simplify the power of $i$.     $i^{83} = i^{80} \cdot i^2 \cdot i^1 = (1)(-1)(i) = -i$      Any power of $i$ that is a multiple of 4 is equal to 1. | 11. Simplify the power of $i$.     $i^{52}$ |

### Concept 3: Definition of a Complex Number

| Video 6: Definition of a Complex Number (2:52 min) |
|---|

▶ Play the video and work along.

**Definition of a Complex Number**
• A **complex number** is a number in the form $a + bi$, where $a$ and $b$ are real numbers and $i$ is the imaginary unit, the square root of -1.

• For a complex number, $a + bi$, the value $a$ is the **real part** and $b$ is called the **imaginary part**.

- If $b = 0$, then the complex number $a + bi$ is a real number.

- If $b \neq 0$, then $a + bi$ is an **imaginary number**.

- The complex numbers $a + bi$ and $a - bi$ are **conjugates**.

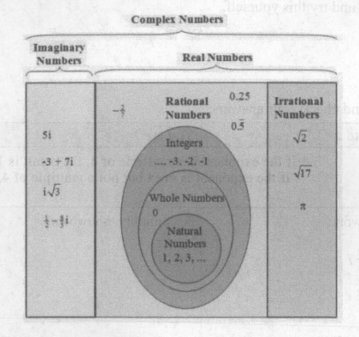

| Review and complete | You try |
|---|---|
| 12. True or False. | 13. True or False. |
| a. $5 + 8i$ is a complex number.<br>True | a. $-2 + 8i$ has imaginary part 8. |
| b. $5 + 8i$ has real part 5.<br>True | b. $-2 + 8i$ has a conjugate of $-2 - 8i$. |
| c. $5 + 8i$ is a real number.<br>False. Complex numbers are not real numbers. | c. $-2 + 8i$ is an imaginary number. |

## Concept 4: Addition, Subtraction, and Multiplication of Complex Numbers

### Video 7: Adding and Subtracting Complex Numbers (3:12 min)

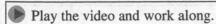 Play the video and work along.

Perform the indicated operation. Write the answers in the form $a+bi$.

$(4-5i)+(8+9i)$     $(1.2+4.7i)-(6.5-3.6i)$

$$\left(\frac{4}{3}-\frac{3}{2}i\right)-\left(\frac{1}{6}+\frac{5}{4}i\right)+\left(\frac{2}{3}+i\right)$$

| Review and complete | You try |
|---|---|
| 14. Perform the indicated operation. Write the answer in the form $a+bi$.<br><br>$(1.6-2i)-(2+4.2i)$<br>$\quad=1.6-2i-2-4.2i$<br>$\quad=(1.6-2)+(-2-4.2)i$<br>$\quad=-0.4-6.2i$ | 15. Perform the indicated operation. Write the answer in the form $a+bi$.<br><br>$(21-i)+(3.7+2i)$ |

### Video 8: Multiplying Complex Numbers  (3:13 min)

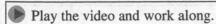 Play the video and work along.

Perform the indicated operation. Write the answers in the form $a+bi$.

$(3-8i)(5+2i)$     $(5-2i)(5+2i)$     $(4+3i)^2$

| Review and complete | You try |
|---|---|

16. True or False? $(5+i)^2 = 25+(-1) = 24$

17. Perform the indicated operation. Write the answer in the form $a+bi$.

$$(3-i)(3+i) = (3)^2 - (i)^2$$
$$= 9-(-1)$$
$$= 10$$

> The product of conjugates is a difference of squares.

18. Perform the indicated operation. Write the answer in the form $a+bi$.

$$(3-2i)^2$$

## Concept 5: Division and Simplification of Complex Numbers

**Video 9: Dividing Complex Numbers (2:37 min)**

▶ Play the video and work along.

Divide. Write the answer in the form $a+bi$.   $\dfrac{2+4i}{3-5i}$

| Review and complete | You try |
|---|---|

19. When we divide complex numbers, we **do not** just divide the real parts and the imaginary parts. We multiply the numerator and denominator by the _____ of the denominator.

20. Divide. Write the answer in the form $a+bi$.

$$\dfrac{4-i}{2+9i}$$

21. Divide. Write the answer in the form $a+bi$.

$$\dfrac{5-2i}{10+i}$$

$$= \frac{4-i}{2+9i} \cdot \frac{(2-9i)}{(2-9i)}$$

Multiply by the conjugate of the denominator.

$$= \frac{8-36i-2i+9i^2}{4-81i^2}$$

$$= \frac{8-38i+9(-1)}{4-81(-1)}$$

Replace $i^2$ with -1.

$$= \frac{-1-38i}{85}$$

$$= -\frac{1}{85} - \frac{38}{85}i$$

## Video 10: Simplify a Radical Expression (2:05 min)

▶ Play the video and work along.

Simplify.  $\dfrac{15 + \sqrt{-75}}{10}$

| Review and complete | You try |
|---|---|
| 22. Simplify. | 23. Simplify. $\dfrac{25 + \sqrt{-150}}{70}$ |

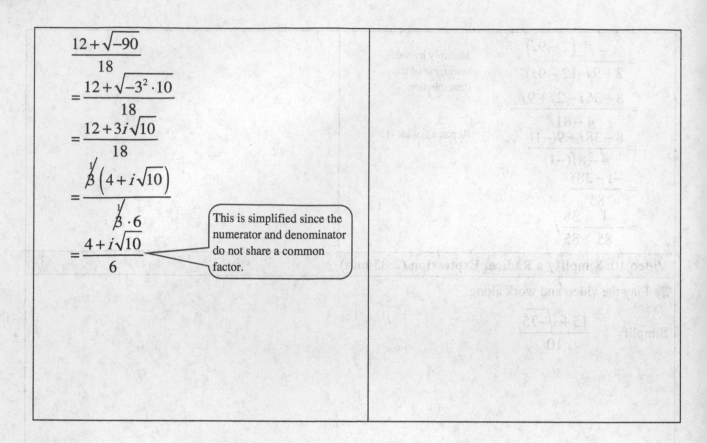

$$\frac{12+\sqrt{-90}}{18}$$

$$=\frac{12+\sqrt{-3^2\cdot 10}}{18}$$

$$=\frac{12+3i\sqrt{10}}{18}$$

$$=\frac{\cancel{3}\left(4+i\sqrt{10}\right)}{\cancel{3}\cdot 6}$$

$$=\frac{4+i\sqrt{10}}{6}$$

This is simplified since the numerator and denominator do not share a common factor.

**Answers**: Section 6.1

1. $b^2 = a$; two

3a. 7 and $-7$

3b. no real number

3c. 4 and $-4$

5a. 100

5b. $-\dfrac{1}{2}$

5c. 0.5

6. odd

8. $-5$

10a. $-1$

10b. 10

10c. $-0.2$

11. nonnegative

13a. $|y|$

13b. $-4$

13c. $|m+n|$

15a. $2a^3$

15b. $y^4$

17. 26 cm

18. positive

20. 20 mi.

22.

| $x$ | $f(x)$ |
|----|----|
| $-1$ | 1 |
| 0 | 2 |
| 1 | 3 |
| 2 | 3.26 |

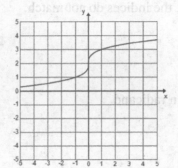

24. Domain $(-\infty, \infty)$; No, this is not the graph.    25. nonnegative

27. Domain $[3, \infty)$

| $x$ | $f(x)$ |
|----|----|
| 2 | undefined |
| 3 | 0 |
| 4 | $-1$ |
| 5 | $-1.41$ |

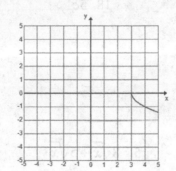

**Answers**: Section 6.2

1. $\sqrt[n]{a}$

3a. 2

3b. $\dfrac{1}{2}$

4. base

6. $-243$

8a. $5\sqrt{a}$

8b. $4x^{1/7}$

10a. $y^{5/2}$

10b. $x^{58}$

10c. $\dfrac{a^4}{b^2}$

11. 1

13a. $\dfrac{1}{2}$

13b. 6

15. $x$

17. $\dfrac{x^3 y^{3/2}}{27}$

19. $\dfrac{b^8}{a^{19/3}}$    21a. 512    21b. $\dfrac{1}{16}$    21c. 16

**Answers:** Section 6.3

1. index    3. No, the indices do not match.    5a. fraction in radicand

5b. perfect square factor in radicand    5c. radical in denominator of fraction

6. positive    8a. $c^3$    8b. $m^5\sqrt{m}$    10. $mn^2\sqrt{mp}$    12. $5\sqrt{10}$

14. $3x^3y^4\sqrt{5y}$    16. $10x^2y^3\sqrt[3]{10xy}$    18. $5c^4$    20. $\dfrac{11\sqrt{11}}{6}$

**Answers:** Section 6.4

1. index; radicand    3. $-\sqrt{7}$    5. $8\sqrt[3]{x}+5\sqrt{x}$    6. radicands

8. $\sqrt{3t}$    10. $7\sqrt{5}+9\sqrt{10}$    11. Yes, to preserve the order of operations

13. $(9y+8xy)\sqrt{5y}$

**Answers:** Section 6.5

1. index     3. $7\sqrt{2x}$     5. $30x^3\sqrt{6}$     7. $6t - t\sqrt{3} + 6\sqrt{t}$

9. $-7\sqrt{30}$     10. distributive     12. $6 + \sqrt{x} - 2\sqrt{y} - x - \sqrt{xy}$

14. $3x + 67$     15. False     17. $9m - 24\sqrt{m} + 16$     18. False

20. $x + 10 - 6\sqrt{x+1}$     22. 99     24. $\sqrt[4]{20}$

**Answers**: Section 6.6

1. rationalizing     3. $5\sqrt{m}$     5. $\dfrac{4\sqrt{y}}{5x^2}$     6. index     8. $7y\sqrt{y}$

10. $\dfrac{x\sqrt[3]{49}}{7}$     11. False     13. $4\sqrt{2}$     15. $\dfrac{\sqrt{30y}}{12}$     17. $\dfrac{2\sqrt[3]{3xy}}{x^3}$

19. $\dfrac{\sqrt[5]{27x^4y^4z^3}}{xyz}$     20. Conjugate     22. $-\sqrt{15} - 4$     24. $\dfrac{11 - 2\sqrt{10}}{9}$

**Answers**: Section 6.7

1. square     3. $\{9\}$     5. $\{3\}$     6. three     8. $\left\{\dfrac{17}{3}\right\}$

10. no solution     11. False     13. $\left\{-\dfrac{10}{3}\right\}$     15. $\{-1\}$     17. $\{3\}$

19. $\left\{\dfrac{3}{2}\right\}$     21. $h^2 = \dfrac{r^2}{\pi^2} - r^2$ or $h^2 = \dfrac{r^2 - \pi^2 r^2}{\pi^2}$

23a. $s(100) = 32$ mph ; The car was traveling 32 mph.     23b. The skid marks would be 625 ft.

**Answers**: Section 6.8

1. $-1$       3a. $i\sqrt{7}$   3b. $4i\sqrt{3}$     4. $i$        6a. $-4\sqrt{2}$

6b. $\sqrt{2}$      8. $i$     9. False       11. 1        13a. True

13b. True      13c. True     15. $24.7 + i$       16. False      18. $5 - 12i$

19. conjugate      21. $\dfrac{48}{101} - \dfrac{25i}{101}$      23. $\dfrac{5}{14} + \dfrac{\sqrt{6}}{14}i$

## Concept 1: Solving Quadratic Equations Using the Square Root Property

| Video 1: Introduction to the Square Root Property  (2:26 min) |
| --- |
| ▶ Play the video and work along. |
| Solve the equation.<br>$x^2 = 25$ |

| Review and complete | You try |
| --- | --- |
| 1.  The square root property states that the solutions to an equation of the form $x^2 = k$, where $k$ is a real number, are _____ and _____. | |
| 2.  Solve the equation.<br><br>$x^2 = 49$<br>$x = \pm\sqrt{49}$<br>$x = \pm 7$<br><br>The solution set is $\{-7, 7\}$ or $\{\pm 7\}$. | 3.  Solve the equation.<br><br>$x^2 = 81$ |

| Video 2: Solving Quadratic Equations Using the Square Root Property (2:56 min) |
| --- |
| ▶ Play the video and work along. |
| Solve the equations.<br><br>$x^2 = 17 \qquad\qquad\qquad\qquad y^2 = 50$ |
| ⏸ Pause the video and try this yourself.<br><br>$z^2 + 3 = 0 \qquad\qquad\qquad\qquad 2w^2 + 98 = 0$ |

▶ Play the video and check your answer.

| Review and complete | You try |
|---|---|
| 4.   Solve the equation.<br><br>$x^2 = -90$<br><br>$x = \pm\sqrt{-90}$ 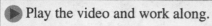 Apply the square root property.<br><br>$x = \pm\sqrt{-1 \cdot 9 \cdot 10}$<br><br>$x = \pm i \cdot 3\sqrt{10}$<br><br>$x = \pm 3i\sqrt{10}$ | 5.   Solve the equation.<br><br>$x^2 - 45 = 0$ |

### Video 3: Solving Quadratic Equations Using the Square Root Property (2:59 min)

▶ Play the video and work along.

Solve the equations.

$(t - 4)^2 = 15$ 

$m^2 + 2m + 1 = 12$

| Review and complete | You try |
|---|---|
| 6.   The square root property was used in the second example above since the left side was a perfect square _____, which we factored as the square of a binomial. | |

7.  Solve the equation.

$$(m+4)^2 = 40$$

> Apply the square root property.

$$(m+4) = \pm\sqrt{40}$$
$$m+4 = \pm 2\sqrt{10}$$
$$m = -4 \pm 2\sqrt{10}$$
$$\{4+2\sqrt{10}, 4-2\sqrt{10}\}$$

8.  Solve the equation.

$$m^2 + 10m + 25 = 7$$

## Concept 2: Solving Quadratic Equations by Completing the Square

### Video 4: Practice Completing the Square (4:14 min)

▶ Play the video and work along.

Perfect square trinomial

$$m^2 + 2m + 1$$

Factored Form

$$(m+1)^2$$

Perfect square trinomial

$$x^2 - 14x + 49$$

Factored Form

$$(x-7)^2$$

A trinomial of the form $x^2 + bx + n$ is a perfect square trinomial if the constant term, $n$, is equal to the square of one-half the middle coefficient.

$$x^2 + bx + n$$

$$n = \left(\frac{1}{2}b\right)^2$$

Determine the value of n so that the trinomial is a perfect square trinomial. Then factor the result.

$$x^2 + 18x + n$$ $$y^2 - 24y + n$$

⏸

Video and Study Guide to accompany Intermediate Algebra, 5th ed., Miller/O'Neill/Hyde
Copyright © 2018 McGraw-Hill Education

Pause the video and try this yourself.

$$t^2 + 13t + n$$

$$z^2 - \frac{5}{3}z + n$$

▶ Play the video and check your answers.

| Review and complete | You try |
|---|---|
| 9. Determine the value of $n$ so that the trinomial is a perfect square trinomial. Then factor the result.<br><br>$x^2 + 9x + n$<br><br>$n = \left(\frac{1}{2} \cdot 9\right)^2 = \left(\frac{9}{2}\right)^2 = \frac{81}{4}$<br><br>$x^2 + 9x + \frac{81}{4} = \left(x + \frac{9}{2}\right)^2$ | 10. Determine the value of $n$ so that the trinomial is a perfect square trinomial. Then factor the result.<br><br>$x^2 + 3x + n$ |

**Video 5: Solving a Quadratic Equation by Completing the Square and Applying the Square Root Property (3:16 min)**

▶ Play the video and work along.

Solve the equation by completing the square and applying the square root property.

$$x^2 + 8x + 24 = 0$$

**Step 1**: Divide both sides by the leading coefficient.

**Step 2**: Isolate the variable terms on one side of the equation.

**Step 3**: Add the square of one-half the coefficient of the linear term to both sides.

Factor the trinomial.

**Step 4**: Apply the square root property and solve the resulting equation.

**Video 6: Solving a Quadratic Equation by Completing the Square and Applying the Square Root Property (3:33 min)**

▶ Play the video and work along.

Solve the equation by completing the square and using the square root property.

$4x^2 = 20x - 14$

| Review and complete | You try |
|---|---|
| 11. Solve the equations by completing the square and using the square root property.<br><br>a. $x^2 + 10x + 7 = 0$<br>$\qquad x^2 + 10x = -7$<br>$\qquad x^2 + 10x + 25 = -7 + 25$<br>$\qquad (x+5)^2 = 18$<br>$\qquad x + 5 = \pm\sqrt{18}$<br>$\qquad x + 5 = \pm 3\sqrt{2}$<br>$\qquad x = -5 \pm 3\sqrt{2}$<br><br>*25 completes the square. Make sure to add 25 to both sides.* | 12. Solve the equation by completing the square and using the square root property.<br><br>a. $x^2 + 12x - 1 = 0$ |
| b. $4a^2 + 12a - 5 = 0$ | b. $2x^2 + 10x - 3 = 0$ |

$$\frac{4a^2}{4} + \frac{12a}{4} - \frac{5}{4} = \frac{0}{4}$$

$$a^2 + 3a - \frac{5}{4} = 0$$

$$a^2 + 3a = \frac{5}{4}$$

$$a^2 + 3a + \frac{9}{4} = \frac{5}{4} + \frac{9}{4}$$

$$\left(a + \frac{3}{2}\right)^2 = \frac{14}{4}$$

$$\left(a + \frac{3}{2}\right)^2 = \frac{14}{4}$$

$$a + \frac{3}{2} = \pm\sqrt{\frac{14}{4}}$$

$$a = -\frac{3}{2} \pm \frac{\sqrt{14}}{2}$$

## Video 7: Approximating the Solutions to a Quadratic Equation (3:20 min)

▶ Play the video and work along.

Use a calculator to approximate the solutions to the equation.

$$4x^2 = 20x - 14 \qquad\qquad x = \frac{5 \pm \sqrt{11}}{2}$$

Determine the $x$-intercepts of the graph of $f(x) = 4x^2 - 20x + 14$. Then graph the function on a graphing calculator.

| Review and complete | You try |
|---|---|
| 13. Use a calculator to approximate the solutions to the equation $x^2 - 5x - 12 = 0$. Determine the | 14. Use a calculator to approximate the solutions to the equation $x^2 - 2x - 10 = 0$. Determine |

x-intercepts of the graph of $f(x) = x^2 - 5x - 12$. Then graph the function on a graphing calculator.

$$x^2 - 5x - 12 = 0$$

$$x^2 - 5x = 12$$

$$x^2 - 5x + \frac{25}{4} = 12 + \frac{25}{4}$$

$$\left(x - \frac{5}{2}\right)^2 = \frac{73}{4}$$

$$x - \frac{5}{2} = \pm\frac{\sqrt{73}}{2}$$

$$x = \frac{5 \pm \sqrt{73}}{2}$$

$$\frac{5 + \sqrt{73}}{2} \approx 6.77, \frac{5 - \sqrt{73}}{2} \approx -1.77$$

The x-intercepts of the function are the solutions above, approximately $(6.77, 0)$ and $(-1.77, 0)$.

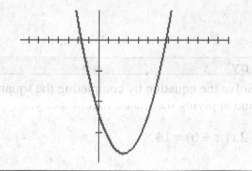

the x-intercepts of the graph of $f(x) = x^2 - 2x - 10$. Then graph the function on a graphing calculator.

| Video 8: Solving a Quadratic Equation by Completing the Square and Applying the Square Root Property (4:09 min) |
| --- |

▶ Play the video and work along.

Solve the equation by completing the square and applying the square root property.

$$2y(y-2)=3+y$$

| Review and complete | You try |
| --- | --- |
| 15. Solve the equation by completing the square and applying the square root property. | 16. Solve the equation by completing the square and applying the square root property. |

**15.**

$$(2x+5)(x-1)=2$$
$$2x^2-2x+5x-5=2$$
$$2x^2+3x-7=0$$
$$\frac{2x^2}{2}+\frac{3x}{2}-\frac{7}{2}=0$$
$$x^2+\frac{3}{2}x+\frac{9}{16}=\frac{7}{2}+\frac{9}{16}$$
$$\left(x+\frac{3}{4}\right)^2=\frac{65}{16}$$
$$x+\frac{3}{4}=\pm\frac{\sqrt{65}}{\sqrt{16}}$$
$$x=-\frac{3}{4}\pm\frac{\sqrt{65}}{4}$$

**16.**

$$2x(x+6)=14$$

## Concept 3: Literal Equations

**Video 9: Using the Square Root Property in an Application (2:17 min)**

▶ Play the video and work along.

The distance $d$ (in feet) that an object falls in $t$ second under acceleration due to gravity $g$ is given by:

$$d = \frac{1}{2}gt^2, \text{ where } t \geq 0$$

Solve the equation for $t$. Do not rationalize the denominator.

Determine the amount of time required for an object to fall 256 ft under the acceleration due to gravity on the Earth $g = 32$ ft/sec$^2$ .

| Review and complete | You try |
|---|---|
| 17. The volume of a can that is 6 in. tall is given by the equation $V = 6\pi r^2$. | 18. The volume of a storage barrel that is 3.2 ft tall is given by the equation $V = 3.2\pi r^2$. |
| a. Solve the equation for $r$. Do not rationalize the denominator.<br><br>$$V = 6\pi r^2$$<br><br>$$\frac{V}{6\pi} = \frac{\cancel{6\pi} r^2}{\cancel{6\pi}}$$<br><br>$$r^2 = \frac{V}{6\pi}$$<br><br>$$r = \sqrt{\frac{V}{6\pi}}$$ | a. Solve the equation for $r$. Do not rationalize the denominator. |
| b. Use the equation to determine the radius of a can that has volume equal to 117.75 in$^3$. Use 3.14 for $\pi$ .<br><br>$$r = \sqrt{\frac{V}{6\pi}} = \sqrt{\frac{117.75}{6 \cdot 3.14}} = \sqrt{6.25} = 2.5 \text{ in.}$$ | b. Use the equation above to determine the radius of a storage barrel that has volume equal to 15.7 ft$^3$. Use 3.14 for $\pi$ . |

**Concepts 1 and 2: Derivation of the Quadratic Formula and Solving Quadratic Equations by Using the Quadratic Formula**

| Video 1: Introduction to the Quadratic Formula  (2:58 min) |
|---|

▶ Play the video and work along.

Solve the equation by using the quadratic formula.

$$2x^2 + 5x - 4 = 0$$

| Review and complete | You try |
|---|---|

1.  Given a quadratic equation, $ax^2 + bx + c = 0$ $(a \neq 0)$, the solutions are:

2.  True or False? The quadratic formula can be used to solve any quadratic equation.

| Review and complete | You try |
|---|---|
| 3.  Solve the equation by using the quadratic formula.<br><br>$x^2 - 6x + 4 = 0$<br><br>$a = \underline{\quad}, b = \underline{\quad}, c = \underline{\quad}$<br><br>$x = \dfrac{-(-6) \pm \sqrt{(-6)^2 - 4(1)(4)}}{2(1)}$<br><br>$x = \dfrac{6 \pm \sqrt{36 - 16}}{2}$<br><br>$x = \dfrac{6 \pm \sqrt{20}}{2} = \dfrac{6 \pm 2\sqrt{5}}{2} = \dfrac{\cancel{2}(3 \pm \sqrt{5})}{\cancel{2}}$ | 4.  Solve the equation by using the quadratic formula.<br><br>$2m^2 + 3m - 7 = 0$<br><br>$a = \underline{\quad}, b = \underline{\quad}, c = \underline{\quad}$ |

The solution set is $\left\{3+\sqrt{5}, 3-\sqrt{5}\right\}$.

## Video 2: Solving a Quadratic Equation Using the Quadratic Formula (3:53 min)

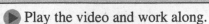 Play the video and work along.

Solve the equation by using the quadratic formula.

$3x(x-6) = x-20$

$$x = \frac{-b \pm \sqrt{b^2 - 4ac}}{2a}$$

| Review and complete | You try |
|---|---|
| 5. When the solutions to a quadratic equation are rational numbers that means the equation can also be solved by _____. | |
| 6. Solve the equation using the quadratic formula. $6x(x-3) - x = 7$ | 7. Solve the equation using the quadratic formula. $5m(m-1) - 2m = -2$ |

$$6x^2 - 19x = 7$$

$$6x^2 - 19x - 7 = 0$$

$$x = \frac{-(-19) \pm \sqrt{(-19)^2 - 4(6)(-7)}}{2(6)}$$

$$x = \frac{19 \pm \sqrt{361 - (-168)}}{12}$$

$$x = \frac{19 \pm \sqrt{529}}{12}$$

$$x = \frac{19 \pm 23}{12}$$

$$x = \frac{19 + 23}{12} = \frac{42}{12} = \frac{7}{3} \text{ or } x = \frac{19 - 23}{12} = \frac{-4}{12} = -\frac{1}{3}$$

The solution set is $\left\{\dfrac{7}{3}, -\dfrac{1}{3}\right\}$.

## Concept 3: Using the Quadratic Formula in an Application

### Video 3: Using the Quadratic Formula in an Application Involving the Pythagorean Theorem (4:47 min)

▶ Play the video and work along.

Martha and Will attend the same college. Martha's house is due north of the school and Will's house is due east of the school. Martha's house is 3 mi further from the school than Will's house. If the distance between Martha's house and Will's house is 9 mi, determine how far Martha lives from the school and how far Will lives from the school.

| Review and complete | You try |
|---|---|
| 8. One leg of a right triangle is 2 in. longer than the other leg, and the hypotenuse is 4 in. Find the lengths of the legs. Round to one decimal place.    $x+2$   4   $x$ | 9. One leg of a right triangle is 4 in. more than the other. The hypotenuse is 12 in. Find the lengths of the legs. Round to one decimal place. |

$$x^2 + (x+2)^2 = 4^2$$

$$x^2 + (x^2 + 4x + 4) = 16$$

$$2x^2 + 4x - 12 = 0$$

$$\frac{2x^2}{2} + \frac{4x}{2} - \frac{12}{2} = 0$$

$$x^2 + 2x - 6 = 0$$

Apply the quadratic formula with $a = 1$, $b = 2$, $c = -6$.

$$x = \frac{-2 \pm \sqrt{(2)^2 - 4(1)(-6)}}{2(1)}$$

$$x = \frac{-2 \pm \sqrt{4 + 24}}{2}$$

$$x = \frac{-2 \pm \sqrt{28}}{2}$$

$$x = \frac{-2 + \sqrt{28}}{2} = 1.7 \text{ or } x = \frac{-2 - \sqrt{28}}{2} = \cancel{3.6}$$

The legs are 1.7 in and 3.7 in.

---

## Video 4: Using the Quadratic Formula in an Application Involving an Object in Free Fall (3:09 min)

▶ Play the video and work along.

A child tosses a ball straight upward with an initial velocity of 60 ft/sec. The height of the ball can be modeled as a function of time by the function,

$$h(t) = -16t^2 + 60t + 4 \quad \text{where } h(t) \text{ is the height and } t \text{ is the time in seconds}$$

Determine the times at which the ball is at a height of 36 ft.

| Review and complete | You try |
|---|---|

10. A child tosses a ball straight upward with an initial velocity of 60 ft/sec. The height of the ball can be modeled as a function of time by the function $h(t) = -16t^2 + 60t + 4$ where $h(t)$ is the height and $t$ is the time in seconds. Determine the time at which the ball is at a height of 2 ft.

$$2 = -16t^2 + 60t + 4$$

$$\frac{16t^2}{2} - \frac{60t}{2} - \frac{2}{2} = 0$$

$$8t^2 - 30t - 1 = 0$$

$$a = 8, b = -30, c = -1$$

$$x = \frac{-(-30) \pm \sqrt{(-30)^2 - 4(8)(-1)}}{2(8)}$$

$$x = \frac{30 \pm \sqrt{932}}{16}$$

$$x = \cancel{-0.3}, x = 3.78$$

The ball is at a height of 2 ft after 3.78 sec.

11. On the moon, an astronaut throws a rock from the deck of a spacecraft at an initial velocity of 2.4 m/sec. The height in meters $t$ seconds after he throws it is $h(t) = -0.8t^2 + 2.4t + 8$. Find the time at which the height of the rock is 6 m. Round to two decimal places.

## Concept 4: Discriminant

**Video 5: Introduction to the Discriminant (1:05 min)**

 Play the video and work along.

---

**FORMULA  The Discriminant**

Given a quadratic equation, $ax^2 + bc + c = 0$ where $a \neq 0$ and $a$, $b$, and $c$ are real numbers. The radicand within the quadratic formula is called the discriminant.

$$x = \frac{-b \pm \sqrt{b^2 - 4ac}}{2a}$$    The **discriminant** is $b^2 - 4ac$.

---

**SUMMARY  Using the Discriminant to Determine the Number and Type of Solutions to a Quadratic Equation**

**Case 1**: If $b^2 - 4ac > 0$ there will be two real solutions.
   a.  If $b^2 - 4ac$ is a perfect square, the solutions will be rational.
   b.  If $b^2 - 4ac$ is not a perfect square, the solutions will be irrational numbers.

**Case 2**: If $b^2 - 4ac < 0$, there will be two imaginary solutions.

**Case 3**: If $b^2 - 4ac = 0$, there will be one rational solutions.

---

| Review and complete | You try |
|---|---|

Video and Study Guide to accompany Intermediate Algebra, 5th ed., Miller/O'Neill/Hyde

| 12. A quadratic equation has a discriminant of –10. Describe the solutions to the quadratic equation. | 13. A quadratic equation has a discriminant of 100. Describe the solutions to the quadratic equation. |
|---|---|
| Since $b^2 - 4ac = -10 < 0$, the equation has two imaginary solutions. | |

## Video 6: Using the Discriminant to Determine the Number and Type of Solutions to a Quadratic Equation (4:14 min)

▶ Play the video and work along.

**Case 1:** If $b^2 - 4ac > 0$ there will be two real solutions.

a. If $b^2 - 4ac$ is a perfect square, the solutions will be rational numbers.

Equation
$2x^2 - 5x - 12 = 0$

Discriminant
$b^2 - 4ac$

Solutions:

$$x = \frac{-(-5) \pm \sqrt{(-5)^2 - 4(2)(-12)}}{2(2)}$$

$$x = \frac{5 \pm \sqrt{121}}{4} = \frac{5 \pm 11}{4} \quad \left\{4, -\frac{3}{2}\right\}$$

b. If $b^2 - 4ac$ is not a perfect square, the solutions will be irrational numbers.

Equation
$2x^2 - 4x + 1 = 0$

Discriminant
$b^2 - 4ac$

Solutions:

$$x = \frac{-(-4) \pm \sqrt{(-4)^2 - 4(2)(1)}}{2(2)}$$

$$x = \frac{4 \pm \sqrt{8}}{4} = \frac{4 \pm 2\sqrt{2}}{4} \quad \left\{\frac{2 + \sqrt{2}}{2}, \frac{2 - \sqrt{2}}{2}\right\}$$

**Case 2:** If $b^2 - 4ac < 0$, there will be two imaginary solutions.

Equation:
$\frac{1}{2}x^2 - 2x + 3 = 0$

Discriminant
$b^2 - 4ac$

Solutions:

$$x = \frac{-(-2) \pm \sqrt{(-2)^2 - 4(\frac{1}{2})(3)}}{2(\frac{1}{2})}$$

$$x = \frac{2 \pm \sqrt{-2}}{1} = 2 \pm i\sqrt{2} \qquad \left\{ 2 + i\sqrt{2}, \; 2 - i\sqrt{2} \right\}$$

**Case 3**: If $b^2 - 4ac = 0$, there will be one rational solutions.

Equation:
$$x(x + 6) = 9$$

Discriminant
$$b^2 - 4ac$$

Solutions:

$$x = \frac{-(6) \pm \sqrt{(6)^2 - 4(1)(9)}}{2(1)}$$

$$x = \frac{-6 \pm \sqrt{0}}{2} = \frac{-6 \pm 0}{2} = \frac{-6}{2} = -3$$

| Review and complete | You try |
|---|---|
| 14. (a) Write the equation $3x(x+15)=0$ in the form $ax^2+bx+c=0$. (b) Find the value of the discriminant. (c) Use the discriminant to find the number and type of solutions. <br><br> (a) $\begin{aligned} 3x(x+15) &= 0 \\ 3x^2+45x &= 0 \end{aligned}$ <br><br> (b) $\begin{aligned} b^2-4ac &= 45^2-4(3)(0) \\ &= 2025 \end{aligned}$ <br><br> (c) The discriminant is a perfect square, so there are two rational solutions. (**Case 1a**). | 15. (a) Write the equation $3x^2=7$ in the form $ax^2+bx+c=0$. (b) Find the value of the discriminant. (c) Use the discriminant to find the number and type of solutions. |

**Video 7: Determining the Number of x-intercepts for a Parabola (Introduction) (1:03 min)**

▶ Play the video and work along.

Determine the x-intercepts of the given function.

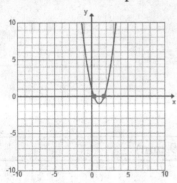

$f(x)=2x^2-4x+1$

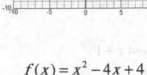

$f(x)=x^2-4x+4$

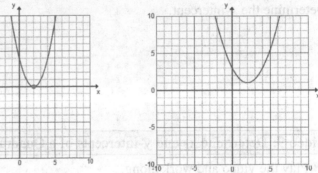

$f(x)=\dfrac{1}{2}x^2-2x+3$

| Review and complete | You try |
|---|---|
| 16. How many x-intercepts are possible for the graph of a quadratic function? Circle all that apply. <br><br> <div align="center">0   1   2   3</div> | |

17. We find the $x$-intercept(s) of a function by setting the $y$-variable equal to _____.

18. The number of real-valued solutions to the quadratic equation that results from setting the $y$-variable equal to zero tells us the number of _____ the graph will have.

## Video 8: Determining the x- and y-intercepts of a Quadratic Function (2:08 min)

▶ Play the video and work along.

For the function defined by $f(x) = \dfrac{1}{2}x^2 - 2x + 3$,

Determine the $x$-intercepts.

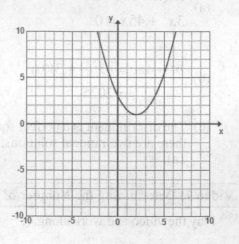

Determine the $y$-intercept.

## Video 9: Determine x- and y-intercepts of a Quadratic Function (3:39 min)

▶ Play the video and work along.

For the function defined by $f(x) = 2x^2 - 4x + 1$,

Determine the $x$-intercepts.

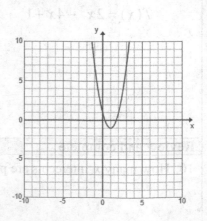

Determine the $y$-intercept.

| Review and complete | You try |
|---|---|
| 19. For the function defined by $g(x) = -x^2 + x - 1$, find the $x$-intercepts and the $y$-intercept. | 20. For the function defined by $g(x) = -5x^2 + x - 15$, find the $x$-intercepts and the $y$-intercept. |

**Review and complete (19):**

$x$-intercepts:

$$0 = -x^2 + x - 1$$

$$x = \frac{-1 \pm \sqrt{(1)^2 - 4(-1)(-1)}}{2(-1)}$$

$$x = \frac{-1 \pm \sqrt{1 - 4}}{-2}$$

$$x = \frac{-1 \pm \sqrt{-3}}{-2}$$

No $x$-intercepts.

$y$-intercept:

$$g(0) = -\left(0\right)^2 + \left(0\right) - 1 = -1 \quad (0, -1)$$

## Concept 5: Mixed Review: Methods to Solve a Quadratic Equation

**Video 10: Solving a Quadratic Equation in which the Coefficients are Fractions (3:52 min)**

▶ Play the video and work along.

Solve the equation by using the quadratic formula.

$$\frac{2}{5}x^2 = \frac{3}{5}x - \frac{3}{10}$$

| Review and complete | You try |
|---|---|
| 21. Solve the equation by using the quadratic formula. | 22. Solve the equation by using the quadratic formula. |

**21.** Solve the equation by using the quadratic formula.

$$\frac{1}{2}x^2 + \frac{2}{3}x = \frac{1}{4}$$

$$12\cdot\left(\frac{1}{2}x^2 + \frac{2}{3}x - \frac{1}{4}\right) = 0\cdot12 \quad \text{(Clear the fractions.)}$$

$$6x^2 + 8x - 3 = 0$$

$$a = 6, \ b = 8, \ c = -3$$

$$x = \frac{-8 \pm \sqrt{(8)^2 - 4(6)(-3)}}{2(6)}$$

$$x = \frac{-8 \pm \sqrt{64 - (-72)}}{2(6)}$$

$$x = \frac{-8 \pm \sqrt{136}}{2(6)}$$

$$x = \frac{-8 \pm 2\sqrt{34}}{12}$$

$$x = \frac{-4 \pm \sqrt{34}}{6}$$

The solution set is $\left\{\dfrac{-4+\sqrt{34}}{6}, -\dfrac{4-\sqrt{34}}{6}\right\}$.

**22.** Solve the equation by using the quadratic formula.

$$\frac{2}{3}x^2 - \frac{1}{6}x + \frac{1}{2} = 0$$

---

**Video 11: Summary of Techniques to Solve a Quadratic Equation (Problem Recognition Exercises (5:00 min)**

▶ Play the video and work along.

**SUMMARY Methods to Solve a Quadratic Equation**

1. **Factor and use the zero product rule.**

   Example: $x^2 - x - 12 = 0$    Example: $5x^2 - 3x = 0$

**2.    Use the square root property. Complete the square if necessary.**

- Good choice if the equation is in the form $x^2 + bx + c$ where $b$ is even.

    Example: $x^2 - 10x + 3 = 0$

- Good choice if the equation is in the form $ax^2 + c = 0$. (middle term is zero)

    Example: $4x^2 + 7 = 0$

**3.    Apply the quadratic formula.**

Example: $3x^2 + 7x - 4 = 0$

| Review and complete | You try |
|---|---|
| 23. Solve by any method. $$(2y+3)^2 = 9$$ The square root property is appropriate here. $$(2y+3)^2 = 9$$ $$2y+3 = \pm\sqrt{9} = \pm 3$$ $$2y+3 = 3$$ $$y = 0$$ $$2y+3 = -3$$ $$y = -3$$ $$\{0, -3\}$$ | 24. Solve by any method. $$5h^2 - 125 = 0$$ |

## Concept 1: Solving Equations by Using Substitution

### Video 1: Identifying the Characteristics of an Equation in Quadratic Form (3:41 min)

▶ Play the video and work along.

| Equation | Substitution | Quadratic Equation |
|---|---|---|
| $\left(2+\dfrac{3}{x}\right)^2 + 10\left(2+\dfrac{3}{x}\right)+9=0$ | $u =$ | |
| | $u =$ | |
| $(x^2-23)^2 + 6(x^2-23)-7=0$ | | |

Solve.  $(x^2-23)^2 + 6(x^2-23)-7=0$

### Video 2: Solving an Equation in Quadratic Form (2:32 min)

▶ Play the video and work along.

Solve.  $w^{2/3}-6w^{1/3}=27$

| Review and complete | You try |
|---|---|

1. One clue that an equation is quadratic in form is that the exponent on the first variable factor is
_____ the exponent on the second variable factor.

2. Solve the equation by using substitution.

$$(x^2 - 4)^2 + 5(x^2 - 4) - 14 = 0$$
$$u = x^2 - 4$$
$$u^2 + 5u - 14 = 0$$
$$(u + 7)(u - 2) = 0$$

$$u = -7 \qquad\qquad u = 2$$
$$x^2 - 4 = -7 \qquad x^2 - 4 = 2$$
$$x^2 = -3 \qquad\qquad x^2 = 6$$
$$x = \pm i\sqrt{3} \qquad\quad x = \pm\sqrt{6}$$

The solution set is $\left\{\pm i\sqrt{3},\ \pm\sqrt{6}\right\}$.

3. Solve the equation by using substitution.

$$p^{2/3} + p^{1/3} - 2 = 0$$

### Video 3: Solving an Equation in Quadratic Form (2:18 min)

▶ Play the video and work along.

Solve the equation.

$$y + 2\sqrt{y} - 35 = 0$$

### Review and complete

4. Solve the equation.
$$2t + 5\sqrt{t} - 3 = 0$$
$$u = \sqrt{t}$$
$$2u^2 + 5u - 3 = 0$$
$$(2u - 1)(u + 3) = 0$$

$$u = \frac{1}{2} \to \sqrt{t} = \frac{1}{2} \to t = \frac{1}{4}$$
$$u = -3 \to \sqrt{t} \neq -3$$

### You try

5. Solve the equation.

$$3m + 5\sqrt{m} = 2$$

The solution set is $\left\{\dfrac{1}{4}\right\}$.

**Concept 2: Solving Equations Reducible to Quadratic**

**Video 4: Solving a Higher Degree Polynomial Equation in Quadratic Form (2:43 min)**

⏸ Before playing the video, try this for yourself.

Solve. $3c^4 + 14c^2 - 5 = 0$

$u = $ _____

▶ Play the video and check your answers.

| Review and complete | You try |
|---|---|
| 6.  Solve. $4m^4 - 9m^2 + 2 = 0$ | 7.  Solve. $c^4 - 7c^2 + 12 = 0$ |
| $\quad 4m^4 - 9m^2 + 2 = 0$ | |
| $\qquad\qquad u = m^2$ | |
| $\quad 4u^2 - 9u + 2 = 0$ | |
| $\quad (4u - 1)(u - 2) = 0$ | |
| $\qquad u = \dfrac{1}{4}, u = 2$ | |

$$u = \frac{1}{4} \qquad u = 2$$

$$\qquad\qquad m^2 = 2$$

$$m^2 = \frac{1}{4} \qquad m = \pm\sqrt{2}$$

$$m = \pm\frac{1}{2}$$

The solution set is $\left\{\pm\dfrac{1}{2}, \pm\sqrt{2}\right\}$.

## Video 5: Solving a Rational Equation that Results in a non-Factorable Quadratic Equation (3:53 min)

▶ Play the video and work along.

Solve. $\dfrac{4}{x+3} + \dfrac{2}{x-1} = \dfrac{2x^2-1}{x^2+2x-3}$

| Review and complete | You try |
|---|---|
| 8. Solve. | 9. Solve. |

**8.** Solve.

$$\frac{x}{3x+2} - \frac{1}{x-1} = 0$$

$$(3x+2)(x-1)\left(\frac{x}{3x+2} - \frac{1}{x-1}\right) = 0 \cdot (3x+2)(x-1)$$

$$x(x-1) - 1(3x+2) = 0$$

$$x^2 - x - 3x - 2 = 0$$

$$x^2 - 4x - 2 = 0$$

Apply the quadratic formula to solve the equation.

**9.** Solve.

$$\frac{x}{x-1} + \frac{4}{x+1} = 4$$

$$x = \frac{-(-4) \pm \sqrt{(-4)^2 - 4(1)(-2)}}{2(1)}$$

$$x = \frac{4 \pm \sqrt{24}}{2}$$

$$x = \frac{4 \pm 2\sqrt{6}}{2}$$

$$x = 2 \pm \sqrt{6}$$

The solution set is $\left\{ 2 \pm \sqrt{6} \right\}$.

## Concept 1: Quadratic Functions of the Form f(x) = x^2+k

| Video 1: Definition of a Quadratic Function  (2:17 min) |
|---|

▶ Play the video and work along.

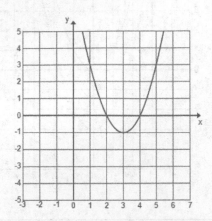

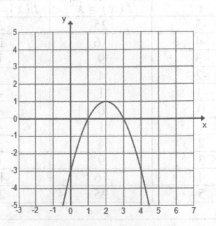

$$g(x) = x^2 - 6x + 8$$

$$h(x) = -x^2 + 4x - 3$$

| Review and complete | You try |
|---|---|

1.  The graph of a quadratic function is a _____ that opens either upward or downward based on the leading _____.

| 2. The function $g(x) = -9x^2 + 2x + 1$ is a quadratic function. The graph of $g(x) = -9x^2 + 2x + 1$ is a parabola which opens downward because $a = -9$. | 3. The function $h(x) = 4x^2 + 1$ is a _____ function. The graph of $h(x) = 4x^2 + 1$ is a parabola which opens _____ because $a = $ _____. |
|---|---|

| Video 2: Investigating the Graphs of Quadratic Functions (vertical shifts) (2:13 min) |
|---|

▶ Play the video and work along.

### Investigating Quadratic Functions

$$f(x) = x^2 + k \qquad k = 2$$

$$f(x) = x^2 + 2$$

| $x$ | $f(x) = x^2$ | $f(x) = x^2 + 2$ |
|---|---|---|
| -5 | 25 | 27 |
| -4 | 16 | 18 |
| -3 | 9 | 11 |
| -2 | 4 | 6 |
| -1 | 1 | 3 |
| 0 | 0 | 2 |
| 1 | 1 | 3 |
| 2 | 4 | 6 |
| 3 | 9 | 11 |
| 4 | 16 | 18 |
| 5 | 25 | 27 |

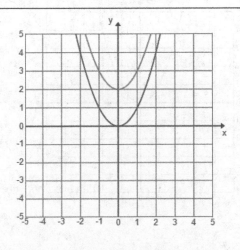

Video and Study Guide to accompany Intermediate Algebra, 5ᵗʰ ed., Miller/O'Neill/Hyde
Copyright © 2018 McGraw-Hill Education

## Investigating Quadratic Functions

$f(x) = x^2 + k \qquad k = -3$

$f(x) = x^2 - 3$

| $x$ | $f(x) = x^2$ | $f(x) = x^2 + 2$ |
|-----|--------------|------------------|
| -5 | 25 | 22 |
| -4 | 16 | 13 |
| -3 | 9 | 6 |
| -2 | 4 | 1 |
| -1 | 1 | -2 |
| 0 | 0 | -3 |
| 1 | 1 | -2 |
| 2 | 4 | 1 |
| 3 | 9 | 6 |
| 4 | 16 | 13 |
| 5 | 25 | 22 |

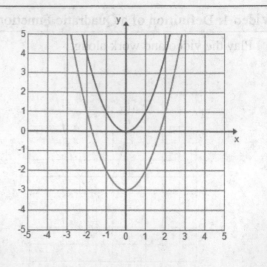

**PROPERTY** Graphs of the form $f(x) = x^2 + k$

- If $k > 0$ then the graph of $f(x) = x^2 + k$ is the same as the graph of $y = x^2$ shifted up $k$ units.

- If $k < 0$ then the graph of $f(x) = x^2 + k$ is the same as the graph of $y = x^2$ shifted down $|k|$ units.

## Video 3: Graphing Quadratic Functions with a Vertical Shift (1:48 min)

▶ Play the video and work along.

$g(x) = x^2 + 5$

$f(x) = x^2 - \dfrac{3}{2}$

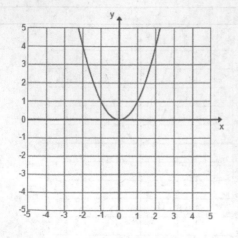

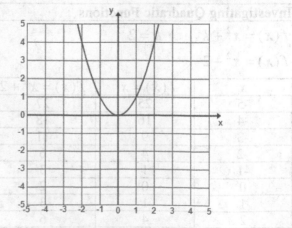

Video and Study Guide to accompany Intermediate Algebra, 5<sup>th</sup> ed., Miller/O'Neill/Hyde

| Review and complete | You try |
|---|---|
| 4. Given the sketch of $y = x^2$, sketch the graph of $f(x) = x^2 + 1$. | 5. Given the sketch of $y = x^2$, sketch the graph of $f(x) = x^2 - 5$. |

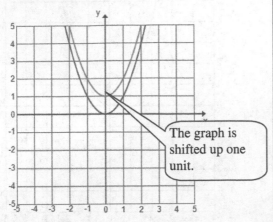

The graph is shifted up one unit.

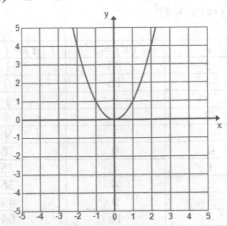

Shift the original parabola _____ 5 units.

## Concept 2: Quadratic Functions of the Form f(x) = (x-h)^2

### Video 4: Investigating the Graphs of Quadratic Functions (horizontal shifts) (2:41 min)

▶ Play the video and work along.

**Investigating Quadratic Functions**

$f(x) = (x - h)^2$    $h = 2$

$f(x) = (x - 2)^2$

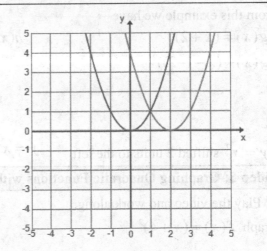

| $x$ | $f(x) = x^2$ | $f(x) = (x - 2)^2$ |
|---|---|---|
| -5 | 25 | 49 |
| -4 | 16 | 36 |
| -3 | 9 | 25 |
| -2 | 4 | 16 |
| -1 | 1 | 9 |
| 0 | 0 | 4 |
| 1 | 1 | 1 |
| 2 | 4 | 0 |
| 3 | 9 | 1 |
| 4 | 16 | 4 |
| 5 | 25 | 9 |

## Investigating Quadratic Functions

$f(x) = (x-h)^2$    $h = -4$

$f(x) = (x+4)^2$

| $x$ | $f(x) = x^2$ | $f(x) = (x+4)^2$ |
|-----|--------------|------------------|
| -5 | 25 | 1 |
| -4 | 16 | 0 |
| -3 | 9 | 1 |
| -2 | 4 | 4 |
| -1 | 1 | 9 |
| 0 | 0 | 16 |
| 1 | 1 | 25 |
| 2 | 4 | 36 |
| 3 | 9 | 49 |
| 4 | 16 | 64 |
| 5 | 25 | 81 |

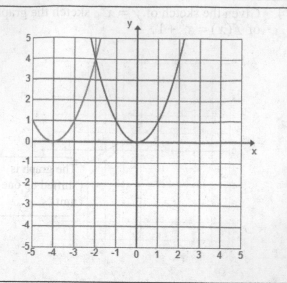

**PROPERTY** Graphs of the form $f(x) = (x-h)^2$

• If $h > 0$, then the graph of $f(x) = (x-h)^2$ is the same as the graph of $y = x^2$ shifted $h$ units to the right.

• If $h < 0$, then the graph of $f(x) = (x-h)^2$ is the same as the graph of $y = x^2$ shifted $|h|$ units to the left.

From this example we have

$g(x) = (x+2)^2$            $h(x) = (x-1)^2$

$g(x) = (x-(-2))^2$

$y = x^2$ shifted 2 units to the left        $y = x^2$ shifted 1 unit to the right

## Video 5: Graphing Quadratic Functions with a Horizontal Shift (1:09)

▶ Play the video and work along.        ⏸ Pause the video and try this yourself.

Graph $f(x) = (x+4)^2$.        Graph $f(x) = (x-2.7)^2$.

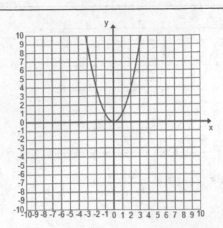

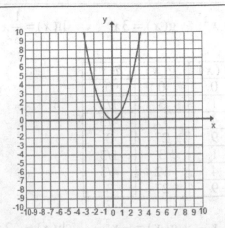

▶ Play the video and check your answer.

| Review and complete | You try |
|---|---|
| 6. Graph $y = x^2$ and $f(x) = (x - 3.5)^2$ on the same set of axes. | 7. Graph $y = x^2$ and $f(x) = (x - 1.2)^2$ on the same set of axes. |

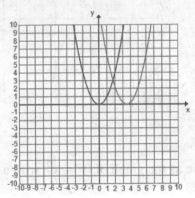

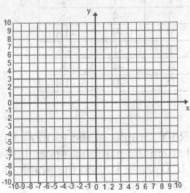

The graph of $y = x^2$ is shifted 3.5 units right to become $f(x) = (x - 3.5)^2$.

## Concept 3: Quadratic Functions of the Form f(x) = ax^2

### Video 6: Investigating the Graphs of Quadratic Functions (vertical stretch and shrink) (3:02)

▶ Play the video and work along.

Graph the functions defined below on the same graph.

$f(x) = x^2$    $g(x) = 3x^2$    $h(x) = \dfrac{1}{3}x^2$

| $x$ | $f(x)$ | $g(x)$ | $h(x)$ |
|-----|--------|--------|--------|
| 0 | 0 | 0 | 0 |
| 1 | 1 | 3 | 1/3 |
| 2 | 4 | 12 | 4/3 |
| 3 | 9 | 27 | 3 |
| -1 | 1 | 3 | 1/3 |
| -2 | 4 | 12 | 4/3 |
| -3 | 9 | 27 | 3 |

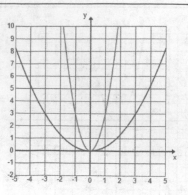

$f(x) = x^2$    $g(x) = -x^2$    $h(x) = -2x^2$

| $x$ | $f(x)$ | $g(x)$ | $h(x)$ |
|-----|--------|--------|--------|
| 0 | 0 | 0 | 0 |
| 1 | 1 | -1 | -2 |
| 2 | 4 | -4 | -8 |
| 3 | 9 | -9 | -18 |
| -1 | 1 | -1 | -2 |
| -2 | 4 | -4 | -8 |
| -3 | 9 | -9 | -18 |

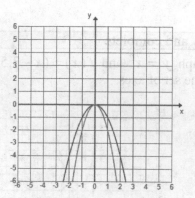

**PROPERTY** Graphs of the form $f(x) = ax^2$

If $a > 0$, then the parabola opens upward. Furthermore,

- If $0 < a < 1$, then the graph of $f(x) = ax^2$ is the same as the graph of $y = x^2$ with a *vertical shrink* by a factor of $a$.
- If $a > 1$, then the graph of $f(x) = ax^2$ is the same as the graph of $y = x^2$ with a *vertical stretch* by a factor of $a$.

If $a < 0$, then the parabola opens downward. Furthermore,

- If $0 < |a| < 1$, then the graph of $f(x) = ax^2$ is the same as the graph of $y = -x^2$ with a *vertical shrink* by a factor of $|a|$.
- If $|a| > 1$, then the graph of $f(x) = ax^2$ is the same as the graph of $y = -x^2$ with a *vertical stretch* by a factor of $|a|$.

**Concept 4: Quadratic Functions of the Form f(x) = a(x-h)^2+k**

**Video 7: Graphing a Quadratic Function and Identifying Key Characteristics 1 (3:15 min)**

▶ Play the video and work along.

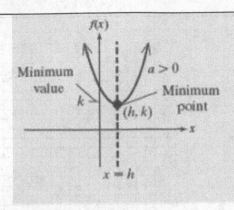 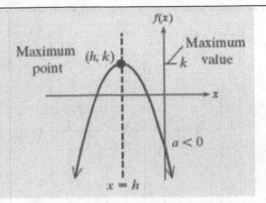

Given the function defined by $k(x) = 2(x+4)^2 + 1$

Vertex form: $k(x) == a(x-h)^2 + k$

a. Identify the vertex.

b. Sketch the function.

c. Determine the axis of symmetry.

d. Is the vertex the minimum point or the maximum point?

e. Write the domain and range in interval notation.

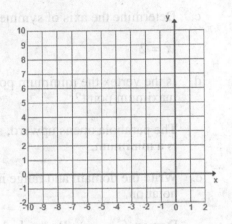

| Review and complete | You try |
|---|---|
| 8. For the function $f(x) = (x-2)^2 - 1$, <br><br> a. Identify the vertex. <br><br> $(2, -1)$ <br><br> b. Sketch the function. | 9. For the function $f(x) = -(x+2)^2 - 1$, <br><br> a. Identify the vertex. <br><br> b. Sketch the function. |

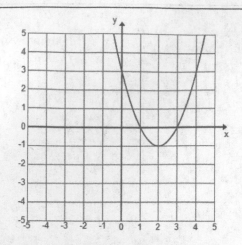

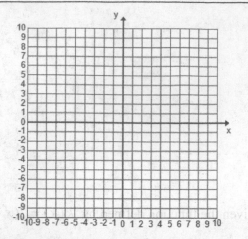

c.  Determine the axis of symmetry.

$x = 2$

d.  Is the vertex the minimum point or the maximum point?

The parabola opens upward, so the vertex is a minimum.

e.  Write the domain and range in interval notation.

Doman $(-\infty, \infty)$ ; Range $[-1, \infty)$

c.  Determine the axis of symmetry.

d.  Is the vertex the minimum point or the maximum point?

e.  Write the domain and range in interval notation.

## Video 8: Graphing a Quadratic Function and Identifying Key Characteristics 2 (2:34 min)

▶ Play the video and work along.

Given the function defined by $k(x) = -3(x-2)^2 + 3$

<u>Vertex form:</u>  $k(x) = a(x-h)^2 + k$

a.  Identify the vertex.

b.  Sketch the function.

c.  Determine the axis of symmetry.

d.  Is the vertex the minimum point or the maximum point?

e.  Write the domain and range in interval notation.

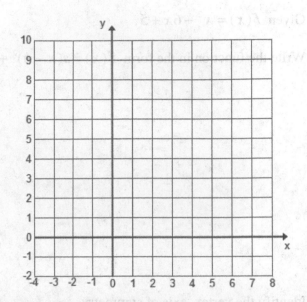

| Review and complete | You try |
|---|---|
| 10. For the function $f(x) = 2x^2 + 3$, | 11. For the function $f(x) = 2(x-3)^2$, |

Review and complete:

10. For the function $f(x) = 2x^2 + 3$,

a.  Identify the vertex.
    (0,3)

b.  Sketch the function.

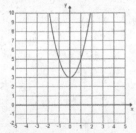

c.  Determine the axis of symmetry.
    $x = 0$

d.  Is the vertex the minimum point or the maximum point?
    The parabola opens upward, so the vertex is a minimum.

e.  Write the domain and range in interval notation.
    Doman $(-\infty, \infty)$  Range $[3, \infty)$

You try:

11. For the function $f(x) = 2(x-3)^2$,

a.  Identify the vertex.

b.  Sketch the function.

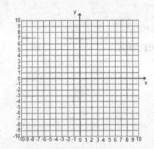

c.  Determine the axis of symmetry.

d.  Is the vertex the minimum point or the maximum point?

e.  Write the domain and range in interval notation.

**Concept 1: Writing a Quadratic Function in the Form f(x) = a(x-h)^2+k**

| Video 1: Completing the Square and Graphing a Quadratic Function  (5:58 min) |
| --- |

▶ Play the video and work along.

Given $f(x) = x^2 - 6x + 5$,

Write the function in the form $f(x) = a(x - h)^2 + k$.

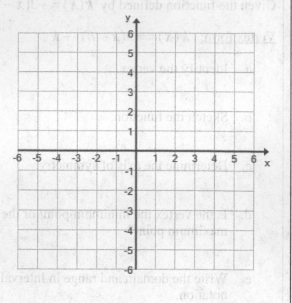

Identify the vertex, axis of symmetry,
and maximum or minimum value.

Determine the $y$-intercept.

Determine the $x$-intercepts.

## Video 2: Completing the Square and Graphing a Quadratic Function (5:20 min)

▶ Play the video and work along.

Given $g(x) = -2x^2 - 16x - 30$,

Write the function in the form $g(x) = a(x - h)^2 + k$.

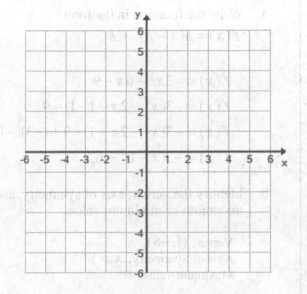

Identify the vertex, axis of symmetry, and maximum or minimum value.

Determine the $y$-intercept.

Determine the $x$-intercepts.

| Review and complete | You try |
|---|---|
| 1. Given $f(x) = -3x^2 + 6x - 9$, | 2. Given $f(x) = -x^2 - 4x$, |

1. Given $f(x) = -3x^2 + 6x - 9$,

   a. Write the function in the form
$f(x) = a(x-h)^2 + k$.

$$f(x) = -3x^2 + 6x - 9$$
$$f(x) = -3(x^2 - 2x + 1 - 1) - 9$$
$$f(x) = -3(x^2 - 2x + 1) - 9 + (-3)(-1)$$
$$f(x) = -3(x-1)^2 - 6$$

   b. Identify the vertex, axis of symmetry, and maximum or minimum value.

      Vertex: $(1, -6)$
      Axis of Symmetry: $x = 1$
      Maximum value: -6

   c. Determine the $y$-intercept.
$$f(0) = -3(0)^2 + 6(0) - 9$$
$$f(0) = -9$$
$$(0, -9)$$

   d. Determine the $x$-intercepts.
$$0 = -3x^2 + 6x - 9$$
$$0 = -3(x^2 - 2x + 3)$$
$$\frac{0}{-3} = \frac{-3(x^2 - 2x + 3)}{-3}$$
$$0 = x^2 - 2x + 3$$
$$x = \frac{2 \pm \sqrt{(-2)^2 - 4(1)(3)}}{2}$$
$$x = \frac{2 \pm \sqrt{-10}}{2}$$
There are no x-intercepts.

   e. Sketch the graph.

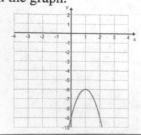

2. Given $f(x) = -x^2 - 4x$,

   a. Write the function in the form
$f(x) = a(x-h)^2 + k$.

   b. Identify the vertex, axis of symmetry, and maximum or minimum value.

   c. Determine the $y$-intercept.

   d. Determine the $x$-intercepts.

   e. Sketch the graph.

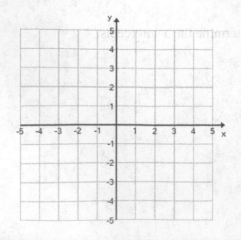

**Concepts 2 and 3: Vertex Formula/Determining the Vertex and Intercepts of a Quadratic Function**

**Video 3: Applying the Vertex Formula and Graphing a Quadratic Function (4:32 min)**

▶ Play the video and work along.

---

FORMULA **The Vertex Formula**

For $f(x) = ax^2 + bc + c$ $(a \neq 0)$, the vertex is given by:

$$\left( \frac{-b}{2a}, \frac{4ac - b^2}{4a} \right) \quad \text{or} \quad \left( \frac{-b}{2a}, f\left( \frac{-b}{2a} \right) \right)$$

---

Given: $f(x) = -3x^2 + 12x - 13$,

Use the vertex formula to find the vertex.

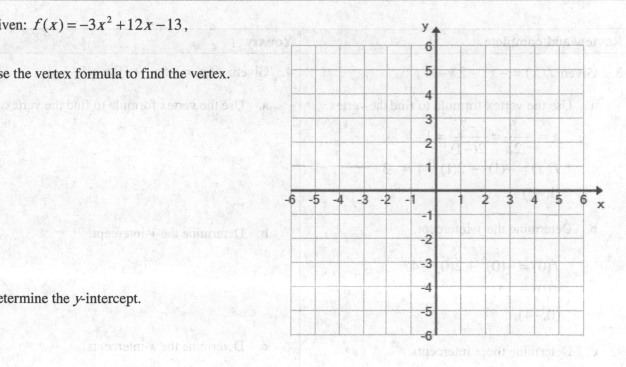

Determine the $y$-intercept.

Determine the $x$-intercept.

| Review and complete | You try |
|---|---|
| 3. Given $f(x) = -x^2 + 2x - 4$, | 4. Given: $f(x) = -5x^2 + 10x - 4$, |
| a. Use the vertex formula to find the vertex.<br><br>$x = -\dfrac{b}{2a} = \dfrac{-2}{2(-1)} = 1$<br><br>$f(x) = -(1)^2 + 2(1) - 4 = -3$<br><br>$(1, -3)$ | a. Use the vertex formula to find the vertex. |
| b. Determine the $y$-intercept.<br><br>$f(0) = -(0)^2 + 2(0) - 4$<br><br>$f(0) = -4$<br><br>$(0, -4)$ | b. Determine the $y$-intercept. |
| c. Determine the $x$-intercepts.<br>$0 = -x^2 + 2x - 4$<br><br>$x = \dfrac{-2 \pm \sqrt{(2)^2 - 4(-1)(-4)}}{2(-1)}$<br><br>$x = \dfrac{-2 \pm \sqrt{-12}}{-2}$<br><br>No $x$-intercepts | c. Determine the $x$-intercepts. |

## Concept 4: Applications and Modeling of Quadratic Functions

### Video 4: An Application of the Vertex Formula: Finding Maximum Height (2:23 min)

Play the video and work along.

A baseball is thrown at an angle of $35°$ from the horizontal. The height of the ball, $h(t)$, (in feet) can be approximated by $h(t) = -16t^2 + 40t + 5$ where $t$ is the number of seconds after release.

How long will it take the ball to reach its maximum height? Round to the nearest tenth of a second.

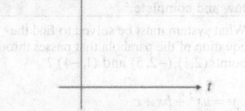

Determine the maximum height. Round to the nearest foot.

| Review and complete | You try |
|---|---|
| 5.  The height of a ball, $h(t)$, (in feet) after it is thrown can be approximated by $h(t) = -16t^2 + 32t + 4$ where $t$ is the number of seconds after release. | 6.  The height of the ball, $h(t)$, (in feet) after it is thrown can be approximated by $h(t) = -16t^2 + 80t + 6$ where $t$ is the number of seconds after release. |
| a.  How long will it take the ball to reach its maximum height? Round to the nearest tenth of a second. $$t = \frac{-b}{2a} = \frac{-32}{2(-16)} = 1 \quad 1 \text{ sec.}$$ | a.  How long will it take the ball to reach its maximum height? Round to the nearest tenth of a second. |
| b.  Determine the maximum height. Round to the nearest foot. $$h(1) = -16(1)^2 + 32(1) + 4 = 20 \text{ ft}$$ | b.  Determine the maximum height. Round to the nearest foot. |

### Video 5: Finding an Equation of a Parabola that Passes through Three Given Points (6:53 min)

 Play the video and work along.

Write an equation of a parabola that passes through the points $(4,7), (1,1)$ and $(3,9)$.

$y = ax^2 + bx + c$

| Review and complete | You try |
|---|---|
| 7. What system must be solved to find the equation of the parabola that passes through the points $(2,1),(-2,5)$ and $(1,-4)$ ? | 8. What system must be solved to find the equation of the parabola that passes through the points $(1,4),(-1,6)$ and $(2,-18)$ ? |

$$y = ax^2 + bx + c$$

$$1 = a(2)^2 + b(2) + c \rightarrow 1 = 4a + 2b + c$$

$$5 = a(-2)^2 + b(-2) + c \rightarrow 5 = 4a - 2b + c$$

$$-4 = a(1)^2 + b(1) + c \rightarrow -4 = a + b + c$$

## Concept 1: Solving Polynomial Inequalities

**Video 1: Solving a Nonlinear Inequality Graphically  (3:39 min)**

▶ Play the video and work along.

Write the solution set.

$$f(x) = x^2 + 2x - 3$$

1.  $x^2 + 2x - 3 = 0$

2.  $x^2 + 2x - 3 < 0$

3.  $x^2 + 2x - 3 \le 0$

4.  $x^2 + 2x - 3 > 0$

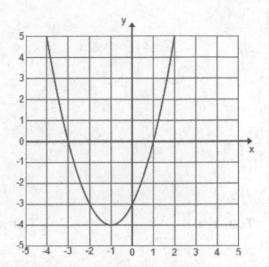

5.  $x^2 + 2x - 3 \ge 0$

| Review and complete | You try |
|---|---|
| 1. Use the graph of $f(x) = x^2 - x - 2$ to find the solution sets of the equations or inequalities. | 2. Use the graph of $f(x) = -x^2 + x + 2$ to find the solution sets of the inequalities. |

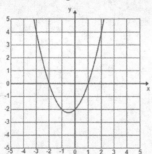

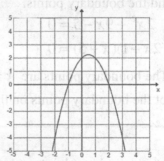

a.  $-x^2 + x + 2 > 0$

a.  $x^2 - x - 2 = 0$

$\{-2, 1\}$ where the graph crosses the $x$-axis.

b.  $x^2 - x - 2 \le 0$

b.  $-x^2 + x + 2 < 0$

$[-2, 1]$ where the graph is below the $x$-axis.

## Video 2: Solving a Quadratic Inequality Using the Test Point Method (4:08)

▶ Play the video and work along.

Write the solution set. $2x^2 + x > 15$

**Step 1:** Solve the related equation and find the boundary points.

**Step 2:** Plot the boundary points on the number line.

**Step 3:** Test a point from each interval to determine if the original inequality is true or false.

**Step 4:** Test the boundary points in the original inequality and write the solution set.

Test $x =$          Test $x =$          Test $x =$

| Review and complete | You try |
|---|---|
| 3. Write the solution set. $2x^2 + 9x - 5 > 0$ | 4. Write the solution set. $m^2 - 12m \le -32$ |

**Review and complete**

3. Write the solution set. $2x^2 + 9x - 5 > 0$

**Step 1:** Find the boundary points.

$$2x^2 + 9x - 5 = 0$$

$$(2x - 1)(x + 5) = 0$$

The boundary points are $x = \frac{1}{2}, x = -5$.

**Step 2:** Plot the boundary points on the number line.

**Step 3:** Test a point in each interval.

$x = -6$   $2(-6)^2 + 9(-6) - 5 = 13 > 0$   True
$x = 0$   $2(0)^2 + 9(0) - 5 = -5 > 0$   False
$x = 6$   $2(6)^2 + 9(6) - 5 = 121 > 0$   True

**Step 4:** Test boundary points and write solution set.

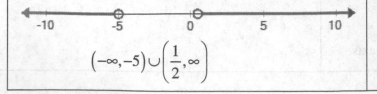

$$\left(-\infty, -5\right) \cup \left(\frac{1}{2}, \infty\right)$$

**You try**

4. Write the solution set. $m^2 - 12m \le -32$

## Video 3: Solve a Polynomial Inequality Using the Test Point Method (1:49 min)

▶ Play the video and work along.

Solve the inequality using the test point method. $x(x-4)(x+2)^2 \le 0$

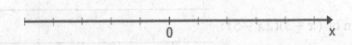

| Test $x=-3$ | Test $x=-1$ | Test $x=2$ | Test $x=5$ |
|---|---|---|---|
| $(-3)(-3-4)(-3+2)^2 \overset{?}{\le} 0$ | $(-1)(-1-4)(-1+2)^2 \overset{?}{\le} 0$ | $(2)(2-4)(2+2)^2 \overset{?}{\le} 0$ | $(5)(5-4)(5+2)^2 \overset{?}{\le} 0$ |
| $(-3)(-7)(-1)^2 \overset{?}{\le} 0$ | $(-1)(-5)(1)^2 \overset{?}{\le} 0$ | $(2)(-2)(-4)^2 \overset{?}{\le} 0$ | $(5)(1)(7)^2 \overset{?}{\le} 0$ |
| $21 \overset{?}{\le} 0$ | $5 \overset{?}{\le} 0$ | $-64 \overset{?}{\le} 0$ | $245 \overset{?}{\le} 0$ |

### Review and complete

5. Solve the inequality using the test point method. $4(x-2)(x+1)^2 \le 0$

  Boundary points: $x=2$ and $x=-1$

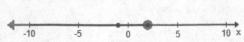

  Test a point in each interval.

  Test $x=-5$:
  $4((-5)-2)((-5)+1)^2 = 4(-7)(-4)^2 \overset{?}{\le} 0$ True

  Test $x=0$:
  $4((0)-2)((0)+1)^2 = 4(-2)(1)^2 \overset{?}{\le} 0$ True

  Test $x=5$:
  $4((5)-2)((5)+1)^2 = 4(3)(6)^2 \overset{?}{\le} 0$ False

### You try

6. Solve the inequality using the test point method. $-2(x-5)(2x+4) > 0$

Solution set: $\left(-\infty, 2\right]$

## Video 4: Solving a Quadratic Inequality by Using a Sign Chart (3:54 min)

▶ Play the video and work along.

Write the solution set.
$2x^2 + x > 15$

Sign of $(x + 3)$:

Sign of $(2x - 5)$:

Sign of $(x + 3)(2x - 5)$:

$-3$       $\frac{5}{2}$

$f(x) = 2x^2 + x - 15$

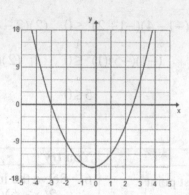

| Review and complete | You try |
|---|---|
| 7. Write the solution set. Use a sign chart. | 8. Write the solution set. Use a sign chart. |

**7.** Write the solution set. Use a sign chart.

$$3x(x + 4) > 10 - x$$
$$3x^2 + 12x - 10 + x > 0$$
$$3x^2 + 13x - 10 > 0$$
$$(3x - 2)(x + 5) > 0$$
$$x = \frac{2}{3}, \ x = -5$$

Sign of $(3x - 2)$:  $-$ | $-$ | $+$
Sign of $(x + 5)$:  $-$ | $+$ | $+$
Sign of $(3x - 2)(x + 5)$:  $+$ | $-$ | $+$

$-5$    $\frac{2}{3}$

**8.** Write the solution set. Use a sign chart.

$$x^2 + 7x < 30$$

$(3x-2)(x+5)>0$   True ┆ False ┆ True

The solution set is $\left(-\infty,-5\right)\cup\left(\dfrac{2}{3},\infty\right)$.

## Video 5: Solve a Polynomial Inequality Using a Sign Chart (2:25 min)

▶ Play the video and work along.

Solve the inequality using a sign chart.   $x(x-4)(x+2)^2 \le 0$

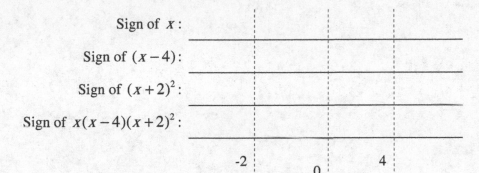

Sign of $x$:

Sign of $(x-4)$:

Sign of $(x+2)^2$:

Sign of $x(x-4)(x+2)^2$:

-2     0     4

| Review and complete | You try |
|---|---|
| 9. Solve the inequality using a sign chart. $x(x+4)(x-1)>0$ | 10. Solve the inequality using a sign chart. $x(x-4)(x+3)\le 0$ |

Sign of $x$:           −  ┆ −  ┆ +  ┆ +
Sign of $(x+4)$:     −  ┆ +  ┆ +  ┆ +
Sign of $(x-1)$:     −  ┆ −  ┆ −  ┆ +
Sign of $x(x+4)(x-1)$:   −  ┆ +  ┆ −  ┆ +

-4   0   1

The solution set is $(-4, 0) \cup (1, \infty)$.

| Video 6: Solving a Quadratic Inequality Using the Test Point Method (3:02 min) |
| --- |

▶ Play the video and work along.

Solve the inequality using the test point method. $x^2 - 3x - 5 \leq 0$

| Review and complete | You try |
| --- | --- |

11. The sign chart method would not be appropriate in the preceding example because the equation is not
_____.

| 12. Solve the inequality using the test point method. | 13. Solve the inequality using the test point method. |
| --- | --- |

**12.**

$2x^2 - x - 5 \geq 0$

$2x^2 - x - 5 \geq 0$

$x = \dfrac{-(-1) \pm \sqrt{(-1)^2 - 4(2)(-5)}}{2(2)}$

$x = \dfrac{1 \pm \sqrt{41}}{4}$

$x \approx 1.9 ; x \approx -1.4$

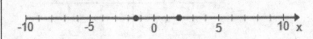

Test $-5$:  $2(-5)^2 - (-5) - 5 = 50 \geq 0$   True
Test $0$ :   $2(0)^2 - (0) - 5 = -5 \geq 0$   False
Test $5$ :   $2(5)^2 - (5) - 5 = 40 \geq 0$   True

**13.**

$3x^2 + 2x - 4 < 0$

Solution set: $\left(-\infty, \dfrac{1-\sqrt{41}}{4}\right] \cup \left[\dfrac{1+\sqrt{41}}{4}, \infty\right)$

## Concept 2: Solving Rational Inequalities

### Video 7: Solving a Rational Inequality Using the Test Point Method (4:00 min)

▶ Play the video and work along.

Write the solution set.    $\dfrac{-2}{x+1} \le 0$

Write the solution set.    $\dfrac{-2}{x+1} \le 0$

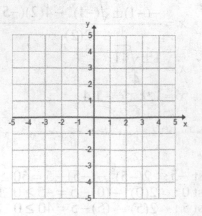

| Review and complete | You try |
|---|---|
| 14. Write the solution set. | 15. Write the solution set. |

14. Write the solution set.

$$\frac{3}{2x-5} \le 0$$

$$\frac{3}{2x-5} = 0$$

$$(2x-5)\left(\frac{3}{2x-5}\right) = (0)(2x-5)$$

$$3 = 0$$

The only boundary point is $x = \frac{5}{2}$. Test one value on the left and one on the right of $x = \frac{5}{2}$.

Test $x = 0$: $\dfrac{3}{2x-5} = \dfrac{3}{2(0)-5} = -\dfrac{3}{5} \overset{?}{\le} 0$    True

Test $x = 5$: $\dfrac{3}{2x-5} = \dfrac{3}{2(5)-5} = \dfrac{3}{5} \overset{?}{\le} 0$    False

Solution set: $\left(-\infty, \dfrac{5}{2}\right]$

15. Write the solution set.

$$\frac{-3}{x+2} \le 0$$

**Video 8: Solve a Rational Inequality with a Sign Chart (2:53 min)**

Video and Study Guide to accompany Intermediate Algebra, 5<sup>th</sup> ed., Miller/O'Neill/Hyde

▶ Play the video and work along.

Write the solution set.   $\dfrac{-2}{x+1} \le 0$

Write the solution set.   $\dfrac{-2}{x+1} \le 0$

Sign of $-2$: _____
Sign of $(x+1)$: _____

Sign of $\dfrac{-2}{x+1}$: _____

$-1$

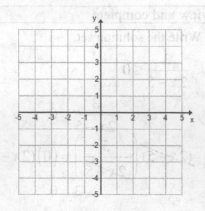

| **Review and complete** | **You try** |
|---|---|
| 16. Write the solution set. | 17. Write the solution set. |
| $\dfrac{9}{x+10} > 0$ | $\dfrac{8}{4x+9} > 0$ |

Sign of $9$:          $+$   $\vert$   $+$
Sign of $(x+10)$:   $-$   $\vert$   $+$

Sign of $\dfrac{9}{x+10}$:   $-$   $\vert$   $+$

$-10$

$-10$ is a restricted value and cannot be included in the solution set.

Solution set $(-10, \infty)$

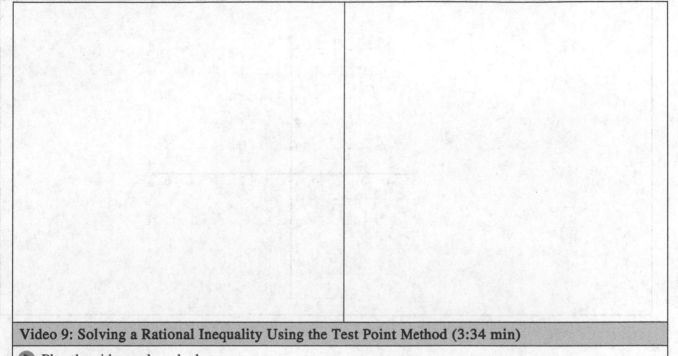

## Video 9: Solving a Rational Inequality Using the Test Point Method (3:34 min)

▶ Play the video and work along.

Write the solution sct.  $\dfrac{x-4}{x-3} \geq 2$

| Test: $x = 0$ | Test: $x = 2.5$ | Test: $x = 4$ |
|---|---|---|
| $\dfrac{(0)-4}{(0)-3} \overset{?}{\geq} 2$ | $\dfrac{(2.5)-4}{(2.5)-3} \overset{?}{\geq} 2$ | $\dfrac{(4)-4}{(4)-3} \overset{?}{\geq} 2$ |
| $\dfrac{4}{3} \overset{?}{\geq} 2$ false | $\dfrac{-1.5}{-0.5} \overset{?}{\geq} 2$ | $0 \overset{?}{\geq} 2$ false |
| | $30 \overset{?}{\geq} 2$ true | |

Write the solution set.  $\dfrac{x-4}{x-3} \geq 2$

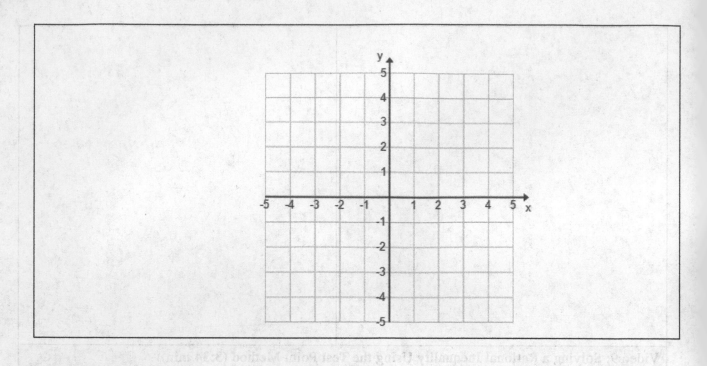

| Review and complete | You try |
|---|---|
| 18. Write the solution set.<br><br>$\dfrac{w-8}{w+6} > 2$ | 19. Write the solution set.<br><br>$\dfrac{x-2}{x+6} \le 5$ |

$(w+6) \cdot \left(\dfrac{w-8}{w+6}\right) = 2 \cdot (w+6)$

$\qquad w - 8 = 2w + 12$

$\qquad w - 8 - w = 2w + 12 - w$

$\qquad\qquad -8 = w + 12$

$\qquad\qquad -8 - 12 = w + 12 - 12$

$\qquad\qquad -20 = w$

The solution $-20$ and restricted value _____ are boundary numbers.

Test points: $x = -25, x = -10, x = 0$

$\dfrac{(-25)-8}{(-25)+6} > 2 \qquad \dfrac{(-10)-8}{(-10)+6} > 2 \qquad \dfrac{(0)-8}{(0)+6} > 2$

$\dfrac{33}{19} \overset{?}{>} 2 \text{ false} \qquad \dfrac{-18}{-4} = \dfrac{9}{2} \overset{?}{>} 2 \text{ true} \qquad -\dfrac{4}{3} \overset{?}{>} 2 \text{ false}$

Solution set $(-20, -6)$

## Concept 3: Inequalities with "Special Case" Solutions

### Video 10: Solving Quadratic Inequalities with "Special Case" Solution Sets (3:34 min)

▶ Play the video and work along.

Write the solution set.
$x^2 - 2x + 1 = 0$

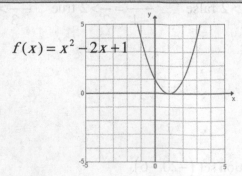

$f(x) = x^2 - 2x + 1$

$x^2 - 2x + 1 < 0$

$x^2 - 2x + 1 \leq 0$

Write the solution set.

$x^2 - 2x + 1 \geq 0$

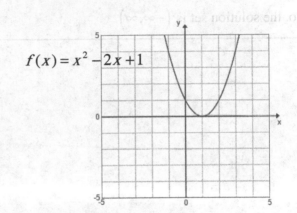

$f(x) = x^2 - 2x + 1$

$x^2 - 2x + 1 > 0$

| Review and complete | You try |
|---|---|

20. If the graph of the parabola for $f(x) = ax^2 + bx + c$ is completely above the $x$-axis, then the solution set to $ax^2 + bx + c > 0$ is _____.

21. If the graph of the parabola for $f(x) = ax^2 + bx + c$ is completely above the $x$-axis, then the solution set to $ax^2 + bx + c < 0$ is _____.

22. Write the solution set.   $2x^2 + 3x + 3 > 0$

Examine the graph of $f(x) = 2x^2 + 3x + 3$.

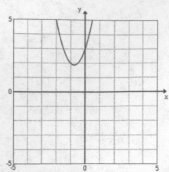

Since the graph of the parabola never drops below the $x$-axis, its function value is always positive.

So, the solution set is $(-\infty, \infty)$.

23. Write the solution set. $2x^2 + 12x + 36 \geq 0$

**Answers**: Section 7.1

1. $\sqrt{k}; -\sqrt{k}$      3. $\{\pm 9\}$      5. $\{\pm 3\sqrt{5}\}$      6. trinomial      8. $\{-5 \pm \sqrt{7}\}$

10. $n = \dfrac{9}{4}; \left(x + \dfrac{3}{2}\right)^2$      12a. $\{-6 \pm \sqrt{37}\}$      12b. $\left\{-\dfrac{5}{2} \pm \dfrac{\sqrt{31}}{2}\right\}$      14. 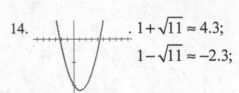 . $1 + \sqrt{11} \approx 4.3$;

$1 - \sqrt{11} \approx -2.3$;

$x$-intercepts: (4.3, 0)

and (−2.3, 0)

16. $\{-7, 1\}$      18a. $r = \sqrt{\dfrac{V}{3.2\pi}}$      18b. 1.25 ft

**Answers**: Section 7.2

1. $x = \dfrac{-b + \sqrt{b^2 - 4ac}}{2a}, x = \dfrac{-b - \sqrt{b^2 - 4ac}}{2a}$      2. True      3. $a = 1, b = -6, c = 4$

4. $a = 2, b = 3, c = -7; \left\{\dfrac{-3 \pm \sqrt{65}}{4}\right\}$      5. factoring      7. $\left\{\dfrac{2}{5}, 1\right\}$      9. 6.2 in., 10.2 in.

11. 3.68 sec      13. two rational solutions      15a. $3x^2 + 0x - 7 = 0$      15b. 84

15c. two irrational solutions      16. 0, 1, and 2      17. zero      18. $x$-intercepts

20. No $x$-intercepts; $y$-intercept: (0, −15)      22. $\left\{\dfrac{1}{8} \pm \dfrac{\sqrt{47}}{8}i\right\}$      24. $\{-5, 5\}$

**Answers**: Section 7.3

1. Exactly double or twice      3. $\{-8, 1\}$      5. $\left\{\dfrac{1}{9}\right\}$      7. $\{2, -2, \sqrt{3}, -\sqrt{3}\}$      9. $\left\{0, \dfrac{5}{3}\right\}$

**Answers**: Section 7.4

1. parabola; coefficient    3. quadratic; upward; 4    5. down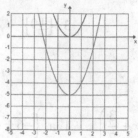

7. 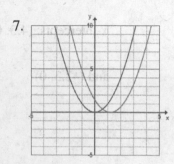    9a. $(-2,-1)$    9b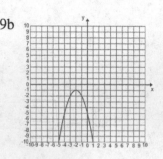

9c. $x=-2$    9d. Maximum    9e. Domain $(-\infty,\infty)$;
                                    Range $(-\infty,-1]$

11a. $(3,0)$    11b. 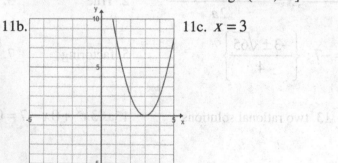    11c. $x=3$

11d. Minimum    11e. Domain $(-\infty,\infty)$;
                    Range $[0,\infty)$

**Answers**: Section 7.5

2a. $f(x)=-(x+2)^2+4$    2b. Vertex $(-2,4)$, axis of symmetry $x=-2$, maximum value $=4$

2c. $(0,0)$

2d. $(0,0),(-4,0)$

2e.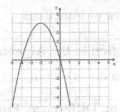

4a. $(1,1)$

4b. $(0,-4)$

4c. $\left(\dfrac{5\pm\sqrt{5}}{5},0\right)$

6a. 2.5 sec

6b. 106 ft

8.   $a+b+c=4$

$a-b+c=6$

$4a+2b+c=-18$

**Answers**: Section 7.6

2a. $(-1,2)$

2b. $(-\infty,-1)\cup(2,\infty)$

4. $\left[4,8\right]$

6. $(-2,5)$

8. $(-10,3)$

10. $(-\infty,-3]\cup[0,4]$

11. factorable

13. $\left(\dfrac{-1-\sqrt{13}}{3},\dfrac{-1+\sqrt{13}}{3}\right)$

15. $(-2,\infty)$

17. $\left(-\dfrac{9}{4},\infty\right)$

18. -6

19. $(-\infty,-8]\cup(-6,\infty)$

20. $(-\infty,\infty)$ All real numbers.

21. $\{\ \}$

23. $(-\infty,\infty)$ All real numbers

## Concept 1: Algebra of Functions

**Video 1: Introduction to the Algebra of Functions (3:03 min)**

▶ Play the video and work along.

**Definition of Sum, Difference, Product and Quotient of Functions**

Given two functions $f$ and $g$, the functions $f+g$, $f-g$, $f \cdot g$ and $\dfrac{f}{g}$ are defined as:

$$(f+g)(x) = f(x) + g(x)$$
$$(f-g)(x) = f(x) - g(x)$$

$$(f \cdot g)(x) = f(x) \cdot g(x)$$
$$\left(\dfrac{f}{g}\right)(x) = \dfrac{f(x)}{g(x)} \text{ provided } g(x) \neq 0$$

Given: $f(x) = 3x$, $g(x) = x^2 - 8x$, $k(x) = 2x + 1$

a. Find: $(f+g)(x)$

b. Find: $(g-k)(x)$

⏸ Pause the video and try this yourself.

c. Find: $(f \cdot k)(x)$

d. Find: $\left(\dfrac{f}{k}\right)(x)$

▶ Play the video and check your answers.

| Review and complete | You try |
|---|---|
| 1. Given $f(x) = x+1$ and $g(x) = x^2$, find the following. $\left(\dfrac{f}{g}\right)(x)$ $= \dfrac{f(x)}{g(x)}$ $= \dfrac{x+1}{x^2}$ provided $x \neq 0$ | 2. Given $m(x) = 4$ and $n(x) = x-4$, find the following. $(m \cdot n)(x)$ |

## Concept 2: Composition of Functions

### Video 2: Composition of Functions (2:52 min)

▶ Play the video and work along.

DEFINITION Composition of Functions

**The composition of $f$ and $g$,** denoted $f \circ g$ is defined by the rule $(f \circ g)(x) = f(g(x))$ provided that $g(x)$ is in the domain of $f$.

**The composition of $g$ and $f$,** denoted $g \circ f$ is defined by the rule $(g \circ f)(x) = g(f(x))$ provided that $f(x)$ is in the domain of $g$.

Given $f(x) = 3x - 5$ and $g(x) = x + 4$, find

a. $(f \circ g)(x)$

b. $(g \circ f)(x)$

### Video 3: Examples with Composition of Functions (2:53 min)

▶ Play the video and work along.

Given $f(x) = x - 3$, $g(x) = x^2$, and $k(x) = \sqrt{x-2}$, find

a. $(f \circ g)(x)$

b. $(g \circ f)(x)$

⏸ Pause the video and try this yourself.

c. $(k \circ f)(x)$

▶ Play the video and check your answers.

| Review and complete | You try |
|---|---|
| 3. Given $m(x) = \sqrt{x}, n(x) = x^4, p(x) = x+2$ find the following.<br><br>  a.  $(m \circ n)(x)$<br>      $= m(n(x))$<br>      $= m(x^4)$<br>      $=$ _____<br>      $=$ _____<br><br>  b. $(p \circ n)(x)$<br>    $(p \circ n)(x) = p(n(x))$<br>       $= p(x^4)$<br>       $=$ _____ | 4. Given $m(x) = \sqrt{x}, n(x) = x^4, p(x) = x+2$ find the following.<br><br>  a.  $(n \circ p)(x)$ Do not expand.<br><br><br>  b.  $(p \circ m)(x)$ |

## Concept 3: Operations on Functions

### Video 4: Practice Combining Functions for Given Values of $x$ (2:39 min)

▶ Play the video and work along.

Given:   $k(x) = x + 6$   and  $p(x) = 3x^2 + x$

  a.  Evaluate $(k \cdot p)(1)$                 b.  Evaluate $\left(\dfrac{p}{k}\right)(-6)$

⏸ Pause the video and try this yourself.

c.  $(p \circ k)(-4)$

▶ Play the video and check your answers.

| Review and complete | You try |
|---|---|
| 5.  To evaluate a composition of two functions for a given value, use the order of operations and evaluate the "inner" function in the expression for that value, and use the output as input in the _____ function. | |

| | |
|---|---|
| 6.  Given $f(x) = 2x - 2$ and $g(x) = \sqrt{x}$, evaluate<br><br>$\left(\dfrac{g}{f}\right)(1)$<br><br>$= \dfrac{g(1)}{f(1)}$<br><br>$= \dfrac{\sqrt{1}}{2(1) - 2}$<br><br>$= \dfrac{\sqrt{1}}{0}$<br><br>undefined | 7.  Given $f(x) = 2x - 2$ and $g(x) = \sqrt{x}$, evaluate<br><br>$(f \circ g)(3)$ |

## Video 5: Determining Function Values from a Graph (3:02 min)

▶ Play the video and work along.

For the functions pictured, find
   a.  $(f + g)(1)$

   b.  $(f \cdot g)(-2)$

   c.  $\left(\dfrac{g}{f}\right)(0)$

   d.  $(g \circ f)(-4)$

   e.  $(f \circ g)(3)$

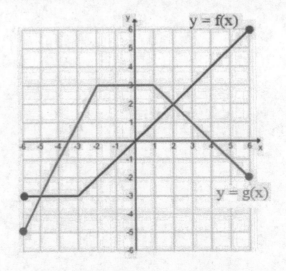

| Review and complete | You try |
|---|---|
| 8. Given the pictured functions, find the value $(f+g)(0)$. <br><br> $(f+g)(0) = f(0) + g(0)$ <br> $= \underline{\quad} + \underline{\quad}$ <br> $= -2$ <br><br> 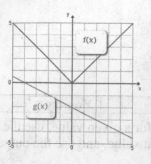 | 9. Given the pictured functions, find the value $(f-g)(2)$. <br><br><br><br> 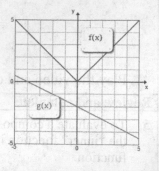 |

## Concept 1: Introduction to Inverse Functions

### Video 1: Introduction to Inverse Functions  (2:43 min)

▶ Play the video and work along.

$f : \{(5,44),\ (7,56),\ (9,68),\ (11,80)\}$

Domain:

Range:

Domain:

Range:

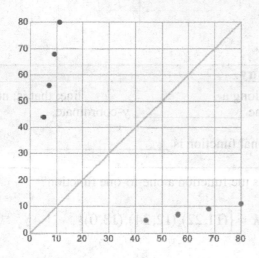

| Review and complete | You try |
|---|---|
| 1. To find the inverse of a relation, we _____ the $x$- and $y$- values in the ordered pairs. | |
| 2. Let $f = \{(1,4),\ (2.5,10),\ (4,16)\}$, write the set of ordered pairs representing $f^{-1}$.<br><br>The $x$- and $y$-values are interchanged to get $f^{-1}$.<br><br>$\quad f^{-1} : \{(4,1),(10,2.5),(16,4)\}$ | 3. Let $f : \{(1,2),(2,3),(3,4)\}$, write the set of ordered pairs representing $f^{-1}$. |

## Concept 2: Definition of a One-to-One Function

| Video 2: Definition of a One-to-One Function (2:13) |
| --- |

▶ Play the video and work along.

$g = \{(3,4),(5,1),(2,4)\}$

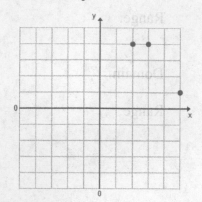
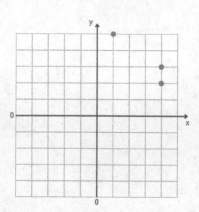

| Review and complete | You try |
| --- | --- |

4.  A **one-to-one function** cannot have two points that lie along a _____ line; that is, no two ordered pairs can have different x-coordinates and the _____ y-coordinate.

5.  A function will have an inverse function only if the original function is _____.

| | |
| --- | --- |
| 6.  Is the function a one-to-one function?<br><br>$m = \{(1,2),(-1,2),(2,-2)\}$<br><br>No, this is a not one-to-one function since two points have the same *y*-coordinate but different *x*-coordinates: $(1,2)$ and $(-1,2)$. | 7.  Is the function a one-to-one function?<br><br>$k = \{(11,22),(12,24),(13,0)\}$ |

| Video 3: Applying the Horizontal Line Test (1:15 min) |
| --- |

▶ Play the video and work along.

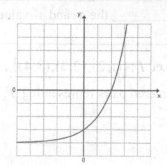
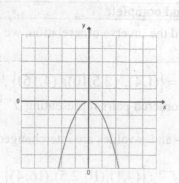

| Review and complete | You try |
|---|---|
| 8.  Does the graph represent a one-to-one function? 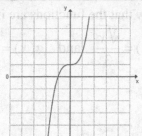 | 9.  Does the graph represent a one-to-one function? 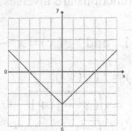 |
| Yes, this is a one-to-one function since no horizontal line intersects the graph in more than one point. 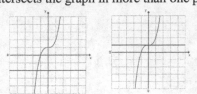 | |

## Concept 3: Definition of the Inverse of a Function

| Video 4: Composition of a Function and its Inverse (2:20 min) |
|---|
| ▶ Play the video and work along.<br><br>Show that the functions are inverses.  $f(x) = 3x - 2$  and  $g(x) = \dfrac{x+2}{3}$ |

| Review and complete | You try |
|---|---|
| 10. The function $g$ is an inverse of $f$ if $(f \circ g)(x)$ and $(g \circ f)(x)$ are equal to _____. | |

| | |
|---|---|
| 11. Show that the functions are inverses.<br><br>$f(x) = \sqrt{x+1}$ and $g(x) = x^2 - 1$<br><br>$\begin{aligned}(f \circ g)(x) &= f(g(x))\\ &= \sqrt{(x^2 - 1) + 1}\\ &= \sqrt{x^2}\\ &= x\end{aligned}$<br>$\begin{aligned}(g \circ f)(x) &= g(f(x))\\ &= \left(\sqrt{x+1}\right)^2 - 1\\ &= x + 1 - 1\\ &= x\end{aligned}$ | 12. Show that the functions are inverses.<br><br>$f(x) = \dfrac{\sqrt[3]{x}}{3}$ and $g(x) = 27x^3$ |
| The functions $f$ and $g$ are inverses. | |

**Concept 4: Finding the Equation of the Inverse of a Function**

| Video 5: Finding the Inverse of a One-to-One Function (1:54 min) |
|---|
| ▶ Play the video and work along.<br><br>Find the inverse.   $f(x) = 3x - 2$ |

| Video 6: Finding the Inverse of a Function (1:41 min) |
|---|
| ▶ Play the video and work along.<br><br>Find the inverse.   $k(x) = \sqrt[3]{2x} - 1$ |

| Review and complete | You try |
|---|---|
| 13. The graphs of inverse functions are _____ | with respect to the diagonal line $y = x$. |

14. Find the inverse.

$$g(x) = \frac{\sqrt[3]{x+1}}{2}$$

$y = \frac{\sqrt[3]{x+1}}{2}$    **Step 1:** Replace $f(x)$ by $y$.

$x = \frac{\sqrt[3]{y+1}}{2}$    **Step 2:** Interchange $x$ and $y$.

$2 \cdot x = \left( \frac{\sqrt[3]{y+1}}{\cancel{2}} \right) \cdot \cancel{2}$

$2x = \sqrt[3]{y+1}$    **Step 3:** Solve for $y$.

$(2x)^3 = \left( \sqrt[3]{y+1} \right)^3$

$8x^3 = y+1$

$y = 8x^3 - 1$

$g^{-1}(x) = 8x^3 - 1$    **Step 4:** Replace $y$ by $f^{-1}(x)$.

15. Find the inverse.

$$g(x) = \sqrt[3]{x} + 2$$

## Video 7: Determining the Inverse Function to a Parabola with a Restricted Domain (2:30 min)

▶ Play the video and work along.

Given the function defined by $k(x) = x^2 + 2$ for $x \geq 0$, find an equation defining $k^{-1}$.

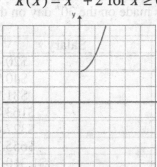

$k(x) = x^2 + 2$ for $x \geq 0$

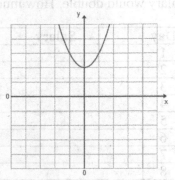

$k(x) = x^2 + 2$

$f(x) = x^2 + 2; x \geq 0$    $k^{-1}(x) = \sqrt{x - 2}$    $y = x$

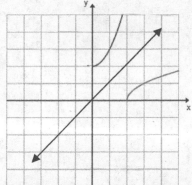

| Review and complete | You try |
|---|---|
| 16. Given $g(x) = x^2 + 5$ for $x \geq 0$, find $g^{-1}(x)$. | 17. Given $f(x) = x^2 - 3$ for $x \geq 0$, find $f^{-1}$. |

16.

$y = x^2 + 5$     Replace g(x) with y.

$x = y^2 + 5$     Interchange x and y.

$x - 5 = y^2$     Solve for y.

$\sqrt{x - 5} = y$     The domain of g is $[0, \infty)$,

$g^{-1}(x) = \sqrt{x - 5}$     so the range of $g^{-1}(x)$ is $[0, \infty)$.

## Concept 1: Definition of an Exponential Function

| Video 1: Introduction to Exponential Functions (3:43 min) |
|---|

▶ Play the video and work along.

Suppose you had a job in which you are paid 2 cents on the first day ($0.02) and every day thereafter, your salary would double. How much money is made on the $30^{th}$ day on the job?

| Day | Salary | Day | Salary | Day | Salary |
|---|---|---|---|---|---|
| 1 | | 11 | $20.48 | 21 | $20,971.52 |
| 2 | | 12 | $40.96 | 22 | $41,943.04 |
| 3 | | 13 | $81.92 | 23 | $83,886.08 |
| 4 | | 14 | $163.84 | 24 | $167,772.16 |
| 5 | | 15 | $327.68 | 25 | $335,544.32 |
| 6 | | 16 | $655.36 | 26 | $671,088.64 |
| 7 | | 17 | $1,310.72 | 27 | $1,342,177.28 |
| 8 | | 18 | $2,621.44 | 28 | $2,684,354.56 |
| 9 | $5.12 | 19 | $5,242.88 | 29 | $5,368,709.12 |
| 10 | $10.24 | 20 | $10,485.76 | 30 | $10,737,418.24 |

Salary in cents      day number

$$y = 2^x$$

| Review and complete | You try |
|---|---|
| 1. An exponential function has a constant as the base and a _____ in the exponent. | |
| 2. Consider the function $f(x) = 0.5^x$.<br>   a. Is $f$ an exponential function?<br>   b. If so, what is the base?<br><br>The function $f(x) = 0.5^x$ is an exponential function with base $0.5$. | 3. Consider the function $h(x) = x^4$.<br>   a. Is $f$ an exponential function?<br>   b. If so, what is the base? |

Video and Study Guide to accompany Intermediate Algebra, $5^{th}$ ed., Miller/O'Neill/Hyde
Copyright © 2018 McGraw-Hill Education

## Concept 2: Approximating Exponential Functions with a Calculator

### Video 2: Evaluating Exponential Expressions on a Calculator (3:01 min)

▶ Play the video and work along.

Given $y = 3^x$, determine the values of $y$ for the given values of x. Round to four decimal places if necessary.

    a.  for $x = 10$

    b.  for $x = -5.2$

    c.  for $x = \pi$

Given $f(x) = \left(\sqrt{5}\right)^x$, determine the function values. Round to four decimal places if necessary.

    a.  $f\left(\dfrac{7}{3}\right)$

    b.  $f\left(\sqrt{11}\right)$

| | You try |
|---|---|
| 4.  Given $f(x) = \left(1.3\right)^x$, determine the function values. Round to four decimal places if necessary.<br><br>    a.  $f\left(\dfrac{7}{3}\right)$<br><br>    b.  $f\left(\sqrt{3}\right)$ | 5.  Given $g(x) = \left(\pi\right)^x$, determine the function values. Round to four decimal places if necessary.<br><br>    a.  $g(-3)$<br><br>    b.  $g\left(\sqrt{7}\right)$ |

## Concept 3: Graphs of Exponential Functions

### Video 3: Graphing Exponential Functions with Base Greater than 1 (6:03 min)

▶ Play the video and work along.

Graph the functions.

$f(x) = 3^x$

| x | f(x) |
|---|------|
| 0 | |
| 1 | |
| 2 | |
| 3 | |
| -1 | |
| -2 | |
| -3 | |

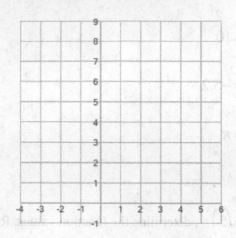

$f(x) = \left(\dfrac{5}{3}\right)^x$

| x | f(x) |
|---|------|
| 0 | |
| 1 | |
| 2 | |
| 3 | |
| -1 | |
| -2 | |
| -3 | |

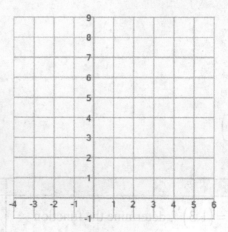

$$f(x) = 3^x \quad g(x) = \left(\dfrac{5}{3}\right)^x$$

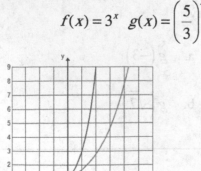

| Review and complete | You try |
|---|---|

**6.** Sketch the graph.    $f(x) = 2^x$

| $x$ | $f(x)$ |
|---|---|
| 0 | 1 |
| 1 | 2 |
| 2 | 4 |
| 3 | 8 |
| -1 | $\dfrac{1}{2} = 0.5$ |
| -2 | $\dfrac{1}{4} = 0.25$ |
| -3 | $\dfrac{1}{8} = 0.125$ |

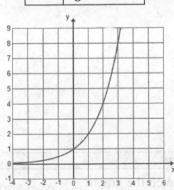

**7.** Sketch the graph.   $f(x) = 1.5^x$

| $x$ | $f(x)$ |
|---|---|
| 0 | |
| 1 | |
| 2 | |
| 3 | |
| -1 | |
| -2 | |
| -3 | |

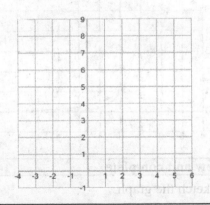

**Video 4: Graph an Exponential Function with Base between 0 and 1   (3:00 min)**

▶ Play the video and work along.

$$h(x) = \left(\frac{1}{2}\right)^x$$

| $x$ | $h(x)$ |
|-----|--------|
| 0   |        |
| 1   |        |
| 2   |        |
| 3   |        |
| -1  |        |
| -2  |        |
| -3  |        |

$$h(x) = \left(\frac{1}{2}\right)^x \qquad y = \left(\frac{1}{5}\right)^x$$

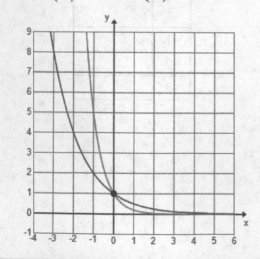

### Review and complete

8. Sketch the graph.

$$b(x) = \left(\frac{1}{4}\right)^x$$

| $x$ | $f(x)$ |
|-----|--------|
| 0   | 1      |
| 1   | $\frac{1}{4} = 0.25$ |

### You try

9. Sketch the graph.

$$f(x) = (0.8)^x$$

| $x$ | $f(x)$ |
|-----|--------|
| 0   |        |
| 1   |        |
| 2   |        |
| 3   |        |

| 2 | $\frac{1}{16} = 0.0625$ |
|----|----|
| -1 | 4 |
| -2 | 16 |

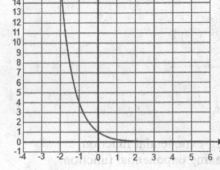

| -1 | |
|----|----|
| -2 | |
| -3 | |

---

### Video 5: Graphs of Exponential Functions: A Summary (1:56 min)

 Play the video and work along.

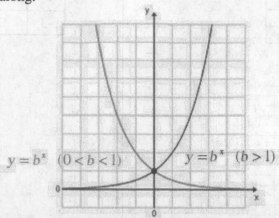

$y = b^x \ \ (0 < b < 1)$   $y = b^x \ \ (b > 1)$

---

**SUMMARY Characteristics of Exponential Functions**

- The graph of $f(x) = b^x$ where $b > 1$, is an **increasing** exponential function.
- The graph of $f(x) = b^x$ where $0 < b < 1$, is a **decreasing** exponential function.
- The graph passes through $(0, 1)$.
- The domain is all real numbers: $(-\infty, \infty)$.
- The range is all positive real numbers: $(0, \infty)$.
- The $x$-axis is a horizontal asymptote.

---

**Concept 4: Applications of Exponential Functions**

---

### Video 6: An Application of an Exponential Function: Radioactive Decay (2:47 min)

 Play the video and work along.

Radioactive iodine is sometimes used in medical treatments such as to treat an overactive thyroid. The half life of radioactive iodine is approximately 8 days. If 1 gram of iodine is present in a sample, then the quantity of iodine, $A(t)$, (in grams) present after $t$ days is given by:

$$A(t) = \left(\frac{1}{2}\right)^{t/8}$$

    a.  Determine the value of $A(8)$ and interpret its meaning in the context of this problem.

    b.  Determine the value of $A(16)$ and interpret its meaning in the context of this problem.

    c.  How much radioactive iodine will be present after 4 weeks? Round to three decimal places.

---

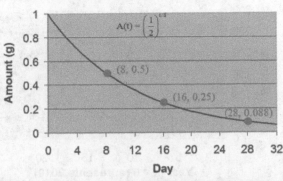

**Amount of Iodine (grams) by Day**

$A(t) = \left(\dfrac{1}{2}\right)^{t/8}$

(8, 0.5)
(16, 0.25)
(28, 0.088)

Amount (g) — vertical axis
Day — horizontal axis

| Review and complete | You try |
|---|---|

**Review and complete**

10. In a sample of 1 g of radium 226, the amount of radium 226 present after t years is given by

$$A(t) = \left(\frac{1}{2}\right)^{t/1260}$$

a.  Find the amount of radium 226 present after 1000 years. Round to three decimal places.

$$A(1000) = \left(\frac{1}{2}\right)^{1000/1260} = 0.577 \text{ g}$$

b.  Find the amount of radium 226 present after 4000 years.

$$A(1000) = \left(\frac{1}{2}\right)^{4000/1260} = 0.111 \text{ g}$$

**You try**

11. The amount of nobelium remaining in a sample that originally contained 1 gram of nobelium is given by

$$A(t) = \left(0.5\right)^{t/10} \text{ where } t \text{ is the time in minutes.}$$

a.  Find the amount of nobelium in the sample after 10 minutes.

b.  Find the amount of nobelium in the sample after one hour.

## Video 7: An Application of an Exponential Function: Population Growth (2:51 min)

▶ Play the video and work along.

In the year 2010, the population of Brazil was approximately 190 million with an annual growth rate of 1.2% per year.

a.  Write a mathematical model that relates the population of Brazil as a function of the number of years since 2010.

b.  If the annual rate of increase remains the same, use this model to predict the population in the year 2025. Round to the nearest million.

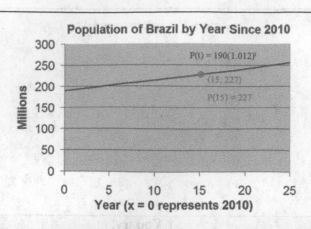

**Population of Brazil by Year Since 2010**

$P(t) = 190(1.012)^t$

(15, 227)

$P(15) = 227$

Year (x = 0 represents 2010)

| Review and complete | You try |
|---|---|

12. In the generic model for population growth $P(t) = P_0(1+r)^t$, the value of $r$ must be expressed as a _____, not a percent.

13. In the year 2000, the population of a certain country was approximately 34 million with an annual growth rate of 0.5% per year.

   a. Write a mathematical model that relates the population of this country as a function of the number of years since 2000.

   $$P(t) = P_0(1+r)^t$$
   $$P(t) = 34(1+0.005)^t$$
   $$P(t) = 34(1.005)^t$$

   b. If the annual rate of increase remains the same, use this model to predict the population in the year 2025. Round to the nearest million.

   $$P(25) = 34(1.005)^{25}$$
   $$\approx 39 \text{ million}$$

14. In the year 2005, the population of a certain state was approximately 3 million, with an annual growth rate of 1.5% per year.

   a. Write a mathematical model that relates the population of this state as a function of the number of years since 2005.

   b. If the annual rate of increase remains the same, use this model to predict the population in the year 2025. Round to the nearest million.

## Concept 1: Definition of a Logarithmic Function

**Video 1: Definition of a Logarithmic Function  (3:41 min)**

▶ Play the video and work along.

**Definition of Logarithmic Function**

If $x$ and $b$ are positive real numbers such that $b \neq 1$, then $y = \log_b x$ is called the logarithmic function with base $b$.

**Definition of Logarithmic Function**

If $x$ and $b$ are positive real numbers such that $b \neq 1$, then

$$y = \log_b x \text{ is equivalent to } b^y = x$$

## Video 2: Converting Between Logarithmic and Exponential Forms (4:36 min)

▶ Play the video and work along.

---

**Definition of Logarithmic Function**

If $x$ and $b$ are positive real numbers such that $b \neq 1$ , then

$$y = \log_b x \text{ is equivalent to } b^y = x$$

---

Write the expression in its equivalent exponential form.

$$\log_3 9 = 2 \qquad\qquad \log_8 1 = 0 \qquad\qquad \log_{1/3}(81) = -4$$

Write the expression in its equivalent logarithmic form.

$$6^2 = 36 \qquad\qquad 10^{-4} = \frac{1}{10,000}$$

| Review and complete | You try |
|---|---|
| 1. Write the expression in its equivalent exponential form.<br><br>$$\log_3 \frac{1}{3} = -1$$<br><br>$$3^{-1} = \frac{1}{3}$$ | 2. Write the expression in its equivalent exponential form.<br><br>$$\log_{10} 100 = 2$$ |
| 3. Write the expression in its equivalent logarithmic form.<br><br>$$5^3 = 125$$<br><br>$$\log_5 125 = 3$$ | 4. Write the expression in its equivalent logarithmic form.<br><br>$$4^6 = 4096$$ |

**Concept 2: Evaluating Logarithmic Expressions**

**Video 3: Evaluating Logarithmic Expressions (1:54 min)**

▶ Play the video and work along.

Evaluate the logarithmic expressions. Assume all variables represent positive real numbers.

$$\log_6 1 \qquad\qquad \log_5 \sqrt[3]{5} \qquad\qquad \log_t\left(t^3\right) \qquad\qquad \log_c c$$

| Review and complete | You try |
|---|---|
| 5. Evaluate the logarithmic expressions. Assume all variables represent positive real numbers. <br><br> a. $\log_5 5 = 1$ because $5^1 = 5$ <br><br> What exponent, applied to base 5, results in 5? <br><br><br> b. $\log_2\left(\dfrac{1}{2^n}\right) = \log_2\left(2^{-n}\right) = -n$ | 6. Evaluate the logarithmic expressions. Assume all variables represent positive real numbers. <br><br> a. $\log_t 1$ <br><br><br> b. $\log_5 125$ |

**Concept 3: The Common Logarithmic Function**

**Video 4: Definition of the Common Logarithmic Function (1:41 min)**

 Play the video and work along.

> **Definition of Common Logarithmic Function**
> The logarithmic function with base 10 is called the **common logarithmic function**. When expressing a common logarithm, the base is often not explicitly written. That is,
> $$y = \log_{10} x \text{ is usually written as } y = \log x$$

Evaluate the logarithmic expressions.

$$\log 100 \qquad\qquad \log 1,000,000 \qquad\qquad \log 0.001$$

| Review and complete | You try |
|---|---|
| 7. When a logarithm is written without a base, the implied base is _____. | |

| | |
|---|---|
| 8. Evaluate the logarithmic expressions.<br><br>a.  $\log 1000 = 3$  ◁ *What exponent, applied to base 10, results in 1000?*<br><br>b.  $\log \dfrac{1}{10^9} = \log 10^{-9} = -9$ | 9. Evaluate the logarithmic expressions.<br><br>a.  $\log 10^{23}$<br><br>b.  $\log \dfrac{1}{10}$ |

## Video 5: Using a Calculator to Approximate Common Logarithms (4:39 min)

▶ Play the video and work along.

Approximate the logarithmic expressions.

$\log 580$                                         $\log(4.3 \times 10^5)$

| Review and complete | You try |
|---|---|
| 10. Approximate the logarithmic expression. Insert consecutive positive integers in the blanks.<br><br>____ $< \log 5600 <$ ____ | Approximate the logarithmic expression. Insert consecutive positive integers in the blanks.<br><br>____ $< \log(9.98 \times 10^{16}) <$ ____ |

## Concept 4: Graphs of Logarithmic Functions

## Video 6: Logarithmic and Exponential Functions as Inverses (5:04 min)

► Play the video and work along.

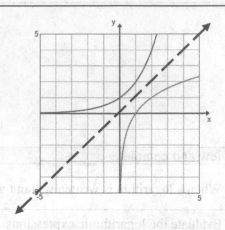

$$y = 2^x \qquad y = x \qquad y = \log_2 x$$

A logarithmic function is the inverse of an exponential function with the same base.

Exponential function        Logarithmic function

$$f(x) = b^x \;\rightarrow\; \text{inverse} \;\rightarrow\; f^{-1}(x) = \log_b x$$

For example:

$$f(x) = 2^x \;\rightarrow\; \text{inverse} \;\rightarrow\; f^{-1}(x) = \log_2 x$$

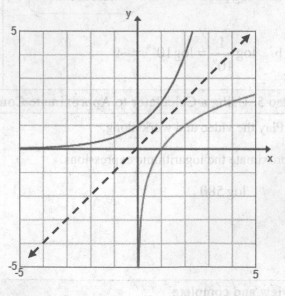

$$y = 2^x \qquad y = x \qquad y = \log_2 x$$

| Exponential Function | Logarithmic Function |
|---|---|
| Domain: $(-\infty, \infty)$ | Domain: $(0, \infty)$ |
| Range: $(0, \infty)$ | Range: $(-\infty, \infty)$ |
| Passes through: $(0,1)$ | Passes through: $(1,0)$ |
| Asymptote: $x$-axis | Asymptote: $y$-axis |

| Review and complete | You try |
|---|---|
| 11. If $f(x) = \log_4 x$, then find an equation for $f^{-1}(x)$. | 12. If $g(x) = (0.9)^x$, then find an equation for $g^{-1}(x)$. |

$$f^{-1}(x) = 4^x$$

The base remains the same.

Video and Study Guide to accompany Intermediate Algebra, 5ᵗʰ ed., Miller/O'Neill/Hyde

## Video 7: Graphing Logarithmic Functions (5:10 min)

▶ Play the video and work along.

Graph the functions.

$y = \log_3 x$

| $x$ | $y$ |
|-----|-----|
|     |     |
|     |     |
|     |     |
|     |     |
|     |     |
|     |     |
|     |     |

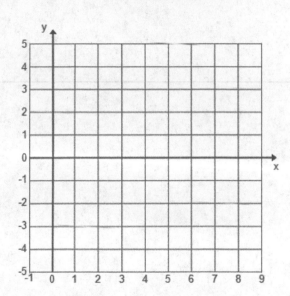

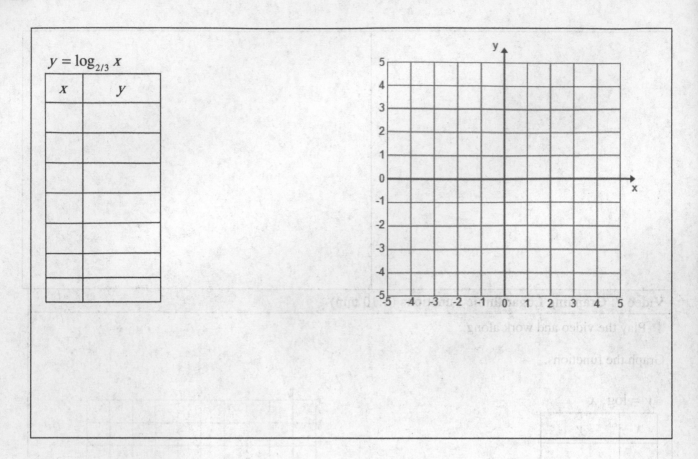

$y = \log_{2/3} x$

| $x$ | $y$ |
|-----|-----|
|     |     |
|     |     |
|     |     |
|     |     |
|     |     |
|     |     |
|     |     |

| Review and complete | You try |
|---|---|
| 13. Graph the function. | 14. Graph the function. |

$y = \log_2 x \rightarrow 2^y = x$

| $x$ | $y$ |
|---|---|
| 1 | 0 |
| 2 | 1 |
| 4 | 2 |
| 8 | 3 |
| $\dfrac{1}{2} = 0.5$ | -1 |
| $\dfrac{1}{4} = 0.25$ | -2 |
| $\dfrac{1}{8} = 0.125$ | -3 |

$y = \log_{1/4} x$

| | |
|---|---|
| | |
| | |
| | |
| | |
| | |
| | |

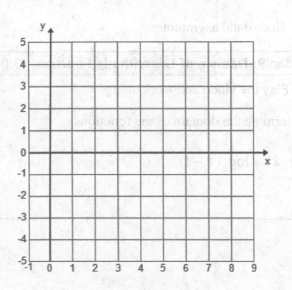

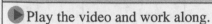

## Video 8: Summary of the Graphs of Logarithmic and Exponential Functions (2:44 min)

▶ Play the video and work along.

Exponential Functions, $y = b^x$

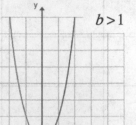

$0 < b < 1$          $b > 1$

Logarithmic Functions, $y = \log_b x$

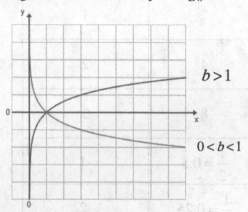

$b > 1$

$0 < b < 1$

| | |
|---|---|
| If $b > 1$, the graph is | If $b > 1$, the graph is |
| If $0 < b < 1$, the graph is | If $0 < b < 1$, the graph is |
| Domain: | Domain: |
| Range: | Range: |
| Passes through: | Passes through: |
| Horizontal asymptote: | Vertical asymptote |

## Video 9: Domain of Logarithmic Functions (3:07 min)

▶ Play the video and work along.

Determine the domain of the functions.

$g(x) = \log_2(x - 1)$                     $f(x) = \log_3(5 - x)$

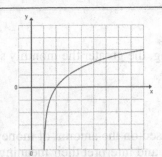

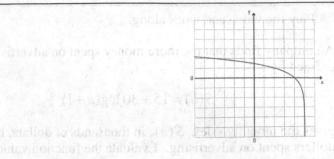

$$g(x) = \log_2(x-1)$$

$$f(x) = \log_3(5-x)$$

**❚❚ Pause the video and try this yourself.**

Determine the domain of the function

$$h(x) = \log_b(2x + 7)$$

**▶ Play the video and check your answer.**

| Review and complete | You try |
|---|---|
| 15. The argument $x$ of $\log_b x$ must be _____. | |
| 16. Determine the domain of the function.<br><br>$$m(x) = \log_4\left(\frac{x+7}{2}\right)$$<br>$$2 \cdot \left(\frac{x+7}{2}\right) > (0) \cdot 2$$<br>$$x + 7 > 0$$<br>$$x > -7$$<br>$$(-7, \infty)$$ | 17. Determine the domain of the function<br><br>$$f(x) = \log_5(5x)$$ |

**Concept 5: Applications of the Common Logarithmic Function**

**Video 10: An Application of a Logarithmic Function (3:52 min)**

▶ Play the video and work along.

A company finds that the more money spent on advertising, the greater the monthly sales. The function defined by

$$S(a) = 15 + 30\log(a+1)$$

gives the monthly sales, $S(a)$, in thousands of dollars, based on the amount of money, $a$, in thousands of dollars spent on advertising. Evaluate the function values and interpret their meaning in the context of this problem.

$S(5)$                    $S(10)$                    $S(15)$

**Monthly Sales versus Advertising Dollars**

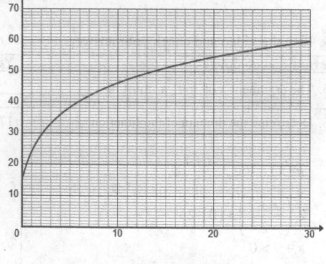

Advertising ($1000)

| Review and complete | You try |
| --- | --- |

18. A company finds that the more money spent on advertising, the greater the monthly sales. The function defined by

$$S(a) = 10 + 20 \log(a+1)$$

gives the monthly sales, $S(a)$, in thousands of dollars, based on the amount of money, $a$, in **hundreds** of dollars spent on advertising. Evaluate the function values and interpret their meaning in the context of this problem.

a.  $S(1) \approx 16$ , so $100 spent on advertising leads to $16,000 in sales.

b.  $S(6) \approx 27$, so $600 spent on advertising leads to $27,000 in sales.

19. A theme park manager finds that the more money spent on prizes for staff training events, the greater the monthly games sales. The function defined by

$$G(a) = 25 + 15 \log(a+1)$$

gives the monthly games sales, $G(a)$, in thousands of dollars, based on the amount of money, $a$, in hundreds of dollars spent on prizes for staff training events. Evaluate the function values and interpret their meaning in the context of this problem.

a.  $G(0)$

b.  $G(30)$

## Concept 1: Properties of Logarithms

| Video 1: Properties of Logarithms  (2:54 min) | | |
|---|---|---|
| ▶ Play the video and work along. | | |
| Property | Explanation | Example |
| $\log_b 1 = 0$ | | |
| $\log_b b = 1$ | | |
| $\log_b b^p = p$ | | |
| $b^{\log_b x} = x$ | | |

| Video 2: Applying the Properties of Logarithms  (1:17 min) |
|---|
| ▶ Play the video and work along. |

Property

Property 1: $\log_b 1 = 0$    Property 2: $\log_b b = 1$    Property 3: $\log_b b^p = p$    Property 4: $b^{\log_b x} = x$

Simplify each expression. Assume all expressions within the logarithms represent positive real numbers.

$$\log_9 9 + \log_4 1 \qquad\qquad 5^{\log_5 (x+4)} \qquad\qquad \log_c c^8$$

| Review and complete | You try |
|---|---|
| 1.  Simplify the expression. Assume all expressions within the logarithms represent positive real numbers.<br><br>$\qquad \log_2 2^5$<br><br>$\qquad \log_2 2^5 = 5$<br><br>*This is an application of property 3.* | 2.  Simplify the expression. Assume all expressions within the logarithms represent positive real numbers.<br><br>$\qquad \log 10^{17}$ |

## Video 3: More Properties of Logarithms (3:40 min)

▶ Play the video and work along.

Property

$$\log_b(xy) = \log_b x + \log_b y$$

Called the **product property** of logarithms.

$$\log_b\left(\frac{x}{y}\right) = \log_b x - \log_b y$$

Called the **quotient property** of logarithms.

$$\log_b x^p = p \cdot \log_b x$$

Called the **power property** of logarithms.

**Summary for Properties of Logarithms**

$$\log_b 1 = 0$$

$$\log_b b = 1$$

$$\log_b b^p = p$$

$$b^{\log_b x} = x$$

$$\log_b(xy) = \log_b x + \log_b y \quad \text{Product Property}$$

$$\log_b\left(\frac{x}{y}\right) = \log_b x - \log_b y \quad \text{Quotient Property}$$

$$\log_b x^p = p \cdot \log_b x \quad \text{Power Property}$$

| Review and complete | You try |
|---|---|
| 3.  True or False?<br><br>$$\log_5(5 \cdot 25) = \log_5(5) \cdot \log_5(25)$$<br><br>False. The logarithm of a product is the sum of the logarithms of the factors, not the product of the logarithms. | 4.  True or False?<br><br>$$\log\left(\frac{6}{7}\right) = \log 6 - \log 7$$ |

## Concept 2: Expanded Logarithmic Expressions

### Video 4: Writing a Logarithmic Expression in Expanded Form (2:08 min)

▶ Play the video and work along.

Write the expression as the sum or difference of logarithms of $x$, $y$, and $z$. Assume all variables represent positive real numbers.

$$\log\left(\frac{x^7}{y^2 z^4}\right)$$

| Review and complete | You try |
|---|---|
| 5.  Write the expression as the sum or difference of logarithms. Assume all variables represent positive real numbers.<br><br>$$\begin{aligned}\log\left(\frac{5}{x^2 y^4}\right) &= \log 5 - \log(x^2 y^4)\\ &= \log 5 - (\log x^2 + \log y^4)\\ &= \log 5 - \log x^2 - \log y^4\\ &= \log 5 - 2\log x - 4\log y\end{aligned}$$ | 6.  Write the expression as the sum or difference of logarithms. Assume all variables represent positive real numbers.<br><br>$$\log\left(\frac{a}{bc^6}\right)$$ |

### Video 5: Writing a Logarithmic Expression in Expanded Form (1:29 min)

Play the video and work along.

Write the expression as the sum or difference of logarithms. Assume that $a$ and $b$ represent positive real numbers.

$$\log_4\left(\frac{64}{\sqrt[3]{a+b}}\right)$$

| Review and complete | You try |
|---|---|

7. The expression $\log(a+b)$ cannot be simplified further, since it is not the log of a _____ or _____.

| | |
|---|---|
| 8. Write the expression as the sum or difference of logarithms. Assume variables represent positive real numbers.<br><br>$\log_2\left(\frac{3\sqrt{x}}{x-1}\right)=\log_2\left(3\sqrt{x}\right)-\log_2\left(x-1\right)$<br>$\qquad=\log_2 3+\log_2\sqrt{x}-\log_2\left(x-1\right)$<br>$\qquad=\log_2 3+\log_2(x)^{1/2}-\log_2\left(x-1\right)$<br>$\qquad=\log_2 3+\frac{1}{2}\log_2 x-\log_2\left(x-1\right)$<br><br>This cannot be simplified any further. | 9. Write the expression as the sum or difference of logarithms. Assume variables represent positive real numbers.<br><br>$\log_b\left(\frac{\sqrt{x}\,y}{z^3}\right)$ |

## Video 6: Writing a Logarithmic Expression in Expanded Form (2:45 min)

 Play the video and work along.

Write the expression as the sum or difference of logarithms. Assume variables represent positive real numbers.

$$\log_b\left(\sqrt[5]{\frac{x^3 z}{y^4}}\right)$$

| Review and complete | You try |
|---|---|
| 10. Write the expression as the sum or difference of logarithms. Assume variables represent positive real numbers.<br><br>$\log \sqrt[3]{\dfrac{4m^2}{n}} = \log\left(\dfrac{4m^2}{n}\right)^{1/3}$<br><br>$= \dfrac{1}{3}\log\left(\dfrac{4m^2}{n}\right)$<br><br>$= \dfrac{1}{3}\left(\log(4m^2) - \log n\right)$<br><br>$= \dfrac{1}{3}\left(\log 4 + \log m^2 - \log n\right)$<br><br>$= \dfrac{1}{3}\left(\log 4 + 2\log m - \log n\right)$<br><br>$= \dfrac{1}{3}\log 4 + \dfrac{2}{3}\log m - \dfrac{1}{3}\log n$ | 11. Write the expression as the sum or difference of logarithms. Assume variables represent positive real numbers.<br><br>$\log\left(\dfrac{v}{w^5}\right)$ |

## Concept 3: Single Logarithmic Expressions

### Video 7: Writing a Sum or Difference of Logarithms as a Single Logarithm (1:41 min)

▶ Play the video and work along.

Write the expression as a single logarithm.

$\log_4 640 - \log_4 5 - \log_4 2$

| Review and complete | You try |
|---|---|
| 12. Write the expression as a single logarithm.<br><br>$\log_2 40 + \log_2 2 - \log_2 5$<br><br>$= \left(\log_2 40 + \log_2 2\right) - \log_2 5$<br><br>$= \left(\log_2 40 \cdot 2\right) - \log_2 5$<br><br>$= \log_2 80 - \log_2 5$<br><br>$= \log_2 \dfrac{80}{5}$<br><br>$= \log_2 16$<br><br>$= 4$ | 13. Write the expression as a single logarithm.<br><br>$\log_5 8 + \log_5 50 - \log_5 16$ |

### Video 8: Writing a Sum or Difference of Logarithms as a Single Logarithm (2:00 min)

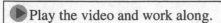 Play the video and work along.

Write the expression as a single logarithm.

$$3 \log x - \frac{1}{3} \log y + \log z$$

| Review and complete | You try |
|---|---|
| 14. Write the expression as a single logarithm. <br><br> $5 \log x + \log y - \frac{1}{2} \log z$ <br><br> $= \log x^5 + \log y - \log z^{1/2}$ <br><br> $= \log x^5 + \log y - \log z^{1/2}$ <br><br> $= \log \left( \dfrac{x^5 y}{z^{1/2}} \right)$ <br><br> $= \log \left( \dfrac{x^5 y}{\sqrt{z}} \right)$ | 15. Write the expression as a single logarithm. <br><br> $2 \log_3 a - \frac{1}{4} \log_3 b + \log_3 c$ |

### Video 9: Using the Properties of Logarithms and Identifying Domain (2:06 min)

▶ Play the video and work along.

When applying properties of logarithms, it is important to take into account domain restrictions.

| Review and complete | You try |
|---|---|
| 16. The argument $x$ of $\log_b x$ must be _____. | |
| 17. Determine the domain of the function.<br><br>$$y = \log_5 x^4$$<br><br>The domain is $(-\infty, 0) \cup (0, \infty)$ since $x^4$ is positive unless $x =$ ____ . | 18. Determine the domain of the function<br><br>$$y = 4 \log_5 x$$ |

## Concept 1: The Irrational Number *e*

### Video 1: Introduction to the Irrational Number e  (2:18 min)

▶ Play the video and work along.

Approximate the value of *e*.

| $x$ | $\left(1+\dfrac{1}{x}\right)^{x}$ |
|---|---|
| 1000 | 2.716923932 |
| 10,000 | 2.718145927 |
| 100,000 | 2.718268237 |
| 1,000,000 | |

### Review and complete

1.  The number *e* is an _____ number, similar to $\pi$. Its value, rounded to three decimal places, is _____.

### Video 2: Graphing the Exponential Function, Base *e*  (2:15 min)

▶ Play the video and work along.

Graph $y = e^{x}$.

| $x$ | $y = e^{x}$ |
|---|---|
| 0 | |
| 1 | |
| 2 | |
| 3 | |
| -1 | |
| -2 | |

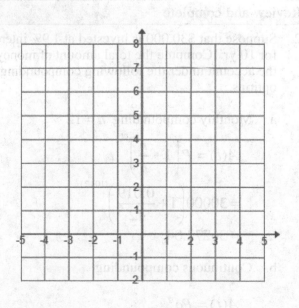

## Concept 2: Computing Compound Interest

| Video 3: An Application of the Exponential Function Base e (Compound Interest) (3:48 min) |
|---|

▶ Play the video and work along.

Compound Interest

$$A(t) = P\left(1 + \frac{r}{n}\right)^{nt}$$

- $P$ is the principal.
- $t$ is the time in years.
- $r$ is the annual interest rate.
- $n$ is the number of compounding periods per year.
- $A(t)$ is the amount in the account after $t$ years.

$$A(t) = Pe^{rt}$$

Used for interest compounded continuously.

Suppose that $10,000 is invested in a tax free municipal bond at 4.5% interest for 20 yr. Compute the total amount of money in the account under the following compounding options:

a. compound monthly

$$A(t) = P\left(1 + \frac{r}{n}\right)^{nt}$$

b. compound continuously

$$A(t) = Pe^{rt}$$

| Review and complete | You try |
|---|---|
| 2. Suppose that $30,000 is invested at 1.9% interest for 10 yr. Compute the total amount of money in the account under the following compounding options. <br><br> a. Monthly compounding $n = 12$ <br><br> $$A(t) = P\left(1 + \frac{r}{n}\right)^{nt}$$ <br> $$= 30000\left(1 + \frac{0.019}{12}\right)^{(10 \cdot 12)}$$ <br> $$\approx 36272.04$$ <br><br> b. Continuous compounding <br><br> $$A(t) = Pe^{rt}$$ <br> $$= 30000e^{(0.019 \cdot 10)}$$ <br> $$\approx 36,276.80$$ | 3. Suppose that $1500 is invested at 1.2% interest for 2 yr. Compute the total amount of money in the account under the following compounding options: <br> a. Annually <br> b. Continuously |

## Concept 3: The Natural Logarithmic Function

### Video 4: Introduction to the Natural Logarithmic Function (3:04 min)

▶ Play the video and work along.

> **Definition of the Natural Logarithmic Function**
> The logarithmic function base $e$ is called the **natural logarithmic function**.
>
> $$y = \log_e x \rightarrow y = \ln x$$

Graph.   $y = \ln x$

| $x$ | $y = \ln x$ |
|-----|-------------|
| 0.5 | |
| 1 | |
| 4 | |
| 8 | |
| 12 | |

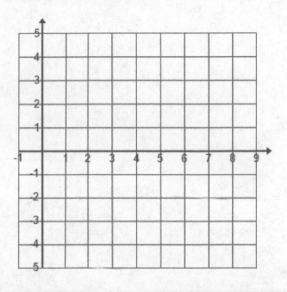

| Review and complete | You try |
|---|---|

4.  The logarithm base 10 is the _____ logarithm. The logarithm base $e$ is the _____ logarithm.

### Video 5: Restating Properties of Logarithms: Using Natural Logarithms (2:37 min)

▶ Play the video and work along.

**Summary of Properties of Logarithms**

$$\log_b 1 = 0 \qquad\qquad \log_b b = 1 \qquad\qquad \log_b b^p = p \qquad\qquad b^{\log_b x} = x$$

Simplify.

$$\ln e^4 \qquad\qquad\qquad 5\ln(1) \qquad\qquad\qquad e^{\ln(2x+5)}$$

| Review and complete | You try |
|---|---|
| 5. The properties of logarithms apply to the natural logarithm function. The base of the natural logarithm is _____. | |
| 6. Simplify.<br>   a. $\ln e = 1$<br><br>   b. $\ln(3-2) = \ln 1 = 0$<br><br>   c. $e^{\ln 6} = 6$ | 7. Simplify.<br>   a. $\ln e^7$<br><br>   b. $\ln(1^6)$<br><br>   c. $e^{\ln 13}$ |

## Video 6: Writing a Sum or Difference of Natural Logarithms as a Single Logarithm (3:34 min)

 Play the video and work along.

**Summary of Properties of Logarithms**

$$\log_b(xy) = \log_b x + \log_b y \qquad\qquad \log_b\left(\frac{x}{y}\right) = \log_b x - \log_b y$$

$$\log_b x^p = p \cdot \log_b x$$

Write the expression as a single logarithm. Assume that all variable expressions within the logarithms represent positive real numbers.

$$3\ln x + \ln(x^2 + x - 12) - \ln(x+4)$$

| Review and complete | You try |
|---|---|
| 8. Write the expression as a single logarithm. Assume that all variable expressions within the logarithms represent positive real numbers.<br><br>$\ln x + 5\ln y - \dfrac{2}{3}\ln z = \ln x + \ln y^5 - \ln z^{2/3}$<br><br>$\qquad = \ln(xy^5) - \ln z^{2/3}$<br><br>$\qquad = \ln\left(\dfrac{xy^5}{z^{2/3}}\right)$<br><br>$\qquad = \ln\left(\dfrac{xy^5}{\sqrt[3]{z^2}}\right)$ | 9. Write the expression as a single logarithm. Assume that all variable expressions within the logarithms represent positive real numbers.<br><br>$2\ln a - \ln b - \dfrac{1}{3}\ln c$ |

### Video 7: Writing a Single Logarithm as the Sum or Difference of Logs  (1:56 min)

 Play the video and work along.

Write the expression as the sum or difference of logarithms of $x$ and $y$. Assume all variables represent positive real numbers.

$$\ln\left(\frac{y^5\sqrt{x}}{e^2}\right)$$

| Review and complete | You try |
|---|---|
| 10. Write the expression as the sum or difference of logarithms of $x$ and $y$. Assume all variables represent positive real numbers.<br><br>$\ln\left(\dfrac{e\sqrt{x}}{y^2}\right) = \ln\left(e\sqrt{x}\right) - \ln y^2$<br><br>$\qquad = \ln e + \ln\sqrt{x} - \ln y^2$<br><br>$\qquad = 1 + \dfrac{1}{2}\ln x - 2\ln y$ | 11. Write the expression as the sum or difference of logarithms of $a$ and $b$. Assume all variables represent positive real numbers.<br><br>$\ln\sqrt{e^2 ab}$ |

## Concept 4: Change-of-Base Formula

### Video 8: Applying the Change-of-Base Formula (5:13 min)

 Play the video and work along.

**Change-of-Base Formula**

Let $a$ and $b$ be positive real numbers such that $a \neq 1$ and $b \neq 1$. Then for any positive real number $x$,

Use a calculator to approximate $\log_4 50$ .

| Review and complete | You try |
|---|---|
| 12. Use the change-of-base formula to rewrite $\log_2 116$. Then use a calculator to approximate its value. Round to four decimal places. | 13. Use the change-of-base formula to rewrite $\log_{3.5} 5000$. Then use a calculator to approximate its value. Round to four decimal places. |

## Concept 5: Applications of the Natural Logarithmic Function

### Video 9: An Application of the Natural Logarithm Function (2:35 min)

▶ Play the video and work along.

Carbon 14 is a radioactive isotope of carbon that has a half-life of 5730 yr. When a living organism dies, it no longer ingests carbon from the environment. As a result the amount of stable carbon 14 remains unchanged from the time of death. But the amount of radioactive carbon 14 begins to decay. As the years go by, there is less and less carbon 14 present in the organism. As a result, scientists can measure the percentage of carbon 14 and determine the age of the specimen. This is given by:

$$A(p) = \frac{\ln p}{-0.000121}$$

$p$ is the percentage of carbon 14 still present.
$A(p)$ is the age of the specimen.

   a.  Determine the age of a tissue sample that has 82% of its carbon 14 remaining.

   b.  Determine the age of a tissue sample that has 35% of its carbon 14 remaining.

| Review and complete | You try |
|---|---|

14. Use the formula presented in Video 9 to determine the age of a tissue sample that has 40% of its carbon 14 remaining.

$$A(p) = \frac{\ln p}{-0.000121} = \frac{\ln 0.4}{-0.000121} \approx 7573 \ years$$

15. Use the formula presented in Video 9 to determine the age of a tissue sample that has 0.8% of its carbon 14 remaining.

## Concept 1: Solving Logarithmic Equations

| Video 1: Introduction to the Logarithmic Equations  (0:51 min) |
|---|

▶ Play the video and work along.

Solve.   $\log_5 x = 3$

| Review and complete | You try |
|---|---|
| 1.  To solve for a variable in a logarithmic equation, the key step is to rewrite the expression in _____ form. | |
| 2.  Solve. $\log_2 x = 5$<br><br>$\log_2 x = 5$<br>$2^5 = x$      Change to exponential form.<br>$32 = x$ | 3.  Solve.  $\log_6 x = -3$ |

| Video 2: Solving a Logarithmic Equation Containing One Logarithm  (3:00 min) |
|---|

▶ Play the video and work along.

Solve.  $-3 + \log_4(3x - 8) = -1$

**Step 1:** Isolate the logarithms on one side of the equation.

**Step 2:** If the equation contains multiple logarithms, combine the logs into a single log.

**Step 3:** Write the equation in exponential form.

**Step 4:** Solve the resulting equation.

**Step 5:** Check the potential solutions in the original equation.

| Review and complete | You try |
|---|---|
| 4.  Solve.  $2 + \log_3(5 + 2x) = 7$<br><br>$2 + \log_3(5 + 2x) = 7$  Subtract 2 from each side.<br>$\log_3(5 + 2x) = 5$  Write in exponential form.<br>$3^5 = 5 + 2x$<br>$243 = 5 + 2x$  Simplify.<br>$238 = 2x$  Subtract 5 from each side.<br>$119 = x$  Divide each side by 2.<br><br>The solution checks. The solution set is {119}. | 5.  Solve. $\log_2(x - 7) + 2 = 7$ |

Video and Study Guide to accompany Intermediate Algebra, 5ᵗʰ ed., Miller/O'Neill/Hyde
Copyright © 2018 McGraw-Hill Education

---

### Video 3: Solving a Logarithmic Equation Containing Multiple Logarithms (4:22 min)

▶ Play the video and work along.

Solve.  $-5 + \log_2 x = -\log_2(x-4)$

**Step 1:** Isolate the logarithms on one side of the equation.

**Step 2:** If the equation contains multiple logarithms, combine the logs into a single log.

**Step 3:** Write the equation in exponential form.

**Step 4:** Solve the resulting equation.

**Step 5:** Check the potential solutions in the original equation.

Check: $x = $ _____

$-5 + \log_2 x = -\log_2(x-4)$

Check $x = $ _____

$-5 + \log_2 x = -\log_2(x-4)$

---

| Review and complete | You try |
|---|---|
| 6.  Solve.  $\log_3 k + \log_3(2k+3) = 2$ | 7.  Solve.  $\log_2(h-1) + \log_2(h+1) = 3$ |
| $\log_3(k \cdot (2k+3)) = 2$ | |
| $\log_3(2k^2 + 3k) = 2$ | |
| $3^2 = 2k^2 + 3k$ | |
| $9 = 2k^2 + 3k$ | |
| $2k^2 + 3k - 9 = 0$ | |
| $(2k-3)(k+3) = 0$ | |
| $k = \dfrac{3}{2}, k = -3$ | |
| The negative answer does not check, since the argument of a logarithm cannot be negative. | |
| The solution set is $\left\{ \dfrac{3}{2} \right\}$. | |

---

**Video 4: Solving a Logarithmic Equation Containing Common Logarithms (2:00)**

▶ Play the video and work along.

Solve. $\log(x-1200)=4.2$

| Review and complete | You try |
|---|---|
| 8. When a logarithm is written without a base, the base is _____. | |
| 9. Solve. $\log(x+42)=3.1$<br><br>$\qquad 10^{3.1}=x+42$<br><br>$\qquad x=10^{3.1}-42\approx 1217$<br><br>The solution set is $\left\{10^{3.1}-42\right\}$. | 10. Solve. $\log(m-3)=-0.2$ |

**Video 5: Solving a Logarithmic Equation Using the Equivalence Property of Logarithms (3:58 min)**

▶ Play the video and work along.

Solve. $\ln(x-4)+\ln(x-2)=\ln(x+2)$

Check  $x=$

$\ln(x-4)+\ln(x-2)=\ln(x+2)$

Check  $x=$

$\ln(x-4)+\ln(x-2)=\ln(x+2)$

| Review and complete | You try |
|---|---|
| 11. Solve. $\log(4m)=\log(m+4)+\log(m-2)$<br><br>$\qquad \log(4m)=\log\big((m+4)(m-2)\big)$<br><br>$\qquad\qquad 4m=m^2+2m-8$<br><br>$\qquad\qquad 0=m^2-2m-8$<br><br>$\qquad\qquad 0=(m-4)(m+2)$<br><br>$\qquad\qquad m=4,\, m=-2$<br><br>Only 4 checks.  So, the solution set is $\{4\}$. | 12. Solve. $\ln(x+5)=\ln(x-1)+\ln(x-5)$ |

## Concept 2: Solving Exponential Equations

### Video 6: Introduction to Exponential Equations (2:23 min)

▶ Play the video and work along.

Solve. $3^{2x+6} = 81$

| Review and complete | You try |
|---|---|
| 13. Solve.  $2^{2x-1} = 32$ | 14. Solve.  $5^{5-x} = 125$ |
| $2^{2x-1} = 2^5$ | |
| $2x - 1 = 5$ | |
| $2x = 6$ | |
| $x = 3$ | |
| The solution set is $\{3\}$. | |

### Video 7: Solving an Exponential Equation (2:28 min)

▶ Play the video and work along.

Solve.  $\left(5^x\right)^{x+4} = \dfrac{1}{125}$

| Review and complete | You try |
|---|---|
| 15. Solve. | 16. Solve.  $\left(4^x\right)^{x+1} = 16$ |
| $\left(5^x\right)^{x+2} = \dfrac{1}{5}$ | |
| $5^{x^2+2x} = 5^{-1}$ | |
| $x^2 + 2x = -1$ | |
| $x^2 + 2x + 1 = 0$ | |
| $(x+1)(x+1) = 0$ | |
| $x = -1$ | |

The solution set is $\{-1\}$.

## Video 8: Solving an Exponential Equation by Taking a Logarithm of Both Sides (4:36 min)

▶ Play the video and work along.

Solve.  $6^x = 87$                      Solve.  $6^x = 87$                      Solve.  $6^x = 87$

| Review and complete | You try |
|---|---|
| 17. Solve.  $7^x = 116$ <br><br> $\log_7 7^x = \log_7 116$ <br><br> $x = \log_7 116$ <br><br> $x = \dfrac{\log 116}{\log 7}$ <br><br> $x \approx 2.44$ | 18. Solve.  $5^x = 602$ . Round the solution to two decimal places. |

## Video 9: Solving an Exponential Equation Containing Base $e$ (2:59 min)

▶ Play the video and work along.

Solve.  $e^{1.8x} = 43{,}500$

| Review and complete | You try |
|---|---|
| 19. Solve  $e^{1.2x} = 78$ <br><br> $\ln\left(e^{1.2x}\right) = \ln(78)$ <br><br> $1.2x = \ln 78$ <br><br> $x = \dfrac{\ln 78}{1.2}$ <br><br> $x \approx 3.63$ | 20. Solve  $e^{0.4x} = 980$ . Round the solution to two decimal places. |

## Video 10: Solving an Exponential Equation Containing Different Bases (3:53 min)

▶ Play the video and work along.

Solve.  $5^{x-2} = 9^x$

| Review and complete | You try |
|---|---|
| 21. Solve. $5^{x-2}=3^x$ <br><br> $\ln 5^{x-2}=\ln 3^x$ <br><br> $(x-2)\ln 5 = x\ln 3$ <br><br> $x\ln 5 - 2\ln 5 = x\ln 3$ <br><br> $x\ln 5 - x\ln 3 = 2\ln 5$ <br><br> $\dfrac{x(\ln 5 - \ln 3)}{(\ln 5 - \ln 3)} = \dfrac{2\ln 5}{(\ln 5 - \ln 3)}$ <br><br> $x = \dfrac{2\ln 5}{\ln 5 - \ln 3}$ <br><br> $x \approx 6.301$ | 22. Solve $4^{3x-4}=3^x$. Round to three decimal places. |

## Concept 3: Applications

### Video 11: An Application of an Exponential Equation (4:07 min)

▶ Play the video and work along.

The population of New Mexico was estimated to be 2 million in 2008. The population growth rate at that time was 1.2%. The function defined by $P(t) = 2(1.012)^t$ represents the population of New Mexico in millions as a function of the number of years since 2008.

    a.  Use the function to estimate the population of New Mexico in 2016.

    b.  Use the function to estimate the number of years since 2008 required for the population of New Mexico to reach 2 million, 480 thousand.

| Review and complete | You try |
|---|---|
| 23. Use the function $P(t) = 2(1.012)^t$ to estimate the number of years since 2008 required for the population of New Mexico to reach 3 million. | 24. Use the function $P(t) = 3(1.04)^t$ to estimate the number of years required for a state that has a population of 3 million to grow to 4 million, at a growth rate of 4%. Round to the nearest whole year. |

$$3 = 2(1.012)^t$$

$$\frac{3}{2} = \frac{\cancel{2}(1.012)^t}{\cancel{2}}$$

$$1.5 = 1.012^t$$

$$\ln 1.5 = t \ln 1.012$$

$$t = \frac{\ln 1.5}{\ln 1.012} \approx 34$$

It will take about 34 years for the population of New Mexico to reach 3 million.

**Answers:** Section 8.1

2. $4x - 16$      3a. $\sqrt{x^4}\,;x^2$      3b. $x^4 + 2$    4a. $(x+2)^4$      4b. $\sqrt{x}+2$

5. outer      7. $2\sqrt{3}-2$      8. 0, -2      9. 5

**Answers:** Section 8.2

1.  interchange      3. $f^{-1}:\left\{(2,1),(3,2),(4,3)\right\}$      4. horizontal; same 5. one-to-one

7. Yes      9. No, it fails the horizontal line test.      10. $x$

12. $(f \circ g)(x) = \dfrac{\sqrt[3]{27x^3}}{3} = x; (g \circ f)(x) = 27\left(\dfrac{\sqrt[3]{x^3}}{3}\right)^3 = x$      13. symmetric      15. $g^{-1}(x) = (x-2)^3$

17. $f^{-1}(x) = \sqrt{x+3}$

**Answers:** Section 8.3

2.  variable      3. Not an exponential function

4a. 1.8445      4b.  1.5753      5a.  0.0323      5b. 20.6697

7.
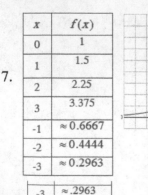

| $x$ | $f(x)$ |
|-----|--------|
| 0 | 1 |
| 1 | 1.5 |
| 2 | 2.25 |
| 3 | 3.375 |
| -1 | $\approx 0.6667$ |
| -2 | $\approx 0.4444$ |
| -3 | $\approx 0.2963$ |

| -3 | $\approx .2963$ |
|-----|--------|

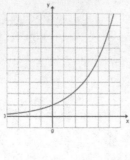

9.

| $x$ | $f(x)$ |
|-----|--------|
| 0 | 1 |
| 1 | 0.8 |
| 2 | 0.064 |
| 3 | 0.512 |
| -1 | 1.25 |
| -2 | 1.5625 |
| -3 | 1.9531 |

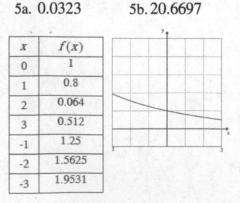

11a. 0.5 g          11b. 0.016 g          12. decimal          14a. $P(t) = 3(1.015)^t$

14b. $P(20) = 4$ million

**Answers**: Section 8.4

2. $10^2 = 100$     4. $\log_4 4096 = 6$     6a. 0          6b. 3          7. 10

9a. 23          9b. -1          10. 3; 4          11. 16; 17          12. $g^{-1}(x) = \log_{0.9} x$

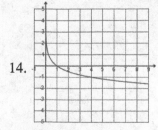

14.                    15. greater than 0          17. $(0, \infty)$

19a. No money on prizes leads to $25,000 sales    19b. $3,000 in prizes leads to $47,000 sales

**Answers**: Section 8.5

2. 17                    4. True        6. $\log a - \log b - 6\log c$        7. product; quotient

9. $\frac{1}{2}\log_b x + \log_b y - 3\log_b z$        11. $\log v - 5\log w$        13. 2

15. $\log_3\left(\dfrac{a^2 c}{\sqrt[4]{b}}\right)$        16. greater than zero        17. 0        18. $(0,\infty)$

**Answers**: Section 8.6

3.  irrational, 2.718        3a. $1536.42        3b. $1536.44        4. common; natural

5. $e$                    7a. 7            7b. 0            7c. 13

9. $\ln\left(\dfrac{a^2}{b\sqrt[3]{c}}\right)$        11. $1 + \dfrac{1}{2}\ln a + \dfrac{1}{2}\ln b$    12. $\dfrac{\ln 116}{\ln 2} \approx 6.8580$    13. $\dfrac{\ln 5000}{\ln 3.5} \approx 6.7987$

15. 39,903 yr

**Answers**: Section 8.7

1.  exponential        3. $\left\{\dfrac{1}{216}\right\}$        5. $\{39\}$        7. $\{3\}$

8. 10                10. $\{10^{-0.2} + 3\}$        12. $\{7\}$        14. $\{2\}$

16. $\{-2, 1\}$ 18. 3.98 20. 17.22 22. 1.812

24. about 7 yr

# Notes

# Notes

# Notes

# Notes

# Notes

# Notes

# Notes

# Notes

# Notes

# Notes

# Notes

# Notes

# Notes

# Notes

# Notes